KU-578-986

PRACTICAL SECURITY

IN COMMERCE AND INDUSTRY

Practical Security in Commerce and Industry

Eric Oliver and John Wilson

Second Edition

Gower Press

First published in Great Britain by Gower Press Limited
Epping, Essex
1968

Second impression 1969
Second edition 1972

ISBN 0 7161 01351

Set in 11 on 13 point Baskerville

Printed in England by
J. W. Arrowsmith Ltd., Bristol

Contents

Contents

Part Three

SECURITY IN OFFICES AND SHOPS

Part Four

SECURITY OF BUILDINGS AND SITES

Part Five

SECURITY AND CONTROL OF ROAD TRANSPORT

Part Six

FIRE AND ACCIDENT PREVENTION

Contents

Part Seven

AIDS TO SECURITY

Part Eight

PROFESSIONAL BODIES

APPENDICES

Preface to second edition

In the first edition of this book we said with some trepidation that it was a challenge to provide practitioners in security and others with a text or reference book which would be found acceptable. After publication we felt that there was much of importance that we had omitted and much which could have been better expressed. However, the appreciation shown by professional colleagues whose expertise we recognised, and the reviews of the Press, restored our confidence. The book (with its companion volume, *Security Manual*) became part of the recommended reading in preparation for the examinations of the Institution of Industrial Security.

When the publishers asked us to update the first edition, we gladly accepted. The changing scene in which security is practised constantly brings new problems, such as those concerning computers and commercial and industrial espionage. The recommended security precautions for computers have been extended and now form a new chapter. The measures against industrial espionage have been developed in the light of experience. New chapters on commercial fraud, planning for emergencies, including bomb scares, and on Scottish law have also been included.

Changes in and new legislation affecting security practices such as the Criminal Damage Act, the Industrial Relations Act and the Fire Precautions Act have been dealt with.

The potential risks to security of drug addicts and their recognition have also been referred to.

Although the law and practice described in this book apply to the United Kingdom, they are not likely to differ in material particulars from those of other countries. Security measures have, of course, international application.

It is our hope this edition will be a useful contribution to the pool of security knowledge and we thank our many friends in the profession for their helpful comments which have been remembered in its preparation.

March, 1972.

Eric Oliver
John Wilson.

PART ONE

THE SECURITY FUNCTION AND ITS PERFORMANCE

I

Forming a security policy and its performance

Security means the protection of property of all kinds from loss through theft, fraud, fire and other forms of damage, and waste. Security duties include factory patrol; gate control; fire prevention, detection and fighting; weighbridge operations; carrying out searches of personnel and vehicles, as well as the enforcement of security procedures in offices and other buildings, including retail and wholesale premises, to prevent losses, and investigations of them when they occur. The protection of secret processes, financial data, sales projects and other information vital to a company's interests against loss through what is described as industrial espionage is an additional responsibility.

SENIOR MANAGEMENT'S ROLE IN COMPANY SECURITY

The first step in developing a sound security programme is to recognise that a company's security is not so much a law enforcement problem, or a problem of social attitudes and education but a management problem. The responsibility rests primarily with its own management and should be approached in the same way as any other management problem.

Analysis and definition—Security has a general application with varied aspects and elements that rarely have equal impact on different company operations. It is management's job to analyse carefully its own particular security situation and to define precisely its own unique problems.

Solution planning—Working within precisely defined areas, management can more easily develop effective remedial and preventive plans, taking into consideration not only the specific security problem but the relationship of one problem to another. Thus a security programme is evolved which puts proportionately greater emphasis on the more important problems.

Programme implementation—Many company security programmes look excellent on paper but fail to work in practice. To have any effect, a programme must be implemented from the management control level down through every level of supervision to all affected personnel. Policies and procedures must be clearly defined, individual responsibilities spelled out and unequivocal ground rules established.

The requirements should be uncomplicated and commonsense in their nature and application, so that they are accepted by all grades of employees without demur as an essential discipline in their daily routine.

Routine follow-up—Security is a continuous problem requiring constant vigilance. It should be a standard procedure to check that the programme is being properly implemented. Controls should be tight enough for minor deviations from the norm to be quickly spotted and investigated. Reporting procedures should be responsible enough to draw appropriate management attention to potential, as well as actual, trouble. Implementation and follow-up functions are more effective if a special official is placed in charge of the overall programme, or given prime responsibility for a specific area. Indeed, the presence of a security specialist benefits any aspect of a security programme.

ROLE OF SUPERVISORY STAFF

The goodwill and cooperation of senior management is fundamental in establishing sound security principles and procedures but day-to-day efficient security practice is more closely dependent on the appreciation and understanding of the principles and procedures amongst the workforce. Shopfloor and office supervisors have an extra responsibility for the protection of their employers' property against loss. There may be an understandable reluctance for them to take action against persons under their charge who may have offended but nevertheless they and not security personnel are the most likely to see or learn of dishonest actions by employees. It is important that they are guided to a realisation that they have this responsibility to the firm and that inaction will be looked upon as impugning their integrity and as indicative of an inability to lead.

Training of supervisory grades

Instruction on security responsibilities and practice should be incorporated in the training of management and supervisory grades. An outline of such a course is given in Appendix 9.

Checklist for supervisors on curbing losses

A good supervisor can do more to stop stealing than the most effective security force or management crash programme. Here are six ways suggested to discourage amateur thieves.

1. Don't let them get started. Stealing is contagious. Once a supervisor lets workers get away with 'borrowing', he has shown them that it is acceptable. They will not be one-time thieves. The stealing virus will spread fast until the whole department is nibbling away at inventories.

2. Remove temptation. A supervisor should not encourage pilfering by leaving broken cases or cartons of material, small tools, hardware, typewriter ribbons, or other items people find useful, lying around unguarded. He should keep unauthorised employees and people from other departments out of sensitive areas by courteously challenging strangers. He should discourage visiting.

3. Control the inventory. In departments where useful, easily concealed parts are used—or in the final assembly and packaging area—a rigid inventory control system is the supervisor's most effective theft preventive. Nothing discourages an industrial thief more than an early detection system.

4. Curb the borrowing. If a company does not permit employees to borrow certain equipment without permission, a supervisor should be sure to enforce the rule. This goes for workers, other supervisors, and himself.

5. Check the hiding places. There are probably many good hiding places around the shop—between bales or boxes, in ventilators, and electric panel boxes. Many a thief stashes his loot away to pick up later when the coast is clear. Except for emergencies, a supervisor should not allow workers to visit parked cars during working hours. He should not hesitate to question suspicious packages or bundles leaving the department.

6. Make tools and supplies hard to get. One of the best ways to impress workers with the value of tools or the small common supplies is to set up a system that makes them hard to get. If a worker has to sign a chit every time he wants a screwdriver, and if a hard-and-fast quota is placed on the perishable items, people will be careful about returning tools and taking supplies for their own use.

Security is the responsibility of individual departments. Each employee in each department is responsible for the protection of the company's assets. The security department should coordinate and integrate the efforts of all to ensure that security policies and practices are effective and followed uniformly.

EMPLOYEES AND SECURITY

The security problems which most concern commercial and industrial organisations are stealing in general, stealing of employers' property by employees in particular, and fraud by employees.

Most employees are subject to some temptations in their employment and it will not be easy for anyone to be dishonest for long before this is noticed by the other employees. It is a primary responsibility of management to ensure that young employees, particularly, are not corrupted through seeing colleagues taking advantage of lax controls. An atmosphere of disapproval must be encouraged so that dishonesty will be brought to the notice of the management if persisted in.

The fact which has to be faced is that no man can be guaranteed to be forever honest, whether he is a leader of the community, a pillar of the church, a holder of a responsible position, or an employee of many years' service.

How prevalent is dishonesty in industry and commerce? What does it cost in loss of profits? How much more has to be sold to account for the raw material or finished goods which have been stolen? Relatively little is known about the precise extent of the losses because only a small percentage is reported to the police and appears in the crime figures published by the Home Office. Many losers are aware only of a general discrepancy and the blame is attributed to bad recording, faulty stock-keeping or a production loss—in effect everything short of unpalatable theft. Quite a lot, however, can be said about the nature of the problem and the means which can be taken to control losses.

Standards of honesty appear to vary with the number of opportunities to steal: in other words a man's or woman's honesty is related to the temptation to which it is exposed. This is recognised by some employers who describe thefts of small articles by their employees as 'pilfering' and not stealing which, it must be emphasised, is what it is. If this goes unchecked it must be expected to increase. Some firms have found that pilfering grows from small beginnings to incredible proportions and the cost of introducing and enforcing satisfactory controls becomes a major exercise involving considerable expense. Incredible though it may be steps to eliminate such losses frequently will be met with real resentment.

The frequency of pilfering will depend, of course, on the nature of the articles concerned. For example, losses in a firm producing heavy iron girders and castings will differ from one where small attractive consumer goods are made. Where the items are valuable and easily disposable the means by which they can be stolen are varied and carefully planned, including collusion between employees and outside persons having means of disposal. No internal control, however sophisticated, can by itself guarantee protection against fraudulent collusion, particularly by persons in authority.

How pilfering (theft) occurs

To pilfer (steal) four factors have to be present: the goods to steal, the pilferer to steal them, theft opportunity situations and the method to move the goods stolen from one place to another.

The goods are always present, the second, human nature being what it is, also is always present to some degree. The theft opportunities are usually caused by inadequate security measures and poor supervisory and management practices. There are those who would attempt to misappropriate anything and everything that can be stolen. The last factor, method of movement, depends on the amount and size of the property which is to be stolen.

We believe that people are basically honest. About 10 per cent of the people would be honest if there are no security measures and regardless of any amount of temptation. About 80 per cent would be honest as long as they are not tempted too far and reasonable security measures and well-defined consequences furnish a deterrent. The remaining 10 per cent are going to lie, cheat and steal regardless of what measures are taken.

THE COST OF SECURITY

Every commercial or industrial organisation insures itself against loss through various causes, and pays to equip its premises with fire-fighting apparatus to the standards required by its insurance company and by the law. These fixed costs are accepted as unavoidable; certain of them may well be allocated to a separate security budget.

Lack of reasonable precautions is no longer tolerated in return for an extra premium and might make a theft risk uninsurable. A risk declined by one insurer is unlikely to be accepted by another.

Further costs of security include, for example, the purchase of safes and anti-thief cash-carrying bags and waistcoats, the installation of burglar-alarms and the fees of security companies for transporting money or patrolling premises to prevent and detect fire and losses from other causes.

These precautions are not mandatory and consequently are critically viewed and reviewed. The costs incurred will probably be justified on the grounds (*a*) of prudence according to the environment and to experience of the degree of risk and (*b*) as preventing an increase in insurance premiums because of repeated losses.

The employment of full-time security staff will depend upon the nature of the business, its loss potentiality and the products manufactured or handled. For example, a firm making highly attractive items, such as cigarettes, spirits or easily disposable foodstuffs, would be likely to find security staff more justifiable in cost than, say, one making industrial chemicals. Further, where there is a high risk of fire, it may be particularly desirable and prudent to supplement fire-detecting equipment by patrolling security staff.

It is, therefore, hardly possible to generalise on the amount of money which should be spent on security as a proportion of company overheads. The cost, however, must be commensurate with the value of what is at risk and the number of persons employed. To assist in deciding on the security protection necessary and therefore the cost involved, advice from someone experienced in evaluating such risks should be obtained. Guidance may be sought from the professional security bodies, commercial security organisations, or the police.

The costs should always be reviewed in the light of any physical changes which may have taken place in the processes or types of product which could have influenced the potential risk. The level of security services maintained should not be overlooked when negotiating premiums for insurance.

RESPONSIBILITY FOR SECURITY

Efforts made to prevent or contain thefts and to deal with other security risks are likely to be ineffective unless their introduction and enforcement is either all or part of the responsibility of a senior manager. On 12 February 1969 in the House of Lords, the Minister of State at the Home Office, Lord Stonham, said, 'I think that in firms of any size at all, even small and middle-sized firms, a director or member of the firm should be partly or wholly engaged in this question of security and protection, not only to protect the employees—though that is vastly important—but also with the vastly greater losses from theft.' This theme is developed in Chapter 2.

COMPANY RULES

Conditions of employment must be agreed to by all employees before engagement. Security will be improved if the conditions include a clause by which the employee agrees to searches of his person, anything he carries or any vehicle he uses, and by which he agrees to procedures that the employer has decided to adopt to prevent and detect theft. It should be made quite clear that the company rules are to be obeyed by everyone, no matter what their status and without exception.

In later chapters the circumstances in which theft is likely to occur and ways by which it can be prevented or lessened will be suggested. These must be economic, effective and weighed against the degree of risk. They must also be of a nature to justify respect. The law affecting the protection of property and the rights and powers of owners to take action against offenders will also be described. The law on searching will be dealt with in Chapter 7.

POLICY ON DISMISSAL AND PROSECUTION

The Industrial Relations Act gives protection to employees in cases of unfair dismissal. There is no doubt that theft would be regarded as a reasonable ground for dismissal, in addition to coming under the general heading of 'misconduct'. Although a successful case in an industrial tribunal would be unlikely, some inconvenience might well be saved by a firm clearly putting its policy into print for the information of employees. If an employee is to be dismissed for misconduct, he has the right to know what actions will be construed as such. It is now extremely important that two similar incidents should not be dealt with in entirely different ways. Once a precedent of not dismissing has been established, this would undoubtedly be quoted in a subsequent case where more stringent disciplinary measures are taken. Theft is a specific instance where a firm directive should be laid down so that everyone knows exactly where they stand. This directive should originate at the highest level so there is no danger of it being watered down for the benefit of particular individuals and relationships.

A suggested form of wording is:

> If any act of theft or other form of dishonesty or criminal damage is admitted or proved to have been committed by an employee in connection with company property, or the property of another employee, or if the unjustifiable use of personal violence against another employee is proved, then the employee who has committed these acts will be dismissed and, where company property is involved, the company will exercise its discretion in whether the matter is to be reported to the police.

This ruling could be put in a list of items that constitute 'misconduct'. This list could include other matters where there should be a right of immediate dismissal, such as clocking offences; being under the influence of drink or drugs during working hours; doing a wilful act which might endanger the health, safety or welfare of a fellow employee or himself; wilful refusal to obey legitimate instructions; and deliberately sleeping during working hours. Whether such a list is compiled is a matter for the personnel manager. There are arguments for and against and, at the time of writing, these have not been resolved. It is clear that discretion should also be available for immediate dismissal on serious but unforeseen grounds.

So far as theft is concerned it is advisable to leave an ultimate discretion as to whether to prosecute which gives room to manoeuvre in certain cases with mitigating circumstances. It may, however, become advisable under the new procedure to prosecute in all instances to resolve the position *vis-à-vis* employment of the individual at the earliest possible time. The police will certainly not be sympathetic to a case being belatedly referred to them because complications over a dismissal for dishonesty have ensued.

It may be advisable to make an inflexible rule where a particular major risk exists. Where there is a prime objective for thieves in a particular industry—for example, copper—it would be understood that anyone who removes it without authority does so at his peril and prosecution will follow at once irrespective of the quantity involved. This is especially essential where large numbers of employees have a constant temptation before them in the form of easily disposable material of any nature.

In matters of dishonesty, trade unions are as responsible in their outlook as management and perhaps even more dogmatic as to how it should be dealt with. Where an employee's personal property is concerned or where he has sustained injury in any way, before any action is taken, particularly involving the police, his agreement must be obtained.

The question of what should be reported to the police is worth considering. Again, the firm's interest is paramount. If the police are brought in, they are unfamiliar with procedures, persons, and places, all of which have to be explained; their presence automatically attracts attention from workers, and men may have to be taken from productive jobs to be questioned. The loss, from the point of view of production, may be out of all proportion to the original theft. A senior member of the security staff should be capable of carrying out most internal inquiries himself. Again, providing the security officers have an accepted status and ability, unions and their members are quite prepared to cooperate in every way.

If intruders are involved in a matter of some consequence, the police might as well be informed from the outset—there is no other avenue of detection and prevention of a recurrence. Where identifiable property has obviously been taken outside the works, again, the police have the only means of recovery. Factors to be taken into account are, of course, value, possibility of identification and recovery, and the inconvenience that police inquiries might occasion.

In connection with prosecution and identification of recovered property, lengthy periods can be spent unprofitably in courts so, if evidence for the firm can be given by a competent junior, it will save the time of a key departmental head. Under the new court procedure of the Criminal Justice Act 1967, personal attendance by witnesses has been much reduced (see pages 143-4).

There are circumstances which can militate against instigating a prosecution that might be merited by the nature of the offence, such as:

1 Will the firm receive adverse publicity by:
(a) A lax system which has thrust temptation in the way?
(b) Having entrusted a low-paid man with too much responsibility?
(c) The ventilation of a grievance in open court, which would show the firm in a bad light?
(d) The compulsory disclosure of matters in court which it would be in the firm's interest to restrict?

2 Will a weakness of the security system be spotlighted and possibly exploited by others?

3 Will the end product of a prosecution be worth the effort and inconvenience? It is not generally recognised that, for trivial offences, the 'sentence' of a court may be a conditional or absolute discharge. (When an accused person is found not guilty at a magistrates' court, he is dismissed.)

4 Considering the age, length of service, and good repute of an employee, would dismissal be a greater penalty than a court might impose?

It must be remembered that malpractices, actual or alleged, by an employer have greater news value than a routine theft by an employee.

If a management is reluctant to lay down a positive ruling that dishonesty incurs dismissal there are several aspects they ought to consider before agreeing to the retention of an offender:

1 Will the leniency extended encourage others to emulate, with the impression that the firm does not regard theft of its property as being of particular importance?

2 When a precedent of leniency has been established, how will the management be able to vary the procedure and then explain their actions if challenged?

3 Should an offender fall from grace again, who will accept the onus of explaining to, say, a board of directors why someone,

who has previously shown lack of honesty, has been given the opportunity to further defraud the firm?

4 Is an offender, knowing that those around him are aware of his dishonesty, likely to be able to work to his full capacity in the firm's interests?

5 How will the fellow workers of the individual regard him thereafter? There will be inevitable suspicion on him when anything goes astray.

6 What is the atmosphere in a department likely to be if a fellow employee's property has been concerned?

Although an inflexible attitude to offenders has its difficulties, reluctant management should have the preceding considerations in mind when arriving at a decision less than dismissal. Above all, even trivial thefts should not be condoned by taking no action. This gives an invitation to others to do likewise.

A person is not guilty until proved so and a firm's action where a person denies the facts alleged against him is well worth prior thought. The issue is straightforward where a person admits an offence and the circumstances put his guilt beyond doubt. An employer, if it is so desired, could dispense with his services forthwith but it would now be advisable to obtain from him a clear written admission of responsibility in an uncontestable form.

Where the theft is denied, even though the management consider they have adequate proof of dishonesty and intend to dismiss in face of strenuous denial, they should think several times before doing so. Lengthy complications could ensue of which the Industrial Relations Act procedure is only one. It would be far more philosophic to take the view that the person has brought this matter on himself and refer it to the police without further qualms.

If the police, as an independent third party, think the issue is self-evident and they are prepared to arrest and charge the person at the time, irrespective of denial, the company could feel justified in suspending him from work until the hearing of the case. Whatever payments should be made in such instances would be for discussion and agreement by the employer and union representatives. Legal opinion is that suspension without pay may only be imposed if the right is clearly written in the employee's terms and conditions of employment.

It is more difficult where the police wish to consider explanations or obtain further evidence before taking legal action. If the employer has deemed it necessary to call in the police, and continues employment of the man during the interim period, this action implies a degree of trust which gives defending counsel a talking point which he will exploit to indicate doubt of guilt. There is little option but to suspend with full pay and urge the police to expedite proceedings as much as they can. This is important because with present-day court congestion, delays of many weeks can occur and a firm could find itself in an invidious position of having paid heavily in wages for no return to a person who could eventually turn up and actually plead guilty to stealing from it. There may be after all cogent reasons for putting a clause into contracts of employment dealing with suspension without pay. It would certainly be wrong to suspend without pay while the police were merely considering whether proceedings should be taken.

In the interests of good industrial relations, and goodwill towards the security department itself, an employee's union representative should be informed at an early stage of action being taken against one of his members—this notification might perhaps be best done by personnel or industrial relations departments, and it would be as well for each firm to establish an agreed practice.

Where a serious contravention of company discipline accompanies the alleged dishonesty or other criminal offence, and is admitted, it may be advisable to deal with this forthwith rather than wait the result of any court proceedings which may be taken. Problems concerning the wages of an employee awaiting trial either at a magistrates' or superior court would be obviated. A decision to dismiss on disciplinary grounds after a court acquittal on the criminal charge would be criticised as 'persecution' even though justified.

The actual mechanics of instigating and carrying through a criminal prosecution are best left entirely to the police whose professional business it is. The law does not preclude any person or body from instituting or carrying out any criminal proceedings in which police will assist but, unless there are very special reasons, there is little point in acting as a private prosecutor. Once notified, the police will ensure that all necessary statements are taken, the case is prepared, and witnesses are warned of the time, date, and place of hearing. Their representative will present the evidence and there can be no suggestion of bias in the conduct of the whole affair. In other words, a minimum of inconvenience and cost is incurred by the firm.

Concealing offences: not reporting thefts

Until the passing of the Criminal Law Act 1967, it would have been 'compounding a felony' to agree with a thief not to prosecute him if he returns the property or the value of it in lieu. 'Misprision of felony' was the offence of concealing the commission of a felony by not reporting the facts to the police. The 1967 Act abolished the classification of offences into felonies and misdemeanours. Hence the offences of compounding, and misprision of, a felony no longer exist. Offences are now classified as arrestable and non-arrestable. Theft and wilful damage are arrestable offences.

Parliament recognised that many thefts, particularly by employees of their employers' property, are not prosecuted and, by section 5(1) of the Act, legalised agreements such as have been described. The wording of the section is:

> Where a person has committed an arrestable offence, any other person who, knowing or believing that the offence or some other arrestable offence has been committed, and that he has information which might be of material assistance in securing a prosecution or conviction of an offender for it, accepts or agrees to accept for not disclosing that information any consideration *other than the making good of loss or injury caused by the offence, or the making of reasonable compensation for that loss or injury,* shall be liable to conviction on indictment to imprisonment for not more than two years.

Assisting anyone to avoid arrest or prosecution

What used to be referred to as being an 'accessory after the fact' (the commission of a felony) is now described in section 4(1) of the Criminal Law Act 1967:

> Where a person has committed an arrestable offence, any other person who, knowing or believing him to be guilty of the offence or some other arrestable offence, does without lawful authority or reasonable excuse *any act* with intent to impede his apprehension or prosecution shall be guilty of an offence.

Failure to inform the police of a theft, in our view, is an omission and not an 'act' as described and therefore no offence would be committed if this were done.

No proceedings may be taken under sections 4 and 5 except by, or with the consent of, the Director of Public Prosecutions.

Offering rewards for return of stolen property

Where any public advertisement of a reward for the return of any goods which have been stolen or lost uses any words to the effect that no questions will be asked, or that the person producing the goods will be safe from apprehension or enquiry, or that any money paid for the purchase of the goods or advanced by way of a loan on them will be repaid, the person advertising the reward and any person who prints or publishes the advertisement shall on summary conviction be liable to a fine not exceeding £100 (Theft Act 1968 section 23).

Power of arrest

It is convenient to refer here briefly to section 2 of the Criminal Law Act 1967, which provides everyone with the power of arrest for theft and attempts to commit the offence. This is developed in Chapter 7.

LOSS FROM OTHER CAUSES

Fire

Another risk which has to be recognised and prepared for is the loss of life, premises, or goods by fire. The increasing seriousness of losses from this cause can be seen from the United Kingdom Fire Statistics 1965, issued by the Ministry of Technology and the Fire Offices Committee, wherein it is pointed out that between 1957 and 1965 fires from all causes in buildings increased by 99 per cent. Incidentally, the report claims that increase in crime over the same period was 108 per cent.

The number of fires started deliberately, under the description of arson and malicious damage, is said to have quadrupled in that time from 50 to 200, and no doubt a number of those described as 'cause unknown' would also have come into that category if all the facts were known. Arson is deliberately and with malice setting fire to premises or property therein (see Chapter 10).

The British Insurance Association claims that in 1958 the value of direct losses from fire was £24 000 000 whereas in 1971 this had risen to £128 700 000. Of course, the changes in the value of money will account for some of the increase.

Ways and means which can be adopted to prevent and detect fire will be referred to. The law on the installation of fire-fighting equipment, means of escape, etc., will be outlined in Chapter 26.

Damage to property from other causes

This is another aspect of the security problem. Damage to valuable and sensitive machinery, for example by the elements, might result in long and unprofitable delays in repairs and replacements. Raw materials, finished goods, stores, and so on, also require to be protected against the effects of bad weather. Provision must be made for the early discovery of defective roofs, windows, gutters, etc., whereby damage may be caused. The stacking of goods so as not to interfere with the operation of sprinklers must also receive attention.

Criminal damage, in some circumstances called sabotage, to property by disgruntled employees and other persons is dealt with in Chapter 10.

Waste

The loss of material due to wasteful treatment or carelessness is another field on which security attention can be focused. Examples are the turning off of valves controlling liquids or steam which have been either carelessly or deliberately left on. Although individually of small consequence, stopping such wastage can represent considerable savings in total. This can even extend to water taps where the usage of water is metered for charging purposes and the switching off of unnecessary lighting or heating.

EMERGENCY OR DISASTER PLANS

The preparation of plans, sometimes described as 'disaster plans', to cope with emergencies arising through natural phenomena, such as floods, with, possibly, accompanying breakdowns in public services, is essential. These should be distributed to all persons required to initiate actions and be in readiness at the designated control centre.

Considerable disruption can also arise following information, usually in the form of a telephone message, that a bomb or some form of explosive device has been placed on premises, which can be controlled if the eventuality has been prepared for. Instructions are required for the telephone operators regarding the action they should take on receipt of such a call. A plan of action regarding evacuation and searching of premises must also be drawn up. Guidelines in this connection will be found in Chapter 28.

COMPANY SECURITY STAFF

Where men and/or women are to be employed as security

officers it is essential that they are of the right calibre and enjoy good status and conditions of service, to accord them the respect that is essential if they are to carry out their duties to the best advantage.

Morale is increasingly important and an efficient security staff must be capable of earning the support and respect of the whole working community while at the same time producing an image of authority which effectively deters dishonesty or disobedience on the part of the minority. This is a demanding role and one in which prevention rather than cure will be of greater importance than in the past. These observations are enlarged upon in Chapter 3.

The duties required of the staff must be committed to writing in the form of standing orders. These are illustrated in detail in Appendix 3. The operation of the duties is described in Chapter 4 and the contribution security staff can make in accident prevention is shown in Chapter 27.

The duties of a modern security officer will be better performed if he receives some formal training in them. The instructional facilities which are available are dealt with in Chapter 5, which gives the subjects contained in the curricula of the training courses.

SECURITY SERVICES COMPANIES

Not every commercial or industrial concern requires the whole-time employment of their own security personnel but circumstances very often require personal attention to be given to security risks—for example, patrolling premises when closed, to prevent and detect fire. Chapter 31 describes some of the services which security companies can supply. These include provision of security personnel to guard or visit premises, investigators to inquire into suspicious stock losses, and so on, store detectives, keyholding and the transport of cash and other valuable property. Precautions which should be taken before and after engaging their services are also mentioned.

CRIME PREVENTION OFFICERS

Management concerned for the security of its property should not neglect taking advantage of the services available free of charge from the crime prevention department of their police force. In recent years the important responsibility of preventing crime has received increasing attention from police authorities. The Home Office Crime Prevention College at Stafford holds instruction courses for all ranks of the Force and for members of other organisations having a special interest in the subject, for example representatives of insurance companies.

Crime prevention officers will be pleased to put their knowledge at the disposal of anyone who requires assistance on the protection of their property against crime.

FIRE PREVENTION OFFICERS

Another source of assistance is the local fire brigade where expert information on prevention of fire is available free of charge on request.

PROFESSIONAL BODIES

Industrial Police and Security Association

This is the only organisation with United Kingdom and overseas coverage for persons employed in the security profession. It is not a negotiating body with respect to conditions of employment nor has it political affiliations. The aims, objectives and membership structure are described in Chapter 34. One of its principal activities is arranging training courses for security personnel, which have been referred to in previous paragraphs.

Institution of Industrial Security

Membership and graduate membership is granted to members of the above Association who have experience in commercial and industrial security for prescribed periods and who fill the requirements of the Governors after they have passed the relevant examination in professional subjects. More about the Institution will be found in Chapter 35 where the by-laws are printed.

2

Responsibility for security

Every commercial concern, no matter how large or small, should have a member of the management who has first responsibility for the security of its property and the safety of its employees and their property whilst engaged at work. Notwithstanding such an appointment it must be emphasised to managers and supervisory personnel primarily concerned with other duties that they, too, have a responsibility for the protection of their employer's property from loss through any means. One company keeps a record of the departments from which property is discovered to have been stolen through the action of the civil police or its own security staff. The person in charge of the department concerned is informed of the discovery and he is expected to take action to prevent a recurrence. Should a theft again occur in that department which reveals that no preventive action has been taken, he is called before senior management to explain what he did on being told of the previous theft and how, in spite of that, this happened again. Where the circumstances call for it, disciplinary action is taken. It is a misconception that the security department is responsible for security. It is there to assist management in all grades in the exercise of that responsibility.

The status of the person with special responsibility for security will depend on the size and importance of the concern and this can range from a member of the board of directors downwards in the operational structure, according to the importance of his responsibilities. For instance, at a factory engaged on government classified contracts it is likely that a director would be responsible with full-time security officers reporting to him from the different units of production. In other circumstances it could be the senior man on the site, such as the factory general manager who would be assisted either by a full-time security officer or a manager who would undertake security duties on a part-time basis.

Whether the person given the responsibilities is untrained in security practices and techniques and carries them out with other duties, which quite understandably he looks upon as more important, or is a fully qualified person in a full-time capacity, depends on a number of factors. These include the vulnerability of the concern to losses through its physical size, location or the type of property which is at risk. For example, the security measures at a small food factory would very likely require to be more comprehensive than, say, a large steel works. Whether the premises have a perimeter fence or wall or are an open site also has a bearing on the type of security required.

The person in charge of security must keep in touch with the *modus operandi* used in recent crimes, through newspaper reports and professional journals. He must keep up to date with the methods of fire prevention applicable to any process which is being carried out and the causes of recent fires. Fires which have occurred in circumstances having a similarity to his employer's should be followed by an inward look at the precautions already taken to discover whether these have weaknesses which require to be strengthened by improved procedures.

Exhibitions of security equipment and technical advances, such as the use of radio and television and specialised electronic devices to prevent crime and fire and to apprehend offenders, should be attended. Sophisticated mechanical aids to security are constantly being invented with the principal intention of replacing expensive manpower without loss of security coverage. Liaison with the local police and fire brigade at a high level must be achieved—this can be beneficial to all parties in the long run.

APPOINTMENT OF SECURITY SPECIALIST

When anyone primarily employed on other work is delegated the security responsibility, the attention he will give to it depends on the time he can spare from his main duties. Under pressure of the principal work, the subsidiary duties can suffer and become neglected. When essential security responsibilities interfere unduly with his effectiveness in his principal occupation, the time has come to consider the appointment of someone qualified for the security responsibilities on a full- or part-time basis. In the latter capacity, other responsibilities having some affinity with security and other duties concerned with the administration of the site can be assumed and these will be referred to again when describing a security services department.

The description given to the holder of the office can be 'security officer', or 'chief security officer' if he is responsible for other personnel engaged on security duties. In some concerns the employment of security staff in different units calls for the appointment of a company or group security officer. Their special and additional responsibilities will be described later in this chapter.

Before making the appointment of a security officer it is advisable to obtain the opinions of the heads of departments with whom he will be likely to come into touch during the course of his duties on the type of man they would find most acceptable. They could also indicate the fields of activity in which they would like to employ his services. From this information—and possibly advice from other concerns employing a security officer—the duties which he will be expected to perform can be decided. What is required before the appointment is a list of his duties, a decision on who he will be responsible to and where he fits in the organisational structure of the concern. Against this job description the suitability of the applicants can be judged and they can have the position fully explained at the time of interview. This will remove the possibility of later misunderstandings.

JOB DESCRIPTION

It is essential that this be prepared in writing. It should describe in broad terms what the employer expects of the security officer or chief security officer. It will have no reference to the salary for the job, which is a personal matter arranged between both parties. It is recommended that before an interview an applicant should be given the opportunity to read the document; if he then wishes to withdraw from consideration for any reason, time would have been saved. The position should be shown as having a 24-hour responsibility and the minimum period of duty expected to be carried out should also be mentioned. The document should explain that he would be expected to visit the premises at all hours at reasonable intervals to ensure that the security precautions are satisfactory.

To ensure the legality of the descriptions of the duties it is recommended they be seen before issue by a legal representative of the employer. An example of a job description is given in Appendix 1.

So that all members of the management are aware of the responsibilities of the security officer, they should see the job description. It must be made clear that no alterations or additions to the documents will be allowed except by the agreement of the person responsible for its compilation. A copy should be given to the security officer for his retention.

Position in the organisation

The person that the security officer reports to will depend on the size of the employing firm and whether there is more than one production or other units. The principle held strongly by those experienced in professional security duties is that the security officer should be directly responsible to the senior member of management in whose area his responsibilities fall. If, however, that person is responsible to someone higher elsewhere in the organisation, the security officer should, when the circumstances justify it, have the authority to approach that person directly. What is clearly undesirable is that the person in charge of security should be required to go through an ascending series of subordinates who, possibly having no experience of security work, may wish to sidetrack or minimise the results of his efforts. Further, he may have to report on and to receive instructions on matters which concern their integrity or that of their friends or associates of which, of course, they should not become aware.

However, for day-to-day matters it is usually convenient for the security officer to report to a less senior manager than has been described and through whom requests for his security services can be channelled. This is often the personnel manager, but before he is chosen it is well to consider the special position he holds in respect of good industrial relations and his welfare responsibilities. Should he be expected to sit in judgment on a breach of security involving an employee and to make recommendations regarding any further action, he could be influenced, consciously or not, by a desire not to be associated with any sanction which might have a bad effect on those relations. Therefore a manager with other responsibilities is in a better position to view such a situation dispassionately.

SALARY

This should be the rate for the job and have a favourable relation to other salaries paid. In arriving at the commencing salary and the maximum which can be reached, sometimes called the maximum job value, no consideration should be given to the possibility that the successful applicant may be in receipt of a pension from a previous occupation. The particular knowledge he is required to possess will probably have been acquired in a superannuated position.

Applicants for the job should be told the commencing salary and that it will be reviewed annually and, if justified, increases will be given up to a maximum figure, but they should not be told that amount. The security officer should, however, be informed when increases have brought his salary to the maximum amount.

STATUS

The status to be given to the security officer must reflect the degree of importance the management attaches to his position. It is recommended that he should be a member of the management structure in a position equated with other management roles, with the same maximum salaries. This will also assist him in his dealings with managers in general. He should enjoy all the privileges accorded to the grade he is placed in, such as membership of a pension scheme, allocation of a company car when this is usual, and so on. He should wear civilian clothes.

In no circumstances should he be called upon to carry out the duties of a subordinate on his staff at times of shortages of security staff. Other means must be adopted to provide the coverage required.

EXPERIENCE

When the duties of the position have been listed, and the status and the salary to be offered have been agreed, the experience and other requirements of the successful applicant have to be decided. The type of security risk which predominates will have a bearing on what is required. However, whilst an otherwise qualified and acceptable candidate for the post may not have sufficient experience in that particular field, before he is rejected on those grounds it should be remembered that training facilities exist for providing instruction in that deficiency. Chapter 5 gives details of courses in security duties and in Chapter 3 are the names and addresses of organisations who provide training in fire duties.

The functions of a security officer can be concerned with the prevention of losses of property through crime (also from waste and damage) and the detection of offenders, and the prevention and fighting of fire. Although those responsibilities frequently are combined in one position, called the security and fire officer, in concerns with a specially high fire risk the responsibilities for the latter duties are often given to a fire officer. His place in the structure of a security services department is shown on the specimen organisation chart in Appendix 2.

A sound knowledge of the law as it affects the protection of property and the rights the owners and their authorised agents have to apprehend and detain offenders is important. Where the risk of fire is present a good knowledge of the sections of the Factories Act 1961 and the Offices, Shops and Railway Premises Act 1963 concerned with the prevention of fire and the protection of employees from it is essential—so is a thorough knowledge of fire-fighting techniques and the equipment available.

The present sources for the recruitment of men with that knowledge and experience are mainly the civil police forces of the United Kingdom and overseas and from the armed forces where experience has been gained in security work. Those with experience of fire usually come from local-authority or industrial fire brigades, although, again, some may come from the forces.

Service in the local police force has advantages, including knowledge of local criminals and persons of bad character, which is useful in the performance of security duties and in excluding from employment undesirable and disruptive elements.

Where the post includes the responsibility for the leadership, control and training of a security staff it is preferable that the holder should have had some earlier experience of a similar nature. In other words, he will have reached a position of command in his previous position. At what level it is preferred will depend on the status and importance of the position it is wished to fill.

PROMOTION

At present there are very restricted promotional opportunities in the security profession as the senior positions are usually taken by experienced men in the older age bracket who remain with the same firm until they reach pensionable age. However, the recognition in recent years by commercial and industrial undertakings of the advantages of a high standard of security should encourage young men with good educational background and other attributes to enter the profession with intent to attain the higher ranks. In the probable absence of sufficient opportunities for promotion with their immediate employers it is to be expected that they will seek advancement with other concerns.

Continuity of employment of security staff of all grades has great advantages, as through it they accumulate a comprehensive knowledge of the procedures and administration and the security risks of their employers. Knowledge of the personnel employed of all grades with their peculiarities and sensitivities, necessary in the efficient performance of security duties, is also acquired. This continuity is more likely to be achieved if older, more mature men are appointed who, if the conditions of employment are satisfactory, will not wish to leave for other positions.

HEAD OF A SECURITY DEPARTMENT

The comments and recommendations made about the duties, status, and salary of the security officer also apply to a chief security officer, company security officer or group security officer. All have the responsibility to advise their employers on security matters and the executive authority to carry out and enforce security measures which have been approved.

Chief security officer

This appointment is usually made where several security personnel are employed who cannot be supervised properly by anyone with a part-time responsibility for security. The holder can also be responsible for ancillary staff carrying out duties having a bearing on the administration of the premises. Examples of these are shown on the organisational chart in Appendix 2. He might also be responsible for fire precautions, but as has already been said it is not unusual for those to be undertaken by another person.

The chief security officer would have the same responsibilities as a departmental manager for the efficient and economic performance of his department. He would be responsible for the preparation of budgetary information to assist the management. He would hold annual reviews with members of his staff and make recommendations when justified for increases in salaries within the scale applicable. He would prepare or approve reports on incidents for submission to higher management. In circumstances where there is a security staff but which do not justify the appointment of a chief they should be answerable to a specific individual in the management structure who is allocated the security responsibility.

Company security officer

This appointment is usually made where a concern has a number of production or other units with or without security staffs. In the former instances there may be a chief security officer or a security officer in charge; to prevent confusion they are usually described as factory or site security officers.

The company security officer is responsible for their supervision and for ensuring with his professional expertise that the security arrangements are satisfactorily carried out, especially when they are not under the care of a full-time professionally qualified person. He is also responsible for coordinating the security arrangements and seeing that there is a consistency in application and performance. Another duty would be to collate the budgetary information from the units for consolidation and submission to the management concerned. He would hold annual interviews with his factory or site security officers and make recommendations on salaries, promotion, etc.

Group security officer

This appointment is to be found where there are a number of concerns with separate corporate identities in a trading group. He has the responsibilities of the company security officer but on a larger scale.

SECURITY ADVISER OR CONSULTANT

Where particularly large companies or groups of companies are concerned these appointments are made to advise on security matters. The holders usually have no executive authority but carry out independent surveys and inspections of existing security measures and the performance of security personnel and prepare reports and recommendations for the central or associated boards.

Their professional advice is sought in any matter involving a security risk.

LIAISON WITH POLICE AND FIRE AUTHORITIES

Everyone responsible for security, whether full or part time, professionally qualified or not, must develop and maintain contacts at a high level with the local police and fire authorities. They will thus have the benefit of advice as necessary in security methods and reciprocally they will keep the police advised of any special temporary risks, such as the holding or expectation of receiving a larger amount of cash or attractive property than usual, and the fire authorities advised of the nature and localities of high fire risks. Crime prevention and fire prevention officers have been appointed to assist commercial and industrial concerns with their particular knowledge.

COMMUNICATIONS AND OFFICES

Office accommodation must be suitable for the position of the person in charge of security and provide some degree of privacy for the interviewing of callers and others. The amount of clerical assistance required will depend on the extent of his responsibilities, but this must always be available as a right. Besides the extension from the general switchboard he may require a separate telephone through which he may make and receive confidential calls. He should be on the telephone at his home so he may be communicated with at all times in order to take charge of any incidents within his responsibility. A contribution should be made by employers towards the rental of the instrument and the cost of outgoing business calls.

SECURITY COMMITTEES

Security arrangements cannot be carried out satisfactorily without the understanding and cooperation of all employees, particularly those in positions of responsibility. To encourage this, whoever is responsible for security should arrange meetings of heads of departments or sections to report on breaches of security and to seek their assistance in the observance of security measures which have been introduced. These will also provide an opportunity for the discussion of grievances of the staff.

TOTAL LOSS CONTROL

This is the description given to the coordination of security, accident prevention, safety, fire prevention, damage control and insurance functions, in commercial and industrial concerns, to prevent and reduce losses. It is sometimes referred to as 'total risk management'. In the larger organisations the various loss prevention responsibilities are coordinated under a full-time total loss controller to whom those responsible for the separate functions described report.

Another way of securing the same result but without the appointment of such a person is to have periodical meetings of those officials under either a permanent chairman or one drawn from the members by rotation.

The permanent inclusion of departmental managers or individual ones as required when a matter affecting their responsibilities is to be discussed has advantages. Where an internal audit department exists the person in charge should be invited to the meetings to collaborate in the assessments of risks and in introducing measures and controls towards the common purpose.

SECURITY SERVICES DEPARTMENT

In large industrial premises a multiple number of services are necessary for administration and the protection of employees and their property, in addition to a security service. These services can be brought together to form a security services department. The services could include:

1 Security, including, where employed, the security staff.

2 Accident prevention.
3 Fire prevention, including the maintenance of fire-fighting equipment.
4 Industrial civil defence.
5 The issue and sale of protective clothing and safety footwear.
6 The laundry and repair of company clothing.
7 The upkeep of notice boards and directional and other notices.
8 The cleaning and maintenance of drains, roofs, paths and roads.
9 Upkeep of gardens and decorative plants.
10 Cleaning and attending lavatories and dressing rooms.
11 Timekeeping.
12 Weighbridge operation.
13 Good housekeeping.
14 Visitors' reception, guides.

The head of the department could be called the 'security manager' with assistant managers or supervisors (foremen) in charge of one or more of the separate services. The organisation of such a department is shown in Appendix 2.

Reports of department

Senior management is not normally in contact with the security force. To keep them informed of its activities so that they are better able to evaluate performance against cost, whoever is in charge should prepare a comprehensive report for their information. The report can be submitted quarterly, half-yearly or annually depending on the extent of the duties performed or the size of the force and should show what services have been performed, and incidents of note which have occurred. Where the costs for security are budgeted, the report could include actual costs for the period concerned for comparison purposes.

3

Employment of security staff

When security staff are being selected, men of mature age who are physically fit and above average height with proportionate build are to be preferred. The older man is more likely to be circumspect in the performance of his duties and thereby command greater respect. Furthermore, he is less susceptible to corruptive influences. Married men with sufficient domestic responsibilities to confirm that they want to stay in a stable job are preferred to single men. Frequent changes in the security staff are to be avoided as much as possible. The men appointed must be tactful, amenable to discipline, capable of exercising discretion within given limits, self-reliant, and able to act decisively in an emergency.

No reflection is intended on the ability of men who have lost limbs or who are severely crippled or infirm to perform security duties of a sedentary nature, but they should not be employed in circumstances where such a disability could have serious consequences, for example, where they are required to exercise physical control over anyone in the course of their duties or in self-defence when under attack. It is unfortunately not uncommon for the security staff, which might consist of one man, in charge of premises to be attacked and overcome.

RECRUITMENT

Through advertisements in national and local newspapers, and the relevant periodicals, vacancies in a security force will reach interested persons. Another possibility is Army, Royal Navy or Royal Air Force depots, which welcome information on opportunities for the employment of men about to leave the forces who may be the right material for the type of work.

Police officers on their retirement are not usually prepared to take up security appointments where shift duty is carried out. After probably many years of that type of work they prefer employment which will give them regular hours.

Transfers

Transfers of members of the general labour force to the security staff are not recommended in principle. Long association with other employees could interfere with their impartial application of security requirements. Exceptions might be made where the man in mind has first-class qualifications for the work and is known to be of the type who would loyally carry out his responsibilities and would not be likely to be affected by previous associations. The advantages of recruitment of such men is that they will have knowledge of the processes and of the workers with, possibly, better acceptance by them. On balance, however, the external recruitment of men who compare favourably with those under consideration from inside is to be preferred. They are less likely to be involved in personal relations which could affect their efficiency.

References

References of applicants must be thoroughly verified before appointments are offered and every effort made to see that only men of unimpeachable character are appointed. The expression 'subject to satisfactory references' qualifying an offer of employment should not be used. If, after the appointment, the references turned out not to be of the standard which would be expected in the special circumstances, an extremely difficult position would have arisen.

A history of frequent changes of occupation without reasonable excuse or unaccounted-for breaks in employment would exclude an applicant from appointment. Risks must not be taken in the employment of security staff, who are to be placed so often in positions of temptation.

Unfortunately, instances have occurred of security officers being charged with criminal offences involving the property of their employers. Such breaches of loyalty are not, of course, confined to men in their positions. Police inquiries, not available to employers, have revealed that the accused have had previous convictions for crime and some have served periods of imprisonment. These disclosures are criticisms of their employers who could not have thoroughly examined their backgrounds before employing them. Such exposures which often get headlines in newspapers bring discredit on the security profession as a whole, which in the great majority of cases is quite undeserved. Instances have occurred of a security officer being known by employees in the same firm to have some discreditable background. In return for favours, which can be imagined, they have promised not to disclose this to their employers.

ATTENDANCE RECORDS

Security officers are required to be of high integrity and worthy of trust and therefore it would be absurd for them to be required to use a time clock to verify the commencement and termination of their periods of duty. The general works labour force who use the time clock would conclude that their employers consider the security staff as being no more worthy of trust than themselves. There should be an attendance register in which the security staff record by signing the times of their starting and finishing a period of duty. Overtime can be calculated from that information.

UNIFORMS

One of the primary questions which has to be decided when

ordering uniforms for the security staff is what colour are they to be. Navy blue is the most popular because of its serviceability and because of its similarity with the colour worn by the forces of the law, the civil police. By this it is hoped that security staff will attract similar respect as symbolising law and order within the areas of their responsibilities.

However, the Police Act 1964, section 52(2), says:

> Any person not being a constable wearing any article of police uniform in circumstances where it gives him an appearance so nearly resembling that of a member of the police force as to be calculated to deceive shall be guilty of an offence and liable to a fine not exceeding £100.

Article of police uniform means 'any article of uniform or any distinctive badge or mark usually issued to police forces or special constable or anything having the appearance of such article, badge, or mark'. 'Calculated' is not free from ambiguity but legal opinion is that it means 'likely'.

This section of the Act was introduced to control the wearing in public places of uniforms having a great similarity to police uniforms due to the increase in the use of uniformed personnel employed by security service companies. When going to and from their assignments they have to pass through the streets and as so many of their uniforms could easily be mistaken for that of a police officer something had to be done to reduce that possibility. One result of the new law is that the companies referred to have made changes in or additions to the uniform of their men so there is now no doubt as to their identity.

With that in mind, if a uniform worn by security staff of a commercial or industrial concern is likely to be mistaken for that of a police officer it must not be worn outside the premises at which the wearers serve. If their duties require this, for example, escorting money to or from a bank or traffic control at factory entrances, the uniform should be of some distinguishing colour—dark green, maroon, or grey are practical as alternatives. If navy blue is still preferred it must bear some identifying badge and/or shoulder flash with the word 'security' on it. A cap badge incorporating the firm's name or its initials would also help in that direction. Stripes of rank should be inverted and for the same purpose metal bars should be substituted for stars on shoulder straps.

One of a security officer's best tools is his uniform. It should be, and well can be, an excellent symbol both of identification and authority and assists him greatly in his dealing with the public, other employees and anyone contravening the law as it affects his employers.

Uniforms must be of good quality and fit so as to encourage the wearers to take a pride in their appearance and be a good advertisement of the firm to callers. The use of second-hand uniforms purchased from outside sources is deplored as having a bad effect on the morale of the staff, and this will obviously reduce efficiency. Uniform should be renewed after specified periods of wear. Trousers for example have a shorter life than overcoats. The older issues of uniform can be worn for night duty with subsequent issues worn on day work when a good appearance is more important.

Standing orders to the security staff (which will be dealt with on page 52) should include instructions respecting the reporting of damage to uniform or equipment and pointing out that the employer reserves the right to assess the wearer for the cost or part cost of replacement or repair of the garment or equipment if it has been caused through negligence.

DESCRIPTION OR TITLE

The titles given to security personnel include:

Security Officer	Works Policeman
Security Guard	Gateman
Security Warden	Gatekeeper
Security Patrol	Watchman

Which one is used can depend on the status accorded the holders and, strangely enough, to some degree on the part of the country where they are employed. For instance, the description 'works policeman' is used more in the north of England than in the south where, in circumstances which justify its use, security officer is more usual. The use of the other names including the word 'security' is not unusual. There is a substantial weight of opinion in the security profession against the use of the word 'police' in the description of security staff on the grounds that it has a connotation which might not be conducive to good relations with other employees. It is interesting to note that a high proportion of the objectors are ex-police officers.

Where the high importance the employing firm attaches to its security staff is reflected by good payment and other conditions of employment the description security officer is most fitting. It properly describes their responsibilities and encourages self-respect. The name 'watchman' should not be used since it is regarded as being particularly derogatory and related to the coke fire by the hole in the road. This description, however, might be given to men, usually aged and infirm, not supplied with uniform whose function is limited to patrolling premises at night and at weekends to prevent and detect fire.

THE STATUS OF THE SECURITY OFFICER

Although security officers, by which name all security personnel will hereafter be described, are not formally of supervisory grade they are in a special position which carries almost supervisory responsibilities. To assist them in carrying out those duties they should be of staff status and enjoy all the relevant privileges. Some reasons are:

1. The security function necessitates delegation of management responsibility beyond that normally given to works employees.
2. It requires a recognised authority which only the staff connotation provides.
3. Security officers on staff conditions are likely to be less involved in issues of general concern to the workforce.
4. The type of man required is more likely to be attracted by staff conditions.

The last reason is especially important because good working conditions can compensate to some extent for the lack, except on rare occasions, of promotional prospects within the security department.

Promotion

Within the security department—Whether there are ranks within the department depends on the number of men employed. Where each shift performs duty in rotation this can mean that usually at night and at weekends they do it without the direct supervision of the chief security officer (CSO) or other person in charge of security. Any decision in circumstances calling for further action is usually taken by the man on duty who is senior in service.

To give each shift someone, irrespective of service, selected for that responsibility, the appointment of senior security officers should be considered. This would provide promotional opportunities for the staff leading to better pay. The rank could be shown on the holder's uniform by a chromium-plated bar on the shoulder straps or by three inverted stripes on the sleeve.

Outside the security department—When selecting men who have shown the right potential for training for supervisory positions the security staff should not be overlooked. A reputation for fairness and consistency in the performance of his duties should make a security officer acceptable to the general workforce in a supervisory capacity after any necessary training.

Trade union membership

In the interests of absolute impartiality, it would perhaps be better for security officers not to be members of a trade union, though this would deprive them of the advantages of various benefits and representation when personal interests were threatened, as would arise in connection with compensation after accidents and unjustified complaints. The nature of a security officer's work gives a clear priority to the interests of his employers and the danger is that union membership, especially where a man became actively involved, might induce him, or perhaps expose him to pressures, to act other than he should. He has almost unrestricted access to offices during quiet periods and may acquire, accidentally or otherwise, classified information management would not wish to be divulged—there must be no clash of loyalties in those circumstances.

However, the advent of the Industrial Relations Act 1971 with its provisions for agency shop agreements gives a clear indication that union membership in some form may well be obligatory in the near future in many organisations. In such cases, membership of a staff union is advisable; this will be in keeping with the status of the function and will maintain a distinction from the general workforce, with whom most daily contact is made.

DUTIES DURING STRIKES

Where security staff have not been union members, no difficulties have arisen in respect of their non-participation at times of industrial action; it has been accepted that they will continue their normal duties but do no additional ones. This is the attitude of responsible unions when the men are in fact members though the exclusion is almost always a verbal agreement at official union level and not necessarily endorsed by local shop stewards at the times of unofficial disputes.

Several acrimonious disputes have followed a return to work when security personnel have not participated in strike action. Their position has been somewhat clarified by paragraphs 21, 22 and 23 of the Code of Practice. These state that some employees have special obligations arising from membership of a profession. While they should respect their union obligations, they should not be called upon by the union to take action which would conflict with the standards of work or conduct laid down for their profession if that action would endanger:

1. Public health or safety
2. The health of an individual needing medical or other treatment
3. The well-being of an individual needing care through the personal social services.

Professional associations, employers and trade unions are asked to cooperate in resolving any conflicts that might ensue.

Security officers, by and large, now have a first-aid responsibility in addition to others relating to safety of premises and individuals, and the term 'professional' can nowadays be applied to their functions. Therefore it seems probable that non-participation in strike action is even less likely than formerly to be condemned.

Strikes are dealt with elsewhere but it is as well to emphasise here that security responsibilities stop at the perimeter of the employer's premises. On no account must security officers become involved in any dispute over the behaviour of pickets towards employees, transport or others beyond that limit.

There is comparatively little law, either in the form of legislation or decided cases, dealing with strikes and there is an understandable reluctance on the part of government to give firm directives. Generally speaking strikes are only illegal in wartime or where there is reasonable cause to believe the probable consequences involve danger to life, serious bodily injury or exposing valuable property to destruction.

Picketing

Peaceful picketing in furtherance of a trade dispute is lawful; this is laid down in section 134 of the Industrial Relations Act 1971. One or more persons, in contemplation or furtherance of a trade dispute, may attend where a person works, or carries on business, or happens to be, if their object in doing so is merely for the purpose of peacefully obtaining or communicating information, or peacefully persuading any person to work or to abstain from working. The wording of the section is such that it would not preclude a group of employers picketing a Working Man's Club, if by doing so they hoped to induce their employees to return to work. The expression 'workman' means all persons employed in a trade or industry, whether or not in the employment of the employer with whom the trade dispute arises. The right the law once gave to peacefully picket a person's home is withdrawn by this section.

Picketing cannot be done on private premises—not even when it is peaceful. If the pickets encroach on land, they may be asked to leave; if they refuse, they become trespassers and like any other trespassers they may be evicted by means of 'reasonable force'—this would of course normally involve calling the police. M'Cusker *v* Smith (1918) is the relevant authority which expressly says that peaceful picketing may still be a matter of trespass.

There are a number of offences stipulated under various Acts but these will be the concern of the police, not the security officer. They include using intimidation or violence to a person, his family or property, hiding tools and persistently following about.

A police officer is justified in limiting the number of pickets to what he deems is reasonable if he thinks there is likely to be a breach of the peace. A refusal to accept this limitation can be deemed obstruction of a police officer. A relevant case is Piddington *v* Bates (1960)—18 pickets to eight workmen with narrow approaches to the place of employment where obviously there would be a chance of trouble.

'Lockout'

There is no statutory definition of a 'lockout'. The most recent under the Trade Disputes and Trade Unions Act 1927, now repealed, gives 'lockout' as the closing of the place of employment, or the stopping of work, or the refusal by the employer to continue the employment of certain persons working for him, in consequence of a dispute and with a view to compelling them to accept terms or conditions affecting their employment. Doing so with the intention of aiding any other employer in compelling persons employed by them to accept terms, etc., is included in the definition. It seems under present conditions that this is a state of affairs which will rarely arise.

ACTION DURING STRIKES

In all strikes that flare up into violence and intimidation, the root cause is likely to be found in an unreasonable attitude or behaviour by one or other of the sides. Recriminations are bound to be on the head of the person or persons who provide the spark and much unnecessary acrimony can be avoided by a flexible, commonsense and reasonable approach from the outset. Picketing is the main danger point, and someone who is knowledgeable on the subject should amicably ensure that the strike leaders know the legal position and the restraints the employers intend to put upon trespass of their property. The whole atmosphere of picketing can be jeopardised by a few unwise words.

Police attendance is not necessary and should not be asked for unless there is serious apprehension of a breach of the peace by pickets acting in contravention of the law by intimidation, violence, trespass or unreasonable physical impedance of traffic on public roads. The police do, however, appreciate formal notification of a strike and impending picketing so that they are aware of a possible source of trouble and know how to regard rumours and hoax telephone calls. They will certainly not intervene nor wish to attend unless it is absolutely essential for them to do so.

The security role is self-evident. While maintaining their employer's rights in respect of trespass, damage, etc., an impartial, courteous and friendly attitude to the strikers can yield dividends in taking the heat out of an immediate situation and yield long-term dividends in goodwill. The person in charge of security should remain readily to hand and keep himself apprised of the whereabouts of those managerial personnel who may be needed at short notice for negotiations or discussions.

TRAINING SECURITY STAFF

General

It is a primary requirement that on appointment a security officer should be taught the organisational structure of his employing firm, the controls instituted for the reception of raw materials, their use and storage, the holding and dispatch of finished goods, the use of transport, petrol and oil, and all general, engineering and canteen stores. Knowledge of the documentation supporting the movement of goods of all kinds, and the extent of the authority of persons concerned, must be thorough so that when spot checks are made they can be carried out with confidence.

Security duties

Formal training in these duties, which is recommended, can now be obtained at courses held in the principal industrial centres of England and Wales by the Industrial Police and Security Association and will be dealt with in Chapter 5. Courses in security duties are now held in certain colleges of further education.

First aid

The courses which have been referred to do not include instruction in first aid to the injured. The Red Cross and St John Ambulance Brigade hold courses of instruction which are available to security personnel. Further, the larger industrial concerns have health departments where similar training can be obtained. Certificates of proficiency issued by the Red Cross and St John, after passing their examinations, are accepted by the factories inspectors as evidence of the holder's training so as to comply with the requirements of the Factories Act 1961, section 61(4). This lays down that a person trained in first-aid treatment must be readily available in factories where more than fifty persons are employed, or such lower number as by regulation may be prescribed, during working hours. In the absence of a trained nurse, usually at night and at weekends, that requirement can be fulfilled by the presence of suitably trained members of the security staff who are constantly on the premises. This qualification sometimes is recognised by the payment of a special grant by the employers; on the other hand it can be a requirement that the security staff shall become qualified in the subject as necessary under the Act as part of their conditions of service before their appointment is confirmed and with no additional payment.

Security staff called to carry out treatment of an employee should be on the alert to detect whether the injury concerned is being falsely claimed or exaggerated to distract them from their duties and provide an opportunity for others to commit some offence against the law or works rules. (For further details on first aid, see Chapter 27.)

Industrial civil defence

The training of security staff should include their incorporation in any scheme for the protection, feeding, and welfare or evacuation of general works personnel in the event of an emergency.

Fire fighting

It is essential that security staff are proficient in the use of fire-fighting equipment and in the elementary techniques used in the handling of fires. If there is not a sufficiently qualified person on the staff to give that training it can usually be obtained through the good offices of the local-authority fire brigade or through membership of the Industrial Fire Protection Association, 36 Ebury Street, London SW1. Also membership of the Fire Protection Association, Aldermary House, Queen Victoria Street, London EC4, will lead to a constant supply of specialised information on experiences with fires and fire-prevention advice.

HOURS OF DUTY AND SHIFT ROTAS

The old concept of security envisaged an almost static role whereby long hours of duty were thought workable without detriment to either man or objective. This view unfortunately persists in some firms who can be easily identified by the appearance and calibre of the staff they employ and by the low efficiency they expect from them. Long unbroken spells of duty are psychologically and physically detrimental to those performing them on a regular basis. The optimum average weekly hours, in conformity with industry in general, is currently accepted as being forty-two, which lends itself to the optimum four-gang three-shift rota, and security staffing should be geared to this.

What is not recommended is a permanent night shift whose members may, by the regularity of the duty, be attracted to part-time day work for another employer which can affect their efficiency in performing their security duties. The arrangement is also undesirable where there is a permanent night shift of the general workforce. This can lead to familiarity which is not a good feature of security practice in any circumstances.

Reserve strength to cater for holidays and sickness will probably be kept to an absolute minimum so that, on occasions, considerable overtime becomes unavoidable. This is economically more desirable than to add extra wages and overheads for men to allow purely for emergencies. During such times, no shift should exceed twelve hours; by then the man's level of performance has sunk to a degree where his capability to deal with unforeseen situations must be suspect.

When the required coverage cannot be achieved without exceeding reasonable hours, the availability of men from the professional security firms on a temporary basis can be investigated. It cannot be expected that they will have the special skills required in some circumstances, but they will have had generalised training, sufficient to bridge an emergency. What must strongly be opposed is any suggestion of temporary transfer of members of the general workforce to security duties. They could thereby have access to information concerning their fellows which should remain confidential. They could gain an insight into the manner in which security patrols and duties are carried out which they could subsequently use to the detriment of the firm. Their only qualification would be familiarity with their surroundings. The fact that it was possible to use them on such a basis lowers the status of the selected and trained man in the eyes of other employees. It could also be anticipated, particularly where this was a staff-grade job, that there could be staff union objections.

Shift duties

These are necessary to provide the continuous coverage required. A proportion of applicants for security employment may not have had experience of the advantages or disadvantages of this form of working. Rather than have a man who finds that it is unacceptable to him when considerable expense has been met in his training and equipment, the full implications of shift working should be made quite clear to him at his initial interview so that he is in no doubt what to expect. This is also true of holidays where he will not be able to coincide with the rest of the community and he will have to accept the possibility of inconvenient overtime at short notice.

FOUR-GANG, THREE-SHIFT DUTIES

This is the normal form of coverage. It can be applied in a number of different ways which are shown in Appendix 4. In every four-week period 168 hours of duty are performed by each man giving an average of 42 hours per week. This requires a minimum of four men or multiples of four men according to the number of staff required. The respective shifts performed over periods of four, eight, twelve or more weeks will balance out so that he does an equal number of each according to the type of rota worked. Where shift premiums are paid for inconvenience and weekend working it is advisable to balance the wages over a complete rota period so that an average weekly wage is paid without the extra work involved in paying a wage fluctuating according to the shifts that have been worked.

The normal roster caters for seven days on early shift (6 a.m./2 p.m.) followed by two days' leave; seven days on late shift (2 p.m./10 p.m.) followed by two days off; then seven days on nights (10 p.m./6 a.m.) followed by three days off; this completes a full cycle. A variant is to work twenty-one days continuously, performing three shifts in sequence followed by a full seven-day leave. Some men will prefer this since holidays preferably should be taken to coincide with the early turn week and this enables, in effect, several holiday fortnights to be taken each year by coinciding a rotational week of leave with seven days' annual leave taken in lieu of the early turn week.

The continental system or 3 × 2 × 2 shift system

An objection to the foregoing three-shift working lies in the complete absence of normal evening social life during the afternoon and night shift periods. By using this variant there are frequent changes of shift, so that some evenings are free each week. An example is shown in Appendix 4 and the common feature of this system is a rotation of 3 morning shifts, 2 afternoon shifts, 2 night shifts; 3 rest days, 2 morning shifts, 2 afternoon shifts; 3 night shifts, 2 rest days, 2 morning shifts; 3 afternoon shifts, and so on.

This system does not usually appeal to the older men who prefer a settled routine which is less likely to cause disturbed sleep and digestive complaints. Longer breaks between shifts are included but with reduced leave days.

All these systems can be varied in respect of the starting day thereby modifying the break periods to coincide with weekends. Whenever it is intended to change a rota system, it is absolutely essential that the staff should have full opportunity to discuss it; if a change is made, all objections have then been ventilated and thoroughly examined so that there can be limited cause for subsequent complaint. With increased unionisation, there is little doubt those bodies will wish to become involved where their members are concerned.

Special arrangements

Conditions may appertain where it is necessary for more men to be on duty at certain times of day than others. The shift rota previously mentioned cannot cater for this and special arrangements must be made, either to devise an alternative acceptable system, or to superimpose upon the shift rota other men carrying out regular duties over the period of greatest demand.

This can particularly apply where a large number of gates are required to be manned during daytime and the necessity lapses when the main body of the workforce leaves, allowing some of the gates to be closed. The same can apply to commercial and retail organisations where there is a necessity for control of members of the public during the day reverting to control of limited staff after normal closing hours. These day staff will not, of course, have wages enlarged by shift premiums but the work will attract men who are averse to shift working and it should be possible to still demand a high standard. Some clerical ability may well be desired and there may be a different job description.

On occasions, however, the desirable extra cover may be needed during the particular period 6 p.m./2 a.m. A variation for this on the normal rotas is also shown in Appendix 4; this utilises ten men as an alternative to employing twelve on a four-gang three-shift basis; cuts overtime in connection with payment of wages on Thursday and Friday of each week and provides a man to cover the 6 p.m./2 a.m. period each day. A further advantage of a system of this nature is that the necessity for this particular duty is progressively reduced during the holiday months so that the men performing it can become holiday reliefs with minimum detriment to the coverage—he is available during the winter months when he is most wanted.

Whenever economies in manpower are desired, the possibility of variations of this nature are well worth considering.

Annual and public holidays

Security staff should enjoy the same holiday privileges as their equivalent grades among other employees. It is quite obvious that they will not be required to be absent at the same time but days in lieu should be logged and taken to mutual convenience of the individual and the requirements of security manning. Where it is possible to allow extra manpower to be off at public holidays, this should be done.

CODE OF PRACTICE

A recommended code of practice for security staff is given in Appendix 32.

4

Duties of security staff

Men employed on security duties need to know their responsibili-ies and these should be described in written form. These are called standing orders and a specimen is given in Appendix 3.

STANDING ORDERS FOR THE SECURITY STAFF

Instructions to the security staff, supplementary to the conditions of employment required for all employees under the Contracts of Employment Act 1972, require to be described in writing in some detail. These standing orders do not take into account all the special duties which the security staff may be called upon to perform when, in the absence of special directions, they would be expected to exercise their training and discretion. Duties required to be carried out in circumstances peculiar to individual firms or industrial practices probably have not been included in the specimen. On the other hand it is possible that no one security staff carries out all the duties which have been outlined.

It is recommended that standing orders be approved by a legal representative before issue. No alterations or additions to the orders are allowed after preparation except by the management responsible for their compilation, and then only in writing. The orders should have 'confidential' printed on them with the caution that they must not be shown to or exposed to the view of anyone outside the security department. All additional instructions to the security staff or requests for their services must be made through the official in charge of them. A copy of the orders should be issued to each member of the security staff for which he must sign. This would be collected from him on his leaving security duties. A copy should always be available in the security department or gatehouse for ready reference by the staff.

It is inevitable that during the course of their duties the security staff may see or hear matters which are confidential. They must be warned against discussing these in places frequented by the general workforce, such as the canteen, lavatories, and amenity rooms. The ability to keep a still tongue is one of the qualifications to be looked for when appointing security personnel.

The duties of some security staff may be restricted by a hard-dying tradition as to their extent and importance. Until fairly recent times 'security' was the backwater where men of advancing years who had become infirm or otherwise unfitted to continue their normal occupation were sent to complete the necessary service before they were pensionable. This is an example of the heart ruling the head. By modern standards their work was extremely limited, unsatisfactorily performed, and consequently a waste of money. The suggestion that they be trained in their duties would have been rejected out of hand as an unjustifiable cost, and further, the facilities were probably not available. Today, however, the recognition of the necessity of good security standards, and the cost, has to be justified by the results which can be sensed rather than seen. Therefore the standard of man employed on those duties has to be high and he must be capable of assimilating training. Training is the subject of Chapter 5.

Against that background, security staff should not be given menial tasks, such as cooking meals for the general workforce when the canteen is closed, stoking boilers, or sweeping roads, which are detrimental to their image and degrading to their self-respect. It is also bad security in so far as advantage can be taken of their presence in a particular place at a known time.

There is no reason, however, why their services should not be utilised to examine recording equipment or experiments and for them to take action as instructed in an emergency.

GATEHOUSE DUTIES

To assist in describing a security officer's duties, they have been divided into those usually carried out at the security office, gatehouse, or lodge near the entrance or entrances of the premises, and those associated with patrolling the premises. A checklist of gatehouse duties is shown in Appendix 21.

Construction

The gatehouse must be strongly constructed with windows providing good visibility in all directions. If this is interrupted by obstructions this can often be overcome by the use of mirrors. The door must be strong with no glass and have a good lock, preferably of the mortise type.

For attending to callers, particularly at night and other vulnerable times, a small window or a hatch in the door should be made so that it can be opened without opening the door. The security officer on duty will then not have to leave the protection of the building. Instances of security officers being overcome by criminals after being drawn out of their buildings by some subterfuge have become too common. When alone they should not leave even at the request of a police officer in uniform unless they are completely satisfied as to his identity. If they have any doubt they should telephone the local police station for confirmation of identity.

The area of the gatehouse and entrance must be well illuminated. An emergency lighting system should be installed to operate automatically should the main electricity supply be interfered with.

In addition to the main gates a drop barrier, like the controversial railway crossing gates, will provide an excellent control on vehicles entering and leaving the premises. They can be operated by hand or by remote control from the gatehouse.

Where premises are completely unattended these barriers can be incorporated in an alarm system which would be activated should they be interfered with. By this means loaded vehicles, particularly those not under the protection of a locked building, cannot be removed from the premises without setting off the alarm.

A lavatory with washing facilities and a changing-room with clothes lockers should be provided. Minimal cooking facilities should be installed for use when the canteen facilities are not available. Where the searching of employees is carried out a room giving some degree of privacy is also required. Notices to the effect that no unauthorised persons may enter should be placed at the entrance and requesting callers and drivers of vehicles to report at the gatehouse on arrival and when leaving.

Registers and other equipment

What will be required is listed below:

Attendance register
Occurrence book
Search register
Property lost and found register(s)
Register of car owners
Telephone
Telephone message pads
Lockable key cupboard
Register of borrowed tools, etc.
Key register
Vehicle register
Standing orders
Phone numbers of senior personnel
Instructions for action in case of fire
Fire extinguishers
Handlamps
Police whistle and truncheon

The purpose of the above will be obvious in some instances. The use of the others will be described in the following paragraphs.

Occurrence (or log) book—This is used to record all events related to security responsibilities which are not entered in a separate register, for example the search register. The occurrence book should be seen daily by the person in charge of security and signed by him. He will be responsible for any further action which might be necessary such as informing the manager of the department concerned of any incident which has received the attention of the security staff and the action which has been taken. This can be by sending a written report or showing the manager the relevant entry in the book and requesting him to initial it as read.

To save repetitious writing of unnecessary detail the occurrence book can be divided into sections, sometimes called 'cuts', for recording recurring events, for example the turning off of electric fires or other apparatus, receiving registered or recorded mail, and so on. As an alternative to the occurrence book where a large security staff is employed, printed report forms can be completed by the security officer concerned in any occurrence and filed in a binder. If a shift of men is under a supervisory member, he must inspect and approve the report of anything which happened during his period of responsibility, adding his signature.

At intervals of not longer than a month the manager responsible for the premises or site should inspect all security registers and reports and afterwards initial them. This has two purposes: firstly, by this means he keeps in touch with what has happened and learns how his security staff has coped with it, and secondly, it shows the staff that their work does not go unnoticed by the top management.

Search register—In this is recorded every instance where an employee or vehicle is searched when either entering or leaving the site. The date and time of the search, the name of the person concerned (and clock number when applicable) and the registered number and owner of any vehicle concerned are shown. The name of the officer requesting the search and the result will be given. If nothing of an incriminating nature is found the attitude of the person searched must be shown.

It is not efficient to carry out searches without making a

record. A written record that searches are carried out is much more useful to management than a verbal claim when dealing with a complaint associated with a search. Research through the records can establish whether complaints have previously been made and assist in placing the current one in perspective.

Property lost and found—Employees must be satisfied that proper attention is paid to their claims that their personal property has been lost or stolen. Records should be made either in a register or on special forms. Appendix 5 is a copy of a suitable form on which the question, 'Have you any further information you wish to give?' is there to allow the opportunity to the loser to reply in the affirmative if he has some suspicions he does not wish to give to the security officer completing the report form. In that event the person in charge of the security staff would see him to listen to his information.

Reports of finding property on the site must be recorded and Appendix 6 shows a copy of a suitable form. When money in any form is handed in, the depositor must be asked to count it and sign the record to the effect that the amount shown is correct. Should a security officer on patrol be handed money or anything containing money which has been found he must record in his notebook the amount of money concerned and ask the depositor to sign his note. This procedure is to protect the security staff against subsequent claims either that less money than that which was deposited has been restored to the loser or has been returned to the finder if it has not been claimed.

Found property which is deposited should have a label attached to it on which essential particulars regarding its finding are entered. An example of a label is given in Appendix 7. The tear-off portion is given to the depositor as a receipt. Nothing is written on the receipt portion to identify the property concerned. Jewellery handed in must be carefully examined and if it has apparently precious stones they must be counted, where it is reasonable to do so, and the number entered in the record. Where found property is claimed the loser's signature must be obtained in acknowledgement of its being returned.

If the property is of some value and has been restored to the finder an indemnification of the company against a later claim by the owner must be obtained and provision for this is shown on the specimen form. The finder of property which is restored to the loser should be informed of this and of the name of the owner. If property found has not been claimed by a specified time after finding, it should be restored to the finder against receipt. As the owner of the premises, etc., on which it was found has next claim after the real owner, some companies prefer that money found is paid into a charity or a welfare fund instead of being returned to the finder.

Car register—The names and departments of the owners of all motor vehicles brought onto the premises are recorded in this register. This is so that the owner can be found in the event of a car being left on the parking area with the engine left running or lights left on, or because a public or factory road is being obstructed. Early information of a flat tyre is very much appreciated.

Borrowed tools register—Some employers permit employees to borrow tools to do work at home. Their authority to take them off the premises is usually given on a form called a 'pass out' signed by an authorised person. This pass out is produced at the gatehouse when the employee leaves and the facts are recorded in this register. Sometimes the pass out is prepared in duplicate, one copy to be given up at the gate and the other to be retained by the employee.

When the employee is returning the tool he reports this and produces it at the gatehouse when he enters and it is noted in the register. Occasionally the register should be checked with the departments of the employees concerned to verify the tools have been returned. The register should also be checked at intervals to see whether tools are outstanding after a reasonable time. If there are and inquiries of the issuing department confirm they have not been returned the necessary inquiries should be made to ensure their return.

Key register—Keys in the gatehouse should be kept in a cupboard which can be locked, the key of which is held personally by the security officer on duty. The list of identifications of the keys must not be visible to anyone entering the gatehouse. There have been instances where important keys hanging on open key-boards or in unlocked cupboards have been removed by unknown persons when the attention of the custodian has been distracted, used for an unlawful purpose, and later returned, again without the security officer knowing.

Only persons authorised to hold keys should be issued with them for a day, a shift, or for a specific purpose. Issues and returns are timed in the register and the entries initialled by the issuer. When it does not unduly interfere, the person receiving the key signs for it.

Requests for keys at unusual times, without previous notice from an authorising person, should be closely questioned and, if possible, the person wanting the key should be accompanied to the place concerned. An instance of when this was not done occurred at a factory when a woman who was known to be a canteen worker called on a Sunday, when the canteen was closed, and asked for the keys of the canteen. She was given them without question and allowed to proceed alone. Because she had not returned the keys after some time a call was made at the canteen where she was found to have gassed herself in an oven. Undue delay in returning keys must be investigated in all cases.

In Chapter 33 a description is given of equipment whereby keys can be separated by distinguishing colour tabs.

Telephone message pads—When a telephone message is received at times when the telephone operator is not on duty it should be entered on message pads in duplicate. The name and initials of the caller, his telephone number, the person for whom the message is intended, and whether it is urgent or not should be included. The message will be completed with the time and date of receipt and the name of the person receiving the message. The original copy is available for delivery and the second copy is filed —this can be useful if the original copy containing important information is lost.

Vehicle register—This is used to record the times of arrival and departure of all non-company vehicles with their registered numbers and owners. Among the useful purposes of this record is ensuring that no vehicle is locked in, and checking by, say, accounts department where there is doubt on the number of journeys or on the arrival of a consignment. Complaints from owners alleging that their vehicles have waited to be loaded or unloaded for unreasonable lengths of time can often be disposed of satisfactorily from information in this register.

Telephone numbers of senior and key personnel—These are necessary to assist the security staff to get in touch with such personnel in an emergency. They must be kept up to date by the persons concerned.

Lifts—The security staff should have available to them the emergency procedure and any equipment or keys to release persons trapped in lifts when electricians are not immediately available.

Passes out

These are forms which can be used to authorise hourly paid employees to leave the premises before their normal finishing time—for medical or dental treatment or on compassionate grounds, for example. The employee hands it to the security officer on gate duty and it is afterwards sent to the personnel or wages department so that any necessary adjustments can be made to his pay.

The form can also be used to authorise the removal from the premises by an employee of small amounts of material of no value to the company, for example, short lengths of conduit pipe, or firewood, for his own use at no charge or on payment of a nominal amount. The pass out must show whether any charge has to be paid and whoever is authorised to receive that payment stamps it accordingly when it has been done.

In some cases the pass out for property is made out in duplicate, one copy to be handed to the security officer at the gate and the other retained by the employee to be produced should he be questioned by police or a member of the security staff. Material passes out should be returned by the security staff to the persons authorising them for them to check for any alterations and to deal with as required. Quantities should be written in words on such forms, figures not being used.

Staff sales

The removal of company products through sales to the staff has to be strictly controlled. The principle to be observed is that employees must not be allowed to take their purchases into their places of work. A staff sales shop close to the exit from the factory has great advantages in that connection. All purchases, which to save time might have been previously ordered and wrapped, must be collected as the purchaser is about to leave the premises.

If the shop is not in such a favourable position one form of control is for the purchaser, having paid for the goods, to be given a ticket bearing the same number as one attached to the wrapping of whatever they have bought. The parcels are later removed by the shop staff to the gatehouse. On surrendering his ticket an employee is given his parcel immediately before leaving the premises.

Suitcases and large parcels

Employees should not be allowed to take such articles into the factory area. They should be offered the facility of leaving them at the gatehouse on entry to be collected when they leave.

PATROLLING DUTIES

Objectives

The objectives in patrolling premises are:

1 To prevent and detect fire.
2 To prevent and detect damage from other causes, and waste.
3 To ensure company rules are observed, for example, no smoking in no-smoking areas.
4 To prevent and detect offences against the company's interest.
5 To prevent accidents.

List of duties for patrolling security officers

The duties to be carried out when patrolling premises have been compiled and are shown below. These should be read in conjunction with the further duties outlined under 'Fire Patrols' in Chapter 26.

1 Examine roofs for holes through which water or sparks from adjoining premises could fall to damage goods or machinery. Any necessary protective action should be taken immediately.
2 Note any other defects in roofs, gutters, walls, or windows through which rain water, smoke, etc., is entering or could enter the premises, and take any necessary action.
3 Check:
(a) locks on warehouses, stores, and offices and mark padlocks to detect any subsequent substitution;
(b) all perimeter fences for breaks, ensuring that goods or other material are not stored directly against them;
(c) all vehicles of employees in car parks to detect lights or engines left switched on, and punctures; inform owners as necessary;
(d) seals on loaded lorries, railway vans, etc.;
(e) insides of empty railway vans waiting to leave the premises, to detect stolen property;
(f) that goods in basements are not stored directly on the floor whereby damage might be sustained from water accumulation;
(g) that any property is not exposed to the weather whereby it is or could be damaged;
(h) that all strong-rooms and safes are locked, removing to safe custody any keys found in locks.
4 Be on alert to notice:
(a) any hazard which might cause or produce an accident and inform appropriate person;
(b) unauthorised persons on premises;
(c) to prevent and detect offences at clocking stations;
5 Turn off all taps or valves through which liquid or steam is escaping.
6 Visit men's dressing-rooms and cloak-rooms to prevent theft and other offences.

7 Visit laboratories and other similar places to check on the operation of any equipment in use, particularly when concerned in an active process at times when those are unattended. (The security department should always be notified in writing of such operations and given directions respecting action to be taken in any emergency. A suggested form of notice to be attached to the equipment in use in such circumstances is shown in Chapter 26.)

Alertness, interest and thoroughness must be displayed. A suspicious mind must be cultivated and anything that appears other than normal must be looked into.

A security officer cannot be expected to acquire more than a superficial knowledge of all the manufacturing and associated processes but it is essential that he should do that as early in his employment as possible. He must learn the identities of the persons in charge of each department and section so as to be able to communicate with them in any circumstances. To call for assistance in any emergency when on patrol the security officer should carry a police whistle and a truncheon to protect himself from attack.

Frequency

The frequencies of the patrols will depend on a number of factors including the length, complexity, and vulnerability to one or more of the risks which patrolling is designed to prevent. In describing the duties of a patrolling officer no distinction will be made between the times when normal production or other processes are being carried out and when the premises are unattended except for the security staff. The times of patrols must be irregular so that the arrival of the officer at any place cannot be anticipated. To prevent any interested person estimating with accuracy where he will be at any time he should retrace his steps occasionally and vary his route.

Clocks and clocking stations

There are divergent views on the requirement that a patrolling security officer should carry a 'watchman's clock' with which to register his attendance at clocking points located on the premises. One view is that if the calibre of the man is of the standard required he will patrol conscientiously and thoroughly and will not require the evidence of the clock to prove it. The other view is that the fact that he can prove his attendance at a given place at a specific time is valuable to him should subsequent events, for example, an undiscovered fire, cast doubts on his claim to have been in the area at a relevant time.

If clocking points are used the number must not be excessive otherwise the officer will be required to spend his time making a round of them to the detriment of his attention to other matters. It is recommended that the first round should be a full one and subsequent ones at the discretion of the officer. Clocking points should be sited near places of high risk—the safe, stores, canteen, distant perimeter fences, and so on—which should receive proportionately more attention than the less vulnerable places. Examination of the tapes of the clocks will show how the patrolling has been carried out and whether a regular pattern of visits to certain areas emerges.

Keys

To save the patrolling officer having to carry a large bunch of keys to premises he is required to visit, the locks to them can be on what is known as a 'suite' which will be opened by one master key (see Chapter 33). Keys to the locks of dry goods and wine, spirits, and tobacco stores should not be held by the patrolling officer nor should they be on a suite of which he has a master key. The reason for this is that should shortages of stock occur suspicion could be attached to him if entry was readily available. However, provision has to be made for him to inspect the interior of stores—for example, through windows—and in the event of an emergency the keys should be within immediate reach. It is suggested they be sealed in a strong envelope and kept under strict security supervision at the security gatehouse. After use the envelope with the key should be produced for resealing.

COMMUNICATIONS

Telephone

Where particularly large areas have to be patrolled this will be done more efficiently if they are divided into named or numbered patrol sections, similar to the beats of the civil police. Where a security officer leaves the gatehouse in these circumstances he should tell any colleague he leaves behind which section or sections he proposes to patrol.

If there is no other SO on the premises, and some process work is being carried on by the general workforce under supervision, he should note in a register where he has gone in case the manager or supervisor on duty requires his assistance. If he has a colleague at base the patrolling officer must telephone him at intervals as a mutual safety precaution. The times should be recorded. If a man fails to communicate or return within a reasonable time, possibly due to accident or illness, the area of search under the circumstances described is much reduced.

Radio

The use of personal two-way radios (transceivers) to provide constant and immediate communication between the gatehouse or other security centre and the patrolling officer has become recognised as a great assistance to good security. (See Chapter 30.) Some of the advantages do not require elaboration but one of them deserves emphasis. This is the ability, where the patrolling officer is out of touch with a telephone or is unable to reach it, to call for assistance in the case of accident or illness to himself or others or when discovering intruders carrying out crime or vandalism.

This facility is illustrated by the experience of one patrolling security officer who found three men on the premises in suspicious circumstances. Before interrogating them, and without their knowing it, he switched on his personal transceiver and through the microphone in his breast pocket his colleague at the security centre heard his questions and their answers. These roused his suspicions so he immediately arranged for assistance to be sent.

The three men were detained and subsequently charged with criminal offences.

One method of calling for a particular person is the radio call system in which a receiver carried in his pocket is operated by radio signal. This emits a bleeping sound and he then telephones the central switchboard to ask who requires him.

Another recent advance is the carrying of a device the size of a cigarette packet which is operated by pressing a button. This gives out a radio signal which operates equipment in the gatehouse to activate a recorded telephone call to police for assistance.

Closed-circuit television

This sophisticated equipment can play a vital part in security and is especially useful in maintaining observation on particular areas from some distance. Its use will be developed in Chapter 30.

Dogs on security duty

This will be dealt with fully in Chapter 32. Suffice it to say here that where a lone security officer is employed on a site where a high risk of theft exists and where he may be in danger of personal attack the employment of a dog to accompany him has great advantages.

Intruders

This description is given to all persons unlawfully or irregularly on the premises. It has already been mentioned that as a means of defence against personal attack, particularly when alone, the SO should carry a truncheon. He must not carry firearms. A security officer guarding an armoured van carrying cash who discharged an automatic pistol at men attempting to steal the money was severely criticised and the incident led to questions in Parliament.

On taking up employment on a security staff a man must accept that this may involve bodily injury in defence of his employer's interests. In circumstances where he feels he cannot cope single-handed with intruders it is better that he should obtain assistance rather than indulge in heroics. A security officer who can identify intruders or thieves who have escaped is more useful than an unconscious one unable to recognise his attackers.

The amount of violence a security officer may use against intruders is the minimum necessary to arrest or to prevent escape. Violence used in self-defence must have a reasonable relationship to that suffered or to that which is reasonably anticipated. Trespassers who have not committed any offence but who refuse to leave the premises when requested may be ejected using as much force as is necessary (see Chapter 10 on 'Trespass').

Vulnerable places

When visiting warehouses, stores and offices where there is a high risk of loss by theft, care must be taken not to give notice of approach by heavy footsteps or the flashing of a torch. A minute or two of silence outside the premises listening for unusual sounds within are worthwhile and rubber soles should be worn.

The need not to announce one's presence is illustrated by the experience of one SO who, without regard for the noise he was creating, entered a building where a safe containing a large sum of money was located. As he entered the office, which was in darkness, he was struck unconscious by a blow on the head from thieves who had forced open the safe and were preparing to leave. He was quite unable to assist the police investigation in any way.

Property found in suspicious circumstances

It is common practice for a thief to hide things he has stolen from his employer on the premises or close to the perimeter fence to be removed later. Sometimes the property is thrown over the wall. Should a security officer come upon such property, which from its nature and the place where it is hidden has clearly been stolen, it must not be moved. Nothing should be done to show anyone who may be watching that the property has been noticed. Observation should be kept on it from a concealed position. No inquiries should be made of anyone outside the security staff.

When anyone is seen taking possession of the property he should be allowed to do so without interference. If this occurs within the premises he should be followed until the opportunity arises for him to be stopped in the presence, preferably, of a member of the managerial or supervisory staff, alternatively another member of the security staff. He should be told what was seen, asked to produce the property and explain his possession of it. If this clearly shows that he has stolen it, the policy of the employers must then be followed. If the property is seen outside the perimeter it should be kept under observation as described and police assistance sought.

Diversionary incidents

Security staff must be on the alert to detect incidents which have been arranged to divert and occupy their attention whilst something unlawful or irregular is being carried out somewhere else. When any incident occurs which requires action from a security officer he should telephone his base and report the matter saying whether he has any suspicions. This is a circumstance where personal radio communication is particularly helpful.

MUTUAL AID SCHEME

An increasing number of security officers, whilst performing duty on their own, have been attacked and incapacitated and in one case killed by intruders in the course of crime. Because of this a scheme whereby they can readily obtain assistance in these circumstances has been devised. It is called a 'mutual aid scheme' and is arranged between the security staff of a number of premises. The details must be kept confidential and the scheme works like this: at a pre-arranged time the security officer (SO) at factory A telephones the SO at factory B, then the SO of B telephones the SO of factory C who afterwards telephones the SO at factory D, he telephones the SO at factory A which completes the cycle. This is repeated at pre-arranged times.

If a security officer fails to answer a call it is repeated after ten minutes and if this is unanswered the SO making the call telephones the police. If a security officer expecting a call does not receive it, he telephones the factory from which he expected the call. If that call and another ten minutes later go unanswered he telephones police. No time is wasted explaining the circumstances to police because all such schemes are known to them. If a security officer has had an accident or has become ill and is unable to reach a telephone, by this arrangement assistance is brought to him within a reasonable time.

If a security officer is under restraint by intruders at the time of a call and they wish him to answer the telephone without disclosing what has happened he would use a previously arranged phrase which would sound quite innocent to the intruders but would inform the caller that assistance is required. The introduction of such a scheme between two or more factories where men are employed in similar circumstances to those described is very comforting to them.

Chief officers of police will not agree to police stations receiving routine reports directly from security officers nor allow telephone calls to be made to them for that purpose. This would unduly engage telephone lines and operators at the stations required for more urgent purposes.

One of the services which can be supplied by security companies is visits to or telephone communication with security custodians of premises and this is dealt with under that heading in Chapter 31.

5

Training security staff

Until 1959 no facilities existed for the formal training of security staff. Training in the duties to be performed depended on the quality and experience of the person responsible for the staff and the time he was able or prepared to give to it. As the required standards of performance of the duties became higher in parallel with the need to justify the expense of security staff, the provision of professional training became imperative.

This was realised by the Industrial Police and Security Association, whose aims and objectives and membership structure are explained in Chapter 34. Training courses which will be described commenced in 1959 and in the following five years 900 security personnel attended them. Their success is shown by the fact that in the four years 1964-7 over 1400 students attended. As the demand from employers for the training of their staff has increased more courses have been arranged.

They are held in commercial and industrial premises of firms supporting the Association, adult education premises, and police buildings usually between the months of September and May. There are three types of course: 'basic', intended for beginners or persons with some experience of security duties and who would benefit from some form of training; 'intermediate', arranged for the more experienced man or as a refresher for those who attended a basic course; and 'advanced', for persons in managerial positions with security responsibilities, for example chief or senior security officers, personnel managers, etc.

The courses have been of three or four days' duration and non-residential, but the recent success of five-day residential courses suggests that there are likely to be more of that pattern in the future.

The lecturers for the basic and intermediate courses include police officers, members of the local authority or industrial fire brigades, former senior police officers now employed in commerce and industry on the relation of the civil and criminal law to security duties, members of management of a variety of firms, factory inspectors, the Royal Society for the Prevention of Accidents, and trade union officials. Demonstrations are given of security equipment such as locks, safes, radio and television by firms supplying such equipment. Visits are arranged to police stations, courts of law, and fire stations. Short plays underlining aspects of the instruction on security duties are staged for the benefit of the students.

At the advanced course lectures are given by senior representatives of organisations having an affinity with security such as the Fire Protection Association, British Insurance Association, and others with the management function to the fore. Group discussions on prepared agendas have proved particularly successful and beneficial.

The subjects on the curricula of the basic and intermediate courses are drawn from the following:

Law of assault
Law of bribery
Law of forgery (elementary)
Law of fraud (elementary)
Law of criminal damage
Law of theft
Law of trespass
Powers of arrest
Powers of search
Factories Act 1961
Offices, Shops and Railway Premises Act 1963
Police Act 1964

Rules of evidence
Notebooks and reporting
Industrial and public relations
Accident prevention
Crime prevention
Dogs on security duty
Liaison with local police and fire brigade
Fire prevention
Use of fire-fighting equipment
General duties of a security officer
Gate control
Key control
Lost and found property

Security Manual, a handbook written by the authors of this book, approved by the National Council of the Industrial Police and Security Association and published by Gower Press, is used as the standard work for the instruction of security personnel at basic and intermediate training courses. It is also recommended reading in preparation for the examinations of the Institution of Industrial Security.

6

Security records and reports

If security officers are to keep notebooks, they must do so in such a manner that their contents will automatically be accepted as accurate, should the need arise for them to be referred to. The police service has learnt by experience that the defence in court proceedings will take any advantage afforded by badly kept notes as a means to challenge and distort the validity of prosecution evidence. For this reason, a simple and logical set of police rules have been laid down which can well be copied by their industrial counterparts.

NOTEBOOKS

Irrespective of criminal matters, there are convincing reasons why security officers should maintain their own personal records of day-to-day duties—in addition to a communal occurrence book. An officer on patrol does not inspire confidence nor impress with efficiency if he has to fumble to find a scrap of paper on which to make an essential note or record an important message.

If used for no other purpose, a notebook would justify its existence purely as an easy means of recording items, no matter how trivial, best not left to fallible memory. It is never possible to tell when a query may arise in the future as to whether a security officer has acted correctly in respect of a matter reported to him—an appropriate brief note could provide an immediate answer either to management or to a court.

Where prosecutions are concerned, it is likely that the conversation between a security officer and an accused person will have a direct bearing upon the issue of a case. What is said by the accused may constitute an admission or an explanation, either consistent with innocence or implying guilt. It is beyond human credence that such conversations could be remembered verbatim for repetition in a court weeks after they have occurred. If such evidence is to be given the only manner in which a court would be fully prepared to accept it would be by the witness referring to notes which the court is prepared to accept as a true record.

Dealing first with the type of book that should be used; obviously it must be of a convenient size for carrying in a pocket, with either stiff or semi-stiff covers to protect the pages. Preferably the pages themselves should be numbered in sequence to preclude any suggestions of tearing out and redrafting entries. For legibility they should be white and lined, and a margin on the left is an advantage for the insertion of times.

Each book taken into use should be date-stamped at the time of issue and have the name of the user inserted. The practice followed in certain police forces of actually numbering the books and keeping a record of allocation is hardly a necessity in the case of security officers.

Entries by the user should be in chronological sequence and an entry made for each day of duty. This should take the form of lining off the previous day's notes and inserting the day and date, followed by the tour of duty for the current day, the actual time of commencing work, and any specific instructions that are given at the time of parading. Incidents of the day should follow in sequence with the corresponding time shown in the margin. Entries should be made as legibly as possible—it must be remembered that the book is the property of the employers who may wish to refer to its contents.

It must not be thought necessary to fill the notebook with trivialities in order to prove that the writer has been active. If there is a spate of incidents to record there is something wrong with the security arrangements—particularly if there is a sequence of thefts or arrests to be shown. The true test of industrial security is whether or not the precautions are adequate enough to strictly limit incidents and dishonesty, a notebook void of everything but dates and times of parading and dismissing may be the best testimonial.

There are certain matters which must be correctly and fully recorded in a notebook. Examples are:

1. In the case of arrests, the full name, age, occupation, and home address of the arrested person must be shown, together with the time of his apprehension and the reason for which he was arrested. Any questions and answers should be clearly set out; it is as important to record the question as the answer and in all matters of consequence the precise words that are used should be quoted.

There is no reason why an abbreviated narrative on the circumstances of an arrest should not be entered, but this should be kept short since, whereas recollection of words may be vague after a period of time, actions of any kind should remain clear and should not be the subject of reference in giving evidence.

2. In the case of lost and found property, full details of time, place, and individual losing or finding must be inserted, as well as a full and identifying description of the property. Where a person has handed over property to the patrolling officer that person should sign the officer's note underneath the description. In the case of found property being claimed during the course of a patrol, the officer must, in addition to establishing ownership beyond all doubt, obtain the signature of the claimant together with details of his department or his home address.

3. Complaints of theft should include full name, address, and means of contacting the complainant; the time the property was last known in order and by whom; when the theft was discovered and by whom; location of the theft; full description of missing articles with their value; and a note of any information bearing on the idenity of the culprit. (See page 57.)

Equally, there are certain things which must not be done in connection with the notebook. Perhaps the most important of these is erasure. No entry should be erased or altered in any way.

There must be no additions written between lines and pages must not be torn from the notebook in order to make fresh entries. Whilst these matters may not be of importance in respect of normal security duties, any signs of this in a notebook produced before a court would result in a most severe criticism of the user and, undoubtedly, the striking out of the evidence to which the entries referred. The correct procedure to follow is to draw a line through the erroneous entry, initial, then record the correct matter.

The question of chronological sequence is of importance when records of conversations are in question. The validity of these entries is sometimes contested on the grounds that they have been made so long after the occurrence that their accuracy must be deemed suspect. All entries therefore must be made as soon as possible or convenient after the happening has taken place. It would be ridiculous to assume that an officer, in the course of making an arrest, stopped to make up his notebook whilst escorting a prisoner to custody!

Where two security officers are acting in unison in the arrest of an individual, there is no objection to their notes on the incident being compiled jointly, always providing, if there is a discrepancy in their recollections of what was said, both will record as they remember.

It is particularly important that the instructions concerning keeping notebooks are observed by those who are engaged in the prevention of theft in stores and other large premises where regular arrests and appearances before a court would be anticipated. Persons so employed will be well advised to meticulously follow police procedures, both in making their notes and in the manner and occasion in which they produce these notes before a court. It should be borne in mind that a defending lawyer cannot demand production and examination of a book unless a witness has referred to it in the course of the court proceedings.

REPORTS

The books of record referred to in Chapter 4 provide only

information in a static form for reference purposes. Reports must be submitted to the management on matters of importance or where the management requires information on which to make decisions. All incidents which necessitate reports should be committed to paper as soon as possible after the actual occurrence. This should be irrespective of whether full details were available at the time the report was made. A brief note of some happening still under investigation could well save embarrassment to senior management, who might otherwise be approached with questions concerning a matter of which they had no previous knowledge. There is no objection to submitting an incomplete report, if it is made clear that a more comprehensive one will follow when fuller information is available.

The object of any report is to convey full and accurate information without ambiguity. The amount of detail which is contained in it will be governed by the known familiarity of the recipient with the matter being reported upon. For example, the head of a department would not need details of the precise function of individuals employed in his department, which might be required by a director who is not conversant with personalities and responsibilities in that department. Similarly, an individual not conversant with a locality or a procedure would need the inclusion of extra details to make the matter comprehensible to him.

Persons for whom a report is intended should be clearly shown on the front sheet in a circulation list, preferably on the left-hand side at the top. It could be advantageous to direct it to a particular person or persons who will have to act on the contents, with copies to those whose status or responsibilities make it advisable that they should be informed. Where the contents are of a confidential nature, or could be construed to reflect upon a named individual, care should be taken to ensure that it is not read in transit by unauthorised persons. In these circumstances, the report itself must be headed confidential and enclosed in an envelope addressed and similarly marked. The onus of deciding what degree of restriction is applied rests with the originator.

No one likes to wade through a quantity of written matter without first knowing what it is all about. Therefore, the first essential of a report is a very brief heading, indicative of the content.

If, for instance, a theft had taken place in Number 2 Warehouse of the Gorbals Works by breaking a window in the supervisor's office and forcing the safe during the evening of Monday 26 February 1968, a suitable heading would be: Office breaking and theft, Number 2 Warehouse, Gorbals Works, 26 February 1968. The reader at once knows what it is all about.

The body of the report should begin without any courteous, old-fashioned preamble by setting out bluntly the time, date and full nature of the occurrence:

> At 21 00 hours on Monday 26 February 1968, security officer Jones found the supervisor's office in Number 2 Warehouse had been entered by breaking the back window and climbing through. The back of the safe had been blown off by explosives; £21 cash was later found to be missing. Glasgow CID were notified and attended; the Fingerprint staff have carried out their examination but ask that there should be no interference with the office and contents until Forensic Science staff have visited—expected at 10 00 hours on 27 February 1968.

Having conveyed the gist of the report more details can follow and everything will fall into perspective without difficulty. The contents of the report should remain terse and to the point. No-one particularly wants to know that Security Officer Jones was patrolling in such and such an avenue examining property, when he saw that a window in Number 2 Warehouse had been broken and he therefore looked inside. The material point is that Security Officer Jones found that the Number 2 Warehouse supervisor's office had been entered by breaking a window.

The language should be plain English of the type which does not allow any misunderstanding; technical terms should not be used unless they are certain to be understood by those reading them. Slang terms must be avoided—they are a confession of a poor vocabulary and reflect adversely on the writer. The report should be as factual as possible and if opinions are quoted they must be soundly based and reasonable and the person expressing them must be sufficiently conversant with the subject to be competent to express such an opinion.

Where statements are attached they should not be recapitulated in the narrative but should be referred to and abbreviated down to the barest detail necessary to convey their content. Naturally, there are a wide variety of reports ranging from those relating to everyday minor occurrences being drawn to the attention of the departmental head to lengthy reports to a senior executive or director outlining or analysing a particular problem and making recommendations for its solution.

With trivia of everyday occurrence a terse stencilled form might lead to clarity and economy in time and writing. Notification of things such as doors left insecure, lights left burning, or electric fires left switched on are all appropriate for a form which would only need the addition of some five or six words and a signature to fully convey what information was required. (See also 'Gatehouse duties' in Chapter 4.) The more lengthy type of document will be compiled almost invariably by chief security officers and their equivalent. Many will be of the type where one need simply state the facts of an incident and if necessary make recommendations based upon that particular set of facts for action to be taken.

It is obviously not feasible to give guidance on all types of report of a complex nature, but where problems are being analysed and recommendations made there are few better ways of doing this than by following the military type of format. No attempts should be made to set out an analysis or complicated report in full in the first instance. If this is done the brain is likely to become immersed in the intricacies of grammatical construction to the omission of points which it was intended to make.

The first step should be the selection of main headings in logical sequence under which detailed rough notes should be prepared, followed by a formally constructed report, after all the implications of what is being proposed have been considered. Unless thoughts are noted down to a pattern it is not possible to review all the factors involved in a complicated situation with complete impartiality. A system of standardised headings, if only in the preparation of notes, facilitates this process and guards the writer against omissions. The following sequence is a logical one for setting out preparatory notes:

1 *The aim* or objective of the report.
2 *The factors* which affect the attainment of the aim or which surround the objective of the report.
3 *The courses* opened which may be followed.
4 *The recommendations* or the proposed action which is suggested.

The whole value of a report depends upon a clear definition of its purpose which must be stated simply and concisely. Care must be taken to avoid any expression of ambiguity and to ensure that the definition of the aim is in accordance with instructions and responsibilities of the writer.

'Factors' are the facts, circumstances, or evidence from which deductions may be subsequently drawn. Those not bearing upon the main purpose of the report should be discarded as being superfluous to it. Each factor will be followed by a deduction and the word 'therefore' is invaluable; for example, in preparing a report recommending an increase in establishment:

> The works area is over 100 acres in extent and, during after-work hours, only one man is available to patrol it *therefore* it is not possible to make an effective patrol of this area more than once every four hours.

A tendency to note down a number of items of information and then draw a composite deduction must be avoided.

The factors to be considered will vary with every situation and it will be found there are usually a few which dominate the others. It is by noting these down and listing deductions from them that a clear and reasonable report will eventually be compiled; minor ones should not be neglected, since a critical examination by a third party may raise these and the answers could then be immediately forthcoming to the credit of the writer.

Only practical courses should be considered with the salient points for and against each. It is valueless to suggest lines of action merely to discard them again as being void of any merit. The course it is intended to recommend should be normally stated last. If recommendations are required they should be clear, unambiguous, and given in sufficient detail for a course of action to be taken upon them.

With any long and complicated report it is advantageous to number paragraphs for easy reference and indeed an index might be helpful. Tables and schedules should be attached and referred to in the body of the report. If statements are attached they should be placed in the order of reference in the text. Maps or plans must have the scale shown upon them and photographs should have a caption to indicate what is shown.

Vehicular accidents on private property

These need rather a specialised form of report. Where there is no negligence on the part of the firm all that is necessary is to effect an exchange of names, addresses, and details of insurance between the parties concerned and briefly note the circumstances as much for information and reference as anything else. Where the firm's transport or interests are directly concerned, full details should be taken, including names of witnesses and statements, if they are prepared to make them. A sketch of the accident could be helpful and where a visiting driver is involved he could be asked his version of the accident if not to give a statement.

Fire and works accident reports are dealt with in Chapter 27 and Appendices 16 and 18. Duplicated forms are suitable for these to ensure complete details are obtained—the same could of course be done for vehicular accidents but it is unlikely that the need would justify a special form.

To sum up, the main essentials of report writing for the average security officer are:

1 Make the report as soon as possible and amplify later if necessary.
2 Indicate clearly for whom the report is intended.
3 Provide brief explanatory heading with the date and time.
4 Be clear, brief, and to the point.
5 Do not pad out the report with superfluous material.
6 Use plain unambiguous English.
7 Limit technical phraseology to those who will understand it.
8 Do not use slang.

PART TWO

LAW AND PRACTICE

7

Arrest and search

Here will be described what constitutes an arrest, what is required before it is carried out, and what has to be done at the time and immediately afterwards. The powers of search of the person and property will also be dealt with. Police action following an arrest by a private person will be given in Chapter 8.

The criminal offences most concerned in arrests are theft, fraud, forgery, criminal damage including arson, and the different types of assaults against the person. There are other powers of arrest which have been provided by statutes and which are sometimes called preventive, in other words they secure the detention of a person reasonably suspected to be about to commit a crime before he does it.

ARREST

An arrest is the taking or restraint of a person from his liberty in order that he shall be forthcoming to answer an alleged crime or offence. It is not necessary to touch or lay hands on a person to arrest him.

Manner of arrest

An arrest should be made as quietly as possible and unnecessary violence avoided. An accused person must be told what he is being arrested for and should always be treated with consideration. Should the arrest subsequently prove unjustified, complaint from the offended party is less likely if he has been treated well.

Once under detention the suspect must not be lost sight of for a moment under any pretext. If he asks to go to the lavatory he must be accompanied—a female, of course, would be escorted only by a woman. Stolen property often has had to be recovered from lavatory pipe bends and drains where it has been disposed of by a suspect to avoid detection. Further, there have been instances of property being hidden in lavatory cisterns to be recovered later, so, when searching, the smallest room must not be neglected. Whilst being detained awaiting the police, suspects have been allowed to visit the lavatory unescorted where they have cut their throat or wrist or used a belt or tie to commit suicide.

Without warrant

The person concerned must be told that he is being arrested ('detained' might be a better word for a private person), on what charge, and that he will be taken to a police station or given into the custody of the police.

By authority of a warrant

An arrest can be with or without a warrant. If the statute concerned with the offence requires a warrant to be issued before a person can be arrested it is applied for either by the police or a private person from a magistrates' court. Statements in writing or proofs of evidence taken from all the relevant witnesses, to prove a *prima facie* case against the person named, are supplied to the magistrate. A nucleus of witnesses sufficient for that purpose appear at the court in person and swear on oath that the contents of their statements are true. On his finding that there is a case to answer the magistrate signs a warrant calling on the police to arrest the accused and bring him before the court where the warrant was issued to answer the charge described thereon. When police find the accused he is told that there is a warrant in existence for his arrest and told the charge which has been made against him. It is not necessary for the arresting police officer to have the warrant in his possession but if he has he shows it to the accused.

Bail

The warrant will state whether or not the accused can be released on bail before appearing at the court concerned. Where bail is to be allowed—the sum of money stipulated will vary with the seriousness of the offence—the accused enters into a written recognisance or undertaking to owe the Crown that sum of money if he fails to attend the court at the required time. If he fails to do so and is subsequently arrested the court can order him to forfeit that sum of money. This is known as estreating bail.

In the more serious cases friends of the accused, known as sureties, may also be required to agree to pay the Crown sums of money should the accused not attend the court and they would similarly lose all or part of their money if he failed to do so. In England and Wales it is not necessary for the accused and/or the sureties to deposit the equivalent amount of money when entering into recognisances.

Where the arrest has been made without a warrant, bail may be allowed at the police station with or without sureties depending on the circumstances, such as the seriousness of the offence and the character and address of the accused.

Common law powers of arrest

Until the passing of the Criminal Law Act 1967 the power of the private person to arrest without a warrant was derived from common law, and had reference only to felonies, which included theft in various forms. The new Act, amongst other things, abolished the distinction between felonies and the other class of offences called misdemeanours but preserved the right of arrest of a private person for what are termed 'arrestable offences'. These are crimes, *and attempts to commit them,* for which a person (not previously convicted) may be sentenced on indictment to imprisonment for five years. These include all forms of theft.

The power of arrest of a private person is now described in section 2(2) as: Any person may arrest without warrant anyone who is, or whom he with reasonable cause suspects to be, in the act of committing an arrestable offence. By section 2(3), where an arrestable offence has been committed any person may arrest without warrant anyone who is, or whom he with reasonable cause suspects to be, guilty of that arrestable offence.

The important requirements which have to be satisfied before a private person decides to arrest anyone are that an arrestable offence *has* been committed and that he has reasonable cause to believe that the person he intends to arrest carried it out. An interesting extension of the power of arrest by a private person is given under the Criminal Justice Act 1967, section 91, whereby anyone who in a public place is guilty, while drunk, of disorderly behaviour may be arrested without a warrant and on conviction can be fined up to £50. This power is not yet effective.

The powers of a police officer extend to arresting a person where he suspects an arrestable offence has been or is about to be committed whether one was committed or not.

Use of force when making an arrest

Section 3(1) of the Act says: 'Any person may use such force as is reasonable in the circumstances in the *prevention* of crime or in effecting or assisting in the lawful arrest of offenders or suspected offenders or of persons at large.' Chapter 10 deals with the use of force by security personnel to evict trespassers.

The Judges' Rules

Persons other than police officers charged with the duty of investigating offences and charging offenders shall, as far as practicable, comply with the Judges' Rules, concerning the interrogation and apprehension of persons, prescribed by the Judges of the High Court. These Rules, primarily directed towards the police, were amended in January 1964 and now for the first time acknowledge that persons other than police officers carry out similar duties and they direct that their actions in connection with interrogations and arrests shall be similarly controlled. The Judges' Rules include how and when an accused or suspected person shall be cautioned and will be dealt with in full in Chapter 11.

Unlawful imprisonment

The possibility of placing oneself in a position where a claim for damages may ensue if a mistake has been made is often raised as a deterrent to those who wish to take advantage of their lawful powers in the prevention of crime or the arrest of offenders. That contingency must be put in perspective: actions in the civil courts for damages for unlawful arrest are relatively rare, taking into consideration the total number of arrests which are made for every sort of offence and those which are not brought to conviction. Some claims, of course, are settled out of court.

If there is *reasonable and probable cause* to believe that the arrested person was responsible for an offence which had been committed he should be detained until he can be handed into the custody of the police. Should their investigations subsequently show that either the offence had not been committed or, if it had been, that the accused person had not committed it a claim for damages will not be likely to succeed. Courts dealing with such claims are sympathetic towards those defendants who honestly believed they were acting in accordance with the law and *without malice*.

Preventive arrests

The Prevention of Offences Act 1851, section 11—This lays down that any person may arrest any person found committing any indictable offence in the night, which is between 9 p.m. and 6 a.m. Indictable offences include all types of theft, criminal damage and offences against the person. Theft is described in Chapter 9 and the other offences are referred to in Chapter 10. Section 12 of the above Act is of considerable interest to security officers as it says that if the person arrested assaults the person arresting he commits an additional offence to the main one for which he may be sentenced to three years' imprisonment.

Vagrancy Act 1824, section 6—The date of this Act is interesting. It was before the establishment of the police force in the form we have become used to and one of its objects was to give householders, property owners, and others the power to apprehend vagrants who are about to commit offences against their interests, and to bring them before a court.

These offenders, according to the seriousness of the offence concerned, were categorised as:

1 Idle and disorderly persons
2 Rogues and vagabonds
3 Incorrigible rogues

The Act is still on the statute book and provides any person, which includes a security officer acting for his principal, with the power to arrest at any time the following:

1. Any person 'found in or upon any dwelling house, warehouse, coach-house, stable, or outhouse, or in any inclosed yard, garden, or area, for any unlawful purpose'. (Yards, etc., must be completely enclosed.)

2. Any 'suspected person . . . frequenting (or loitering about) any river, canal . . . or any quay, wharf, or warehouse near or adjoining thereto, or any place of public resort . . . or any highway or any place adjacent to a street or highway; with intent to commit an arrestable offence'. (Wording amended by the Prevention of Crimes Act 1871, section 15 and the Criminal Law Act 1967, schedule 2, paragraph 2(1)(*b*). 'Arrestable offence is defined on page 87.)

The evidence on which a court will conclude that the intention was unlawful (not trespass—see page 122) will be what he was seen to do, what he said on apprehension and anything he had in his possession which could have been used in the course of committing a crime.

To justify the arrest of a person who is found in the circumstances described and is suspected of being about to commit an arrestable offence, the law requires evidence of two separate acts which were observed. The first which aroused suspicion and the second to show that the accused's intentions were in fact dishonest.

An example having its application to industrial security is where a man is seen in the company car park to try without success to open the door of a parked car which contains some attractive property. He is unsuccessful and passes to other cars where he repeats his action and where he is again unsuccessful. When he is stopped he cannot account satisfactorily for his actions and furthermore has no car in the park of his own. If time permits it is advisable in these circumstances, to enlist the assistance of police in the apprehension of the 'suspected person'.

The Theft Act 1968, section 25—Possession of housebreaking instruments, etc. This says:

1 A person shall be guilty of an offence if, when not at his place of abode, he has with him any article for use in the course of or in connection with any burglary, theft or cheat.
2 A person guilty of an offence under this section shall on conviction on indictment be liable to imprisonment for a term not exceeding three years.
3 Where a person is charged with an offence under this section, proof that he had with him any article made or adapted for use in committing burglary, theft or cheat shall be evidence that he had it with him for such use.
4 *Any person* may arrest without warrant anyone who is, or whom he, *with reasonable cause,* suspects to be, committing an offence under this section.

'Burglary' (section 9) means entering any building as a trespasser and with intent to commit theft, grievous bodily harm, or unlawful damage to the building or anything therein by fire or explosion, or having entered any building or part thereof and committing or attempting to commit any such offence. A building includes an inhabited vehicle or vessel.

This section replaces part of section 4 of the Vagrancy Act 1824, and section 28 of the Larceny Act 1916, which provided powers of arrest for what was loosely called possessing housebreaking instruments with intent to break and enter premises.

THE RIGHT OF SEARCH

No private person, which includes security officers, has the right to search anyone. Consent to a request to do so must be given at the time. The agreement of an employee to a clause in the conditions of employment requiring him or her to submit to a personal search when thefts are suspected does not remove the necessity to obtain consent to the search of the person immediately before doing so.

To put one's hands on the person of another without his consent is an assault and could be followed by a charge of that offence and, although extremely theoretical, a claim in a civil court for damages for trespass of the person (see page 119).

A refusal to be searched is not by itself justification for detaining a person until the police arrive, or attempting to do so. However, a refusal where there is credible evidence that the suspect has stolen property in his possession would be reasonable cause to suspect he was in the act of committing an arrestable offence and he could be lawfully detained until handed to police.

The extent and thoroughness of personal searching depends on the type of property likely to be concerned. The searching of employees, for example, engaged in the refining of precious metal or in the manufacture of jewellery, would not have been efficiently carried out unless it required the stripping of the clothes to the skin. On the other hand where, say, foodstuffs are concerned it might be sufficient to inspect the contents of bags or packages carried and to 'frisk', that is, to run the hands over the outside of the clothes. Cycle clips have been known to support stolen property inside trouser legs.

Conditions of employment

It is generally the practice at production units, particularly where the raw materials or the finished products are attractive and can be secreted on the person, for the conditions of employment to include a clause referring to personal searching. The employer offers to engage an employee providing he agrees to the conditions and, from the employees' viewpoint, he knows what acceptance entails. On the employee accepting the offer a contract has been made between them. If the employee fails to comply with the conditions he breaks the contract and is liable to disciplinary action.

The words of the searching clause differ from firm to firm but in essentials are the same. Some examples are:

1 'The company reserves the right to search all employees leaving *or entering* the premises and to inspect any parcel, package, handbag, or motor vehicle.'
2 'Should pilfering be suspected the company reserves the right to require employees to submit to a search. Any employee found removing company property from the premises without proper authority will be dismissed and may become liable to prosecution.'
3 'When employees join the company they sign an agreement that while on company premises any authorised official of the company may question them concerning company property and may examine any article in their possession or any vehicle used by them and that they will submit if required to a personal search.'
4 'The company may require an employee to submit to being searched and may take such precautionary measures as it considers necessary.'

It may be advisable to add a further clause such as:

> At intervals random checks are made and it should be made clear that selection for search does not imply suspicion and it is hoped that employees will appreciate the need for such checks to be made in the interests of all.

The condition must be shown to apply to management, staff, and labour force alike and without discrimination. At one company the chaffeur of the managing director was using his car to remove company property from the premises. The absence of a searching clause does not preclude the owner of property or his agent from asking anyone to account for the possession of property believed stolen.

The following basic procedure is recommended when proposing to search employees. A woman may be searched only by another woman.

1 Particularly where there is reliable evidence to show that an employee who is about to leave the premises has stolen property in his possession, the searching should not be confined to one person. If the suspicion is confirmed his selection for searching without anyone else could lead to him identifying who exposed him.
2 If selection of those to be searched can be made by some person independent of the security staff it assists in preventing complaints of victimisation. This can be done by arbitrarily picking names or numbers from a figurative hat or by using a random selection unit (see page 97).
3 Having as unobtrusively as possible requested an employee to submit to a search he should be taken to a room giving some privacy (see page 55). There he should be reminded of the condition of employment respecting searching.
4 He should be asked whether he requires a witness to be present and if he does this must be arranged. Another member of the security staff or supervisory grade staff should also attend in these circumstances.
5 The employee should be asked if he has any company property to declare. Note carefully what he says. If it is an admission ask him to produce the property and note from where he took it. Note any explanation he gives. If it is property which can be marked with some distinguishing inscription this must be done and the employee should be asked to write his name on it.
6 If he denies having any property ask if this can be checked by the examination of his clothing. Providing he agrees run the hands over the outside of his clothing and if anything is felt ask for it to be produced.
7 Where company property is found to which the employee has no right, tell him he will be detained until the management is informed.
8 If there is a refusal to be searched, the employee should be reminded of the search clause, if possible in the presence of

a witness to the refusal, and told the facts will be reported to the management.

9 If a search reveals nothing of interest, allow the employee to leave as quickly as possible after thanking him for his cooperation and for acting as an example to others.

10 Record the name and department of the employee, the date and time of the search, the name of the security officer concerned, the result, and whether any complaint was made (see page 56 respecting the search register).

A very firm line must be taken by employers with instances of refusals to be searched, otherwise a mockery will be made of the requirements which, among other things, will encourage more employees to refuse and this would lead to an intolerable situation. Where such breaches of the conditions of employment occur, dismissal is justified and indeed recommended. There has been no occasion, known to the authors, where this has been carried out and objected to by representatives of organised labour.

A complete and thorough examination of motor vehicles, though desirable from time to time, depends on the time available. When any search is carried out the driver and any passengers should be included. Likely places for the secretion of stolen property are the boot, in tool boxes, the spare tyre, behind, in and under seats, under the bonnet, and fixed to the chassis. This problem can be removed or reduced by excluding the cars of employees of all grades from within the premises. If they are admitted, security will be improved if they are required to be parked in a clearly marked area, not on factory roads and as far away from the production and storage areas as possible. The parking area should be well illuminated during the hours of darkness. An additional security measure which might be adopted is to lock the gates of the parking areas except for short periods at the beginning and the end of work periods.

Contractors' vehicles and staff

The agreement in writing should be obtained from building and haulage contractors, whose vehicles and staff enter the premises, to their being subjected to the same searching conditions as company personnel and vehicles. The authorised removal from the premises of the materials and plant of contractors, which have a similarity to company property or of other contractors, can present difficulties. One way to overcome this is to provide the contractors with duplicated forms (a specimen is shown in Appendix 8) to be completed with the name of the firm, the registered number of the vehicle, the date and time of the removal, what is being removed, and the signature of an authorised member of the firm. A copy of all signatures would be supplied to the security department. The pass would be handed to the security officer at the gate and retained.

These forms could provide useful information should claims be received from contractors of the loss of material and plant. It is customary for contractors and other outside persons working on company premises to agree to conditions under which the company is indemnified against common law claims due to the negligent behaviour of their employees. It is suggested that the searching conditions should be included.

Searches at exits

When deciding to carry out searches at the exit from premises these will be more effective if a group of employees is chosen together to be dealt with one by one rather than making individual selections as employees leave. The fact that searches are taking place will soon become known to the later leavers who, if they are carrying stolen property, will try to get rid of it. Therefore after searches have been carried out it is recommended that places where such property could have been hidden on the way to the exit should be examined.

Where small items are likely to be carried, places which might be overlooked are: a rolled or folded newspaper; a container for tea or sugar in a food box; a vacuum flask; a tin of tobacco; the shoulder pads and linings of a coat or jacket, or the waistbands of trousers; strapped or stuck between the shoulders, the backs of calves or thighs; tops of stockings; under the inside sole or the lining of shoes; inside the peak of a cap or the lining; in piled-up hair.

Depending, of course, on the nature of the property at risk from theft, containers of all descriptions and other unexpected places where property can be secreted, such as food boxes, vacuum flasks, and the frame tubes, panniers and handlebars of cycles, should not be neglected when carrying out searches.

Searching women

Some industrial concerns have a high proportion of women employees, either full or part time. The material they are concerned with is often domestically attractive and searching procedures have to be introduced to deter them from stealing. If there is no woman on the security staff it is often the case that the searching of women employees is carried out by a nurse from the health centre or a female member of the personnel staff. This is not a popular duty.

The problem can be overcome by the engagement on a retainer or part-time basis of a woman used to searching females, such as a retired nurse, midwife, policewoman or court matron. When she carries out searches they should be at her discretion within the terms of her engagement. The fact that such an experienced woman has been engaged will be a most effective deterrent to stealing.

Random selection units

To reduce the possibility of complaints in respect of searching procedures a mechanical random selector can be used. This is electrically operated and can be set for a selection rate of, say, 5 of every 100 personnel leaving the premises. This removes from security officers the burden of choosing who to search and who not to. Staveley-Smith Controls Ltd, of Manchester, is one manufacturer of such equipment.

Police procedure

(See also Chapter 8.) It might be as well to mention here that when an accused person is taken into police custody it is the practice approved by the law that he is personally searched for the following reasons:

1 To remove any weapon or anything else by which he may cause injury to himself or others.
2 To prevent damage to property.
3 To find and preserve evidence of the offence concerned and of other offences.

8

Arrest by a private person and police procedure

When police assistance is called for, following an arrest by a private person, the officer attending will ask to be told, *in the presence and hearing of the person detained,* what is the evidence that he has committed an offence. Police procedure may differ in unimportant details from one force to another but basically the actions and requirements are the same. The person responsible for stopping the suspect and who recovered, say, company property from him, would then tell the officer what he saw, what he did, what he said to the suspect, and what the suspect replied. The identification of any property allegedly stolen and its value would be required by the officer from a competent witness.

The officer then will ask the suspect whether, having heard what has been said, he wishes to say anything. He will caution him that he is not required to say anything unless he wishes to do so but what he does say will be taken down in writing and may be given in evidence. The officer will record in writing anything he says. On the assumption that the evidence shows a *prima facie* case against the suspect the officer will then tell him he proposes to arrest him and for what offence and take him to the police station accompanied by the witnesses.

At the police station the station officer on duty will ask to be told the evidence and this will be described by the arresting officer *in the presence and hearing of the accused* or, according to practice, the witnesses will repeat to the station officer what they told the arresting officer. Before accepting the charge, the station officer has to decide what offence appears to have been committed and if there is credible evidence that the accused was responsible for it. He does not have to be convinced of the guilt of the accused.

The offence is written out in plain language on what is known as a charge sheet with a reference to the Act and section which describes it. To ensure that the accused is required to attend court to answer the charge a prosecutor is required. This can be either the owner or his authorised representative who will sign the charge sheet, or the police officer arresting. It has become the practice that where there is clear evidence that the accused person has a case to answer the police officer arresting will sign the charge sheet. The accused is given a form on which the offence with which he is charged is written. This is to give to any legal representative he may wish to instruct.

After this, and when other administrative procedures have been completed and the address of the accused has been verified, he is usually released on bail to appear at the next hearing of the local magistrates' court. (See Chapter 7 for information on the granting of bail.)

Statements or proofs of evidence which witnesses can give are taken down in writing by police officers and signed by the witnesses. They are for the assistance of a lawyer presenting the prosecution or for the clerk of the court in his examination of the witnesses who are called to give evidence. (See Chapter 12 respecting the use of statements instead of the personal attendance of a witness at court.)

At busy courts it can happen that a charge on the list of those to be heard that day is not reached before the court closes. The accused is then readmitted to bail to surrender at the court at a later date. This is called a remand.

Any property involved is retained by the police from the time of arrest until the case is completed, when it is normally returned to the owner. Where, despite a plea of not guilty, the accused is convicted it is customary for the police to retain any exhibits, that is, the property in the case, until the specified time for him to appeal against conviction has passed.

Evidence

The type of evidence which may be given in a court will be described in Chapter 12.

9

Law of theft and related offences

The purpose of this chapter is not to train management and security staff to become lawyers or law enforcement officers. In the performance of security duties, which are fundamentally concerned with the protection of property against being stolen, it is vital to have adequate knowledge of what the law describes as stealing, what evidence is required to justify a prosecution, and what powers exist to arrest the alleged offender.

REPLACEMENT OF OLD LAWS

Following a review of the law of theft (larceny) by the Criminal Law Revision Committee which took seven years to complete, a Bill recommending a number of changes was presented to Parliament. The changes included the repeal or amendment of more than 100 Acts dating from 1275 to 1967. With the exception of minor amendments, the recommendations contained in the Bill were accepted by both Houses of Parliament.

The efforts of the Committee led to the redefinition of theft and the abolition of twenty instances of specialised offences which include obtaining money, chattels, and valuable securities by false pretences with intent to defraud, larceny-bailee, larceny by trick and fraudulent conversion. Those old offences are included in the newly defined offence of 'theft', which will be described later. The Act also incorporates related offences from other Acts—for example, the Falsification of Accounts Act 1875 and the Debtors Act of 1869 (section 13, obtaining credit by fraud) are repealed and the equivalent offences are repeated in the Act. Different terminology and dfinitions are used and new offenes known to the law include blackmail which will be described in section 21 in the extracts from the Act which follow in this chapter. The Act came into effect on 1 January 1969. It has effect only in relation to offences wholly or partly committed on or after that date.

Scotland and Northern Ireland

The Act does not extend to Scotland or, unless separately enacted by the Parliament of Northern Ireland, to that part of the United Kingdom.

Larceny by servant and embezzlement

Some other instances of specialised or aggravated stealing to which police and the security world have become accustomed are not repeated in the Act and probably the most relevant to security duties to disappear is theft and embezzlement by a clerk or servant of the property of his employer.

The legal differences in the offences will soon become historical and space will not be taken up here describing them. Under the old law the offence of stealing from one's employer was viewed as being particularly serious, and a long maximum sentence of imprisonment (fourteen years) could be imposed on conviction.

It is not known whether the Law Revision Committee was influenced in making their recommendations by the alarming increase in theft and embezzlement of recent years, showing that the special relationship that once existed between employees and their employers is not now what it was; however, they decided that breaches of it do not now justify special punishment, and all thefts are recommended to be punishable by up to ten years' imprisonment.

Fraudulent conversion

In the Theft Act the specialised offence of fraudulent conversion is not repeated and offences which come within the old definition will now be treated as theft. Whereas under the old law a warrant was required before an arrest could be carried out, this will not be necessary in future because theft is an offence for which a person may now be arrested without warrant.

Obtaining by false pretences: larceny by trick

These were offences under the old law and one distinction was that for the former offence there was no power of arrest without warrant, whereas in the latter case an arrest could be made without one. The distinction is removed by the new Act and offences which would have been called by either of the names have now been brought under the description of 'criminal deception' (see section 15 which follows) for which under the general provision of the Criminal Law Act 1967 there is a power of arrest without warrant.

EXTRACTS FROM THE THEFT ACT 1968

Section 1. Basic definition of theft

1 A person is guilty of theft if he dishonestly appropriates property belonging to another with the intention of permanently depriving the other of it; and 'thief' and 'steal' shall be construed accordingly.
2 It is immaterial whether the appropriation is made with a view to gain, or is made for the thief's own benefit.
3 The five following sections of this Act shall have effect as regards the interpretation and operation of this section (and, except as otherwise provided by this Act, shall apply only for purposes of this section).

Section 2. 'Dishonestly'

1 A person's appropriation of property belonging to another is not to be regarded as dishonest—

(*a*) if he appropriates the property in the belief that he has in law the right to deprive the other of it, on behalf of himself or of a third person; or

(*b*) if he appropriates the property in the belief that he would have the other's consent if the other knew of the appropriation and the circumstances of it; or

(*c*) (except where the property came to him as trustee or personal representative) if he appropriates the property in the belief that the person to whom the property belongs cannot be discovered by taking reasonable steps.

[*This defines the requirements to substantiate a charge of 'theft by finding'*.]

2 A person's appropriation of property belonging to another may be dishonest notwithstanding that he is willing to pay for the property.

Section 3. 'Appropriates'

1 Any assumption by a person of the rights of an owner amounts to an appropriation, and this includes, where he has come by the property (innocently or not) without stealing it, any later assumption of a right to it by keeping or dealing with it as owner.

2 Where property or a right or interest in property is or purports to be transferred for value to a person acting in good faith, no later assumption by him of rights which he believed himself to be acquiring shall, by reason of any defect in the transferor's title, amount to theft of the property.

Section 4. 'Property'

1 'Property' includes money and all other property, real or personal, including things in action and other intangible property. . . .

[It is not now necessary for a conviction that intangibles should be evidenced by a document which the accused misappropriates as under the old law (see section 4(1)). Consequently a clerk or manager who receives blank signed cheques from his employer with instructions to fill in the names of payees to whom the employer is indebted, will now be guilty of theft if he fills in the names of his own creditors and the cheques are sent to them and cleared, and so are debited to the employer's bank account. Under the old law a clerk or manager committed no offence at all (R. v *Davenport, [1954] 1 All E.R. 602).*

*Similarly a clerk or agent who receives payment of money owed to his principal by means of cheques payable to himself personally and places the cheques into his own bank account and draws against them to pay his private debts, will be guilty of theft, although he too would have escaped punishment under the old law (R.*v *Gale (1876), 2 Q.B.B. 141). This is commendable, but the problem is, how far does the new Act go in including intangibles within its scope? For the wider definition of property may widen the crime of larceny beyond all recognition.]*

Section 5. 'Belonging to another'

1 Property shall be regarded as belonging to any person having possession or control of it, or having in it any proprietary right or interest (not being an equitable interest arising only from an agreement to transfer or grant an interest). . . .

3 Where a person receives property from or on account of another, and is under an obligation to the other to retain and deal with that property or its proceeds in a particular way, the property or proceeds shall be regarded (as against him) as belonging to the other.

4 Where a person gets property by *another's mistake,* and is under an obligation to make restoration (in whole or in part) of the property or its proceeds or of the value thereof, then to the extent of that obligation the property or proceeds shall be regarded (as against him) as belonging to the person entitled to restoration, and *an intention not to make restoration* shall be regarded accordingly as an intention to deprive that person of the property or proceeds. . . .

Section 6. 'With the intention of permanently depriving the other of it'

1 A person appropriating property belonging to another without meaning the other permanently to lose the thing itself is nevertheless to be regarded as having the intention of permanently depriving the other of it if his intention is to treat the thing as his own to dispose of regardless of the other's rights; and a borrowing or lending of it may amount to so treating it if, but only if, the borrowing or lending is for a period and in circumstances making it equivalent to an outright taking or disposal.

2 . . . where a person, having possession or control (lawfully or not) of property belonging to another, parts with the property under a condition as to its return which he may not be able to perform, this (if done for purposes of his own and without the other's authority) amounts to treating the property as his own to dispose of regardless of the other's rights.

Section 7. Theft

A person guilty of theft shall on conviction on indictment be liable to imprisonment for a term not exceeding ten years.
[*All such crimes are therefore arrestable offences, see Chapter* 7.]

Section 8. Robbery

(1) A person is guilty of robbery if he steals, and immediately before or at the time of doing so, and in order to do so, he uses force on any person or puts or seeks to put any person in fear of being then and there subjected to force.

(2) A person guilty of robbery, or of an assault with intent to rob, shall on conviction on indictment be liable to imprisonment for life.

[*To constitute the offence the force used need not have been applied to the person from whom the property is stolen.*]

Section 9. Burglary

(1) A person is guilty of burglary if—

(*a*) he enters *any building or part of a building* as a trespasser and with intent to commit any such offence as is mentioned in subsection (2) below; or

(*b*) having entered any building or part of a building as a trespasser he steals or attempts to steal anything in the building. . . .

(2) The offences referred to in subsection (1)(*a*) above are offences of stealing anything in the building or part of a building in question, of inflicting on any person therein any grievous bodily harm or raping any woman therein, and of doing unlawful damage to the building or anything therein.

(3) References in subsections (1) and (2) above to a building shall apply also to an inhabited vehicle or vessel . . .

(4) A person guilty of burglary shall on conviction on indictment be liable to imprisonment for a term not exceeding fourteen years.

Section 10. Aggravated burglary

(1) A person is guilty of aggravated burglary if he commits any burglary and at the time has with him any firearm or imitation firearm, any weapon of offence, or any explosive; and for this purpose—

(*a*) 'firearm' includes an airgun or air pistol, and 'imitation firearm' means anything which has the appearance of being a firearm, whether capable of being discharged or not; and

(*b*) 'weapon of offence' means any article made or adapted for use for causing injury to or incapacitating a person, or intended by the person having it with him for such use; and

(*c*) 'explosive' means any article manufactured for the purpose of producing a practical effect by explosion, or intended by the person having it with him for that purpose.

(2) A person guilty of aggravated burglary shall on conviction on indictment be liable to imprisonment for life.

[*It is interesting to note that under the old definition of burglary the offence was concerned only with dwelling houses. Now the offence can be committed in* any building *or part of a building, also inhabited caravans and houseboats. An essential element in the old definition was a breaking into or out of the premises concerned, but it is not now necessary to prove this.*]

Section 11. Removal of articles from places open to the public

(1) . . . where the public have access to a building in order to view the building or part of it [or any of its contents], any person who without lawful authority removes from the building or its grounds the whole or part of any article displayed or kept for display to the public in the building or that part of it or in its grounds shall be guilty of an offence. . . .

(4) A person guilty of an offence under this section shall, on conviction on indictment, be liable to imprisonment for a term not exceeding five years.

Section 12. Taking a motor vehicle or other conveyance without authority

(1) Subject to subsections (5) and (6) below, a person shall be guilty of an offence if, without having the consent of the owner or other lawful authority, he takes any conveyance for his own or another's use or, knowing that any conveyance has been taken without such authority, drives it or allows himself to be carried in or on it. . . .

(3) Offences under subsection (1) above and attempts to commit them shall be deemed for all purposes to be arrestable offences within the meaning of section 2 of the Criminal Law Act 1967. . . .

(5) Subsection (1) above shall not apply in relation to pedal cycles; but, subject to subsection (6) below, a person who, without having the consent of the owner or other lawful authority, takes a pedal cycle . . . shall on summary conviction be liable to a fine not exceeding £50.

(6) A person does not commit an offence under this section by anything done in the belief that he has lawful authority to

do it or that he would have the owner's consent if the owner knew of his doing it and the circumstances of it. . . .

Section 13. Abstracting of electricity

A person who dishonestly uses without due authority, or causes to be wasted or diverted, any electricity shall on conviction on indictment be liable to imprisonment for a term not exceeding five years. . . .

Section 15. Obtaining property by deception

(1) A person who by any deception dishonestly obtains property belonging to another, with the intention of permanently depriving the other of it, shall on conviction on indictment be liable to imprisonment for a term not exceeding ten years. . . .

Section 16. Obtaining pecuniary advantage by deception

(1) A person who by any deception dishonestly obtains for himself or another any pecuniary advantage shall on conviction on indictment be liable to imprisonment for a term not exceeding five years. . . .

Section 17. False accounting

(1) Where a person dishonestly, with a view to gain for himself or another or with intent to cause loss to another,—
(*a*) destroys, defaces, conceals, or falsifies any account or any record or document made or required for any accounting purpose; or
(*b*) in furnishing information for any purpose produces or makes use of any account, or any such record or document as aforesaid, which to his knowledge is or may be misleading, false or deceptive in a material particular;
he shall, on conviction on indictment, be liable to imprisonment for a term not exceeding seven years.

(2) For purposes of this section a person who makes or concurs in making in an account or other document an entry which is or may be misleading, false or deceptive in a material particular, or who omits or concurs in omitting a material particular from an account or other document, is to be treated as falsifying the accounts or document. . . .

Section 21. Blackmail

(1) A person is guilty of blackmail if, with a view to gain for himself or another or with intent to cause loss to another, he makes any unwarranted demand with menaces; and for this purpose a demand with menaces is unwarranted unless the person making it does so in the belief—
(*a*) that he has reasonable grounds for making the demand; and
(*b*) that the use of the menaces is a proper means of reinforcing the demand.

(2) The nature of the act or omission demanded is immaterial, and it is also immaterial whether the menaces relate to action to be taken by the person making the demand.

(3) A person guilty of blackmail shall on conviction on indictment be liable to imprisonment for a term not exceeding fourteen years.

Section 22. Handling stolen goods

[*This replaces the offence of receiving stolen property under the Larceny Act 1916 and conviction will not now rely on evidence that the accused actually had physical possession of the stolen property.*]

(1) A person handles stolen goods if (otherwise than in the course of the stealing) knowing or believing them to be stolen goods he dishonestly receives the goods, or dishonestly undertakes or assists in their retention, removal, disposal, or realisation by or for the benefit of another person, or if he arranges to do so.

(2) A person guilty of handling stolen goods shall on conviction on indictment be liable to imprisonment for a term not exceeding fourteen years.

Section 23. Advertising rewards for return of goods stolen or lost

Where any public advertisement of a reward for the return of any goods which have been stolen or lost uses any words to the effect that no questions will be asked, or that the person producing the goods will be safe from apprehension or inquiry, or that any money paid for the purchase of the goods or advanced by way of loan on them will be repaid, the person advertising the reward and any person who prints or publishes the advertisement shall on summary conviction be liable to a fine not exceeding one hundred pounds. . . .

Section 25. Going equipped for stealing, etc.

(1) A person shall be guilty of an offence if, when not at his place of abode, he has with him any article for use in the course of or in conection with any burglary, theft or cheat.
(2) A person guilty of an offence under this section shall on conviction on indictment be liable to imprisonment for a term not exceeding three years.
(3) Where a person is charged with an offence under this section, proof that he had with him any article made or adapted for use in committing a burglary, theft or cheat, shall be evidence that he had it with him for such use.
(4) *Any person may arrest without warrant* anyone who is, or whom he, with reasonable cause, suspects to be, committing an offence under this section.
(5) For the purpose of this section an offence under section 12(1) of this Act of taking and driving away a conveyance shall be treated as theft, and 'cheat' means an offence under section 15 of this Act.

[*Under section 28 of the Larceny Act 1916, the offence of possessing housebreaking implements could be committed only at night, that is between 9 p.m. and 6 a.m. Possession at other times could be prosecuted under the Vagrancy Act 1824, but the punishment was very much lighter. The replacing offence is more comprehensive in its scope than the old law and can be committed at any time.*]

Section 26. Search for stolen goods

(1) If it is made to appear by information on oath before a justice of the peace that there is reasonable cause to believe that any person has in his custody or possession or on his premises any stolen goods, the justice may grant a warrant to search for and seize the same; but no warrant to search for stolen goods shall be addressed to a person other than a constable except under the authority of an enactment expressly so providing. . . .

POWER OF ARREST

For theft and other related offences, *and attempts to commit them* which are offences at common law, the power of arrest without warrant is provided by the Criminal Law Act 1967, section 2. See also Chapter 7.

OFFENCES BY CHILDREN

The age of liability was raised to ten years by the Children and Young Persons Act 1963 section 16(1). However, by a similarly named Act of 1969, section 4, a person shall not be charged with an offence, except homicide, by reason of anything done or omitted while he was a child. A child is a person under the age of fourteen. This restriction has no application to Northern Ireland. The date when this Act comes into effect had not been announced at the time of writing (March 1972).

RESTITUTION AND COMPENSATION IN CRIMINAL PROCEEDINGS

Too often in the past losses from theft and damage have been regarded as matters for recovery through insurance or as part of inevitable wastage. The right to reparations from those responsible is, increasingly, being recognised both by legislators and the judiciary. Claims should be made wherever possible and apart from being a financial recoupment they will provide a salutary experience to the offender and a warning to others. It must be remembered, however, that a court will not automatically make an order—the operative word is 'may' and all contentious issues are likely to be left to civil action.

Restitution of stolen property or its equivalent

The Theft Act 1968 section 28 is the authority most likely to be invoked. Where goods have been stolen and a person is convicted of *any* offence in connection with the theft—this could include the handling of stolen property—a convicting court *may*:

1 Order *anyone* having possession or control of the goods to return them
2 Where the stolen property has been converted into other goods by exchange or otherwise, *on application of the loser or his representative,* order the person convicted to hand over these in lieu of those stolen
3 If, at the time of his arrest, the person convicted had money in his possession, *on application of the loser or his representative,* order money not exceeding the value of the stolen property to be paid out of that money

There is no financial limit imposed in (3). If the goods are in the possession of an innocent purchaser, he too may request an order for repayment of the money he has paid from that found in the possession of the thief at the time of his arrest.

These powers are solely applicable to instances where a person has been convicted for the offence. One which has been 'taken into consideration' is *not* regarded as a conviction and a court can base no orders on it. It is police practice when an individual is responsible for a number of offences to lay substantive charges on perhaps three and allow the accused, if he desires, to ask the court to take others he admits into account when passing sentence. Where a company or anyone else wishes to claim upon the offender, the police should be asked to prefer a charge against him in respect of their particular interests. Neither the procedures of the police nor of the courts are entirely uniform even when ownership is not in dispute. Where the goods have been recovered the police will in most areas automatically hand them over. In a few they or the prosecuting solicitors will make formal application on behalf of the loser under (1) above. If the offender originally pleaded 'not guilty', a period of fourteen days is allowed for the lodging of an appeal and the goods concerned in the charge will be retained by the police until this has lapsed.

Since many cases are now dealt with in the absence of witnesses (see Chapter 12) it is advisable that where a loser desires to claim compensation this should be made clear when the fact of an arrest is known and the application should be incorporated in the statement of the representative who is to give evidence, and for some police forces this is routine practice.

Where the police have come into possession of property in connection with any criminal charge and the ownership is not resolved in court, any claimant for the property can make application to a court for an order to deliver the property to the person appearing to that court to be the owner, or make any other order as they think fit (Police (Property) Act 1897).

Compensation

The Forfeiture Act 1870 section 4 is the authority and the sum that can be awarded is up to a maximum of £400. Compensation will only be awarded upon the application of an aggrieved person and it is not restricted to theft but also can be applied to damage. The application must be made immediately after conviction and to the same court. Applicants are advised to quote total costs of making damage good, not the cost of materials only, and have documents at hand to show a breakdown of the total.

A high percentage of all theft and damages is attributable to children and young persons (under the age of seventeen years). A court may order the parent or guardian of such an offender to be responsible for such fines, damages, costs or compensation as may be awarded.

General advice

1. Where you sustain damage through loss or theft and the offender is traced, always consider the possibility of obtaining compensation from him.
2. Enquire from the police how to lay your claim and indicate your wish to do so at the first available opportunity. Incorporate this request in the statement of an employee who is to give evidence.
3. In the case of damage, itemise and be prepared to supply proof of the cost of making good.
4. If there is any indication that ownership of the stolen goods will be disputed by subsequent purchasers, consult your legal advisers immediately.
5. Whenever possible, and always when an employee is involved, have a representative at the court hearing so as to have first-hand knowledge of the evidence.
6. The foregoing considerations are applicable to English courts, only. Scottish law often differs appreciably and guidance should be sought from the police dealing with the offender when a claim is to be made there.

10

Other offences against security

CRIMINAL DAMAGE

The Criminal Damage Act 1971 has consolidated and made obsolete all previous legislation connected with damage caused deliberately or by completely reckless behaviour with the exception of certain special matters concerning railways and shipping. Its provisions do not apply to Scotland and Northern Ireland.

Where fire is the means employed to cause damage, the offence is called and charged as 'arson'. Where, for example, an employee, impelled by grievance or other motives, damages machinery or destroys records, that is termed 'sabotage' but the charge is not correspondingly named.

CRIMINAL DAMAGE ACT 1971

All offences under this Act make those responsible liable to imprisonment. None of the maximum sentences is less than ten years, indeed, arson and acts likely to endanger life carry a potential life sentence. It follows therefore that all are arrestable offences and therefore any person, which of course includes a security officer and employees generally, may carry out an arrest (see page 87).

The basic offences created by this Act (not verbatim) are:

1 Without lawful excuse, destroying or damaging any property belonging to another intending to do so or being reckless as to whether it was so damaged or destroyed.
2 Without lawful excuse destroying or damaging any property, whether belonging to himself or another, intending to do so, or being reckless as to whether any property would be destroyed or damaged *and* by so doing intending to endanger the life of another or being reckless as to that risk.

To illustrate 2, there could be the deliberate blowing up of premises with an intention to kill someone therein; alternatively, say for insurance purposes, a separately occupied part of larger premises could be set on fire with total disregard of the certain spread to other sections with inevitable risk to the life of those in them.

3 Without lawful excuse, making to another a threat, intending that the other person would fear it would be carried out, to:
(*a*) Destroy or damage any property belonging to that or a third person, or
(*b*) To destroy or damage his own property in a way which he knows is likely to endanger the life of that other or a third person.

It is possible to visualise in embittered industrial disputes that threats under 3(*a*) could be made.

4 Having anything in his custody or under his control intending without lawful excuse to use it or cause or permit another to use it:
(*a*) To destroy or damage any property belonging to some other person or
(*b*) To destroy or damage his own or the user's property in a way which he knows is likely to endanger the life of some other person.

Possession of petrol bombs might come under this section.

'Without lawful excuse'

The following 'lawful excuses' do not apply to any circumstances involving threatening or endangering the life of another. They are statutory excuses peculiar to this Act and in addition to normal defence 'excuses'. They are that:

1 At the time of committing the acts, he believed they had been consented to, or would have been consented to, by those responsible for the property if they had known of the actions and the circumstances.
2 His actions were designed to protect property, rights or interests belonging to himself or another and at the time he honestly believed:
(*a*) That they were in immediate need of protection and
(*b*) The actions used or intended were reasonable having regard to all the circumstances.

These excuses are obviously intended to cater for mistakes and incidents such as breaking down of doors and demolition to stop spread of fire.

Compensation

Where a person is convicted of an offence under section 1 of the Act (actual damage), the court, on being satisfied as to the approximate cost of making good the loss or damage to the property, may order compensation to the loser as it thinks fit—but not exceeding £400 in the case of a magistrates' court. A formal application is not a necessity—but it would be advisable, see pages 113-16.

OFFENCES AGAINST THE PERSON

Assault

Under the Offences Against the Person Act 1861 an assault is constituted by any attempt to apply unlawful force to another. 'Battery' includes every touching and laying hold, however trifling, of another person's clothes in an angry, rude, insolent or hostile manner. If done by accident or with lawful consent it would not constitute an assault and battery. This proviso is referred to in Chapter 7 when dealing with the carrying out of personal searches.

Assaults are separated into common assaults and the more serious ones resulting in some element of grievous bodily harm being sustained. Where a complaint of common assault is made to the police it is usual for the complainant to be recommended to make his own application to a magistrate for a summons against whoever he alleges assaulted him. Where that person also alleges he was assaulted and likewise obtains a summons it is called a cross-summons.

If a weapon has been used in an assault it must be carefully looked after and not cleaned in any way as it may be required to be forensically examined for bloodstains, hair of the injured person, etc. (see Chapter 12 under 'Exhibits'). Where personal injuries are directly attributable to a criminal offence, or sustained when trying to prevent a crime, trying to arrest a suspect or helping the police, notwithstanding that the identity of the person responsible is unknown, the existence of the Criminal Injuries Compensation Board should not be overlooked. Applications for compensation should be addressed to the Board at Russell Square House, 10 Russell Square, London WC1B 5EN.

Power of arrest

For the more serious assaults any person may arrest any person without a warrant (Criminal Law Act 1967). See page 113 respecting persons under the age of fourteen.

Indecent assaults

The Sexual Offences Act 1956 covers this type of assault on a male or female. It should be reported to the police for further action. Consent by a person of sixteen or older nullifies a charge of indecent assault upon him or her.

BRIBERY

When this subject is mentioned it appears that most persons

believe it concerns only an employee in a government or other public body or the police. The Prevention of Corruption Act 1906, section 1, extends to persons in all descriptions of employment and says: 'Any agent who corruptly accepts from any person any gift or consideration as an inducement or reward for doing or forbearing to do any act in relation to his principal's affairs or business, or any person who corruptly gives, agrees to give, or offers such gift or consideration for so doing shall be guilty of an offence.' 'Agent' includes any person employed by or acting for another. 'Consideration' includes valuable considerations of any kind.

If any offence comes to light the first step is to report it to the police. No prosecution, however, may be instituted without the consent of the Attorney-General or the Solicitor-General.

Security officers are especially open to corruption by interested parties and should at once report to their superior any approach which appears to be made with that intention, taking care not to show in any way to the person making the approach that this is going to be done.

FORGERY

This means altering a genuine document or making a false document so as to appear genuine with intent to defraud. That definition extends beyond merely the forging of a signature. The actual use of a forged document is called 'uttering' and is an offence additional to the forgery. The Forgery Act 1913 is the relevant statute.

Documents controlling the loading of goods or the passage of them through any gate control are susceptible to being altered with intent to get stolen goods through. An example is where the practice was to summarise the individual deliveries of goods on a lorry onto a loading sheet to assist the loaders. Following shortages of stock an investigation showed that some of the summaries had been forged by adding the figure 1 in front and 0 behind others. This had led to the loads being increased over the correct totals and the drivers, who were parties to the fraud, sold the excess goods. It is recommended that if a document is used to authorise the loading or dispatch of goods it should be typewritten in duplicate and one copy retained by the authorising department. Spot checks should be made on the loading bays to see that the original compares with the copy.

Obtaining stores of any description by falsely increasing the amount to be drawn or adding additional items with intent to steal them is obtaining by means of a forged instrument. Requisitions should be checked from time to time with the authorisers to see whether any alterations have been made. The fact that requisitions are subject to check is a deterrent to altering them.

Power of arrest

Any person may arrest without warrant any person who has committed the offences which have been described (Criminal Law Act 1967). See page 113 respecting persons under the age of fourteen.

PROSECUTION

To assist in the prosecution of offences against the various Acts which have been mentioned, written statements, sometimes called 'proofs of evidence', are required from the witnesses of fact. These may be obtained by the person in charge of security if he is qualified to do so or by the police. Taking statements is dealt with in Chapter 12.

TRESPASS

Although this is not now a criminal offence unless specifically laid down by an Act of Parliament, such as trespassing on the property of railways and airfields and in factories and stores containing explosives, it would be as well if a little attention were given to the subject to bring the powers of a security officer into perspective in relation to trespassers. Trespass constitutes in law an act of interference with the possession of land and is actionable in a civil court without proof of damage.

An order of the court may be made prohibiting a repetition. If the order is disobeyed damages may be awarded.

Notices bearing the words 'trespassers will be prosecuted' are put up by owners of land and premises who know they are empty threats but nevertheless hope that the readers of them will be impressed and keep out.

Eviction

If a trespasser refuses to leave property when requested he may lawfully be removed by the use of only such force as is necessary to eject him. A security officer employed by the owner of the land or premises concerned is his agent and can exercise all the powers of the owner. Chapter 7 contains some observations on the use of violence in self-defence and other circumstances and should be read in this connection.

In the majority of cases the trespasser has quite innocent intentions and will leave immediately he is told he is trespassing. However, should his actions suggest he is there for a sinister purpose the police should be informed and in the meantime, if he is unaware his presence has been discovered, he should be kept under observation. If, on the other hand, he knows he has been discovered and his anwers to questions are unsatisfactory he should be detained and the police called.

The name and address of a trespasser with whom the police will not be concerned should be requested, but no further action is available in law should this be refused.

OFFICIAL SECRETS ACT

Where work of any nature is being carried out which is described as 'classified', the government usually require the security measures to protect information about it getting into the wrong hands to be to their satisfaction. Divulgence of such information to unauthorised persons is an offence under the Official Secrets Act. The measures recommended and introduced to prevent such leakages are of course confidential and cannot be dealt with here.

DRUGS

The effects of taking drugs can include a diminished sense of responsibility for the protection of employers' property against loss and for the observance of safety rules.

Drugs create an increasing reliance on them and the addict often becomes unable, from normal resources, to pay for them. He is then compelled to obtain money from some other source, for example, stealing cash and company property which is convertible into cash with which to purchase what is required. But for the craving, such actions would be unthinkable to the addict.

While under the effect of a drug the addict is unstable and his behaviour could vitally harm the interest of his employer or himself.

It is therefore essential that indications of drug abuse must be identified as early as possible for action to be taken. These include:

1 A dramatic increase in pilferage.
2 Empty pill or medicine bottles and used syringes in rubbish bins or wastebaskets.
3 The smell of burning rope in washrooms or hallways.
4 Sudden behaviour changes in otherwise stable employees.
5 Erratic changes in emotion or activity by formerly stable employees.
6 Degeneration of performance level.

Some of the physical symptoms that supervisors can be trained to recognise may include:

1 The drug user may wear sunglasses at inappropriate times to cover dilated or constricted pupils, and wear long sleeves at all times, even on hot summer days, to cover marks caused by intravenous injection of drugs.
2 The abuser of depressants such as the barbiturates usually shows most of the signs of alcohol intoxication—slurred speech, staggering, etc.—with one exception: there is no odour of alcohol on the breath. This type of abuser may fall asleep on the job or appear confused.

3 Abuse of amphetamines (stimulants) such as diet pills or 'pep' pills brings about excessive activity, excitability or irritability, and heavy perspiration. The pupils may be dilated, even in bright light. These drugs also may cause unusual bad breath unlike that resulting from garlic or alcohol, and chapped or cracked lips caused by drying of the mucous membranes. The amphetamines also cause complexion problems. Finally, excessive nervousness may lead to itching, chain smoking or talkativeness.

4 The most obvious symptoms of abuse of heroin are injection marks and constricted pupils. Sometimes the addict may make frequent or prolonged trips to his locker or to the restroom for privacy to inject the drug. He may become excessively secretive about his belongings. After injection the user may become drowsy and lethargic.

5 Marihuana cigarettes rolled in brownish paper and crimped at the ends are usually smoked while the users are gathered in groups. The smoke of marihuana is recognisable by its odour, like that of burning rope. Smoking marihuana may produce talkativeness, bursts of hilarity, and distortions of time and space. There is some loss of coordination, and the eyes may become bloodshot. As the drug wears off, drowsiness and lassitude comes on.

6 Although it is unlikely that an employee will use hallucinogens at work, it is possible. Under the influence of hallucinogens like LSD, the individual appears to be in a trance or dream-like state. The most common symptom is dilation of the pupils. Mental effects are unpredictable; the user may experience exhilaration, an urge to self-destruction, or sheer panic. 'Flashback' drug episodes may occur weeks later, perhaps while on the job, without further ingestion of the drug.

Supervisors should be certain that there are no other explanations for the symptoms before concluding that the employee is a drug user. Simple fatigue, for example, can cause a number of symptoms similar to those listed.

11

Questioning and the Judges' Rules

It is difficult to visualise any set of circumstances in which a security officer can act without indulging in some form of questioning. For the most minor breach of works regulations, where he may have to request the name and works number of the employee and point out the nature of the breach, his mode of approach may be the subject of complaint by the person to whom he is speaking unless he behaves in a reasonable manner.

It is unfortunately a characteristic of persons in trouble that they are apt to find fault with the individual who is responsible for their predicament. This is sometimes seen in its ultimate phase at the criminal courts in the form of making unfounded allegations against the conduct of prosecution witnesses, with a view to diverting the attention of the jury or magistrates away from the essence of the matter—the guilt of the accused. Apart from the precautionary nature of so doing it is common sense to adopt the reasonable attitude which one would expect if the positions were reversed.

No excessive use of authority should be made or even implied. However, many offenders regard attack as being the best form of defence, so neither should there be any question of backing down in the face of an aggressive approach from someone who is in the wrong. Politeness coupled with firmness soon makes such a person see that he is making a fool of himself to no purpose.

In contrast to the police, the only security officers who are likely to need to develop a genuine technique of questioning are a limited number of chief security officers, whose firms prefer to conduct their own internal inquiries into major defalcations. For the ordinary security officer, questioning should be kept to a minimum, consistent with establishing that no mistakes have been made and that the individual being spoken to has had every opportunity to clarify the circumstances.

The Judges' Rules, which are dealt with in Chapter 7 and later in this chapter, were made to regulate the conduct of questioning by the police but rule 6 deals with their application to other persons:

> Persons other than police officers charged with the duty of investigating offences or charging offenders shall, so far as may be practicable, comply with these Rules.

It is evident that a security officer, who is employed to safeguard his employers' interests and apprehend those who attempt to steal their property, is a person who has such a duty. This is again particularly applicable to the actions of store detectives. See Chapter 21 for a High Court interpretation of the Rules where a shop manager questioned an alleged shoplifter.

The foundation stone on which the Judges' Rules are based is set out in Appendix *A*, subsection *E* of the Judges' Rules and Administrative Directions to the Police 1964:

> That it is a fundamental condition of the admissibility in evidence against any person, equally of any oral answer given by that person to a question put by a police officer and of any statement made by that person, that it shall have been voluntary, in the sense that it has not been obtained from him by fear of prejudice or hope of advantage, exercised or held out by a person in authority, or by oppression.

Appendix *A* terminates with the following sentence:

> Non-conformity with these Rules *may* render answers and statements liable to be excluded from evidence in subsequent criminal proceedings.

It should be noted that the operative word here is 'may'; this places on the court itself the responsibility of deciding whether a statement or answer shall or shall not be admissible in the course of proceedings.

Security officers are likely to be regarded more leniently by a court than would be their police equivalents for contraventions of these Rules. The circumstances will be the deciding factor. There will be instances where what has been said is so vital to the prosecution that the defence may well try to have it erased completely from the proceedings. There is little doubt that this would be successful where an employee who was questioned by a senior security officer later claimed that he was not aware of his rights and had answered because he thought he had to, in view of the status of the security officer with his employers. If a court decides that the security officer should have cautioned the person to whom he was speaking, it will simply not allow that part of his evidence to be given or will disregard it if, by chance, it already has been spoken.

Suffice it to say then, that if a matter under inquiry is discovered to be of some gravity and the replies to questions being put are vitally important to the truth of the matter in issue, it may be best to administer a caution to ensure that the subsequent conversation is admissible in a court. It is as well, therefore, for anyone in authority, who may interrogate in connection with a criminal offence, to know the essential points of the Judges' Rules. In outlining these later, some with no security interest have been omitted or abbreviated, others with limited application have been included for information and to show the trend of thought of the legislators.

It is true to say that facts speak for themselves. Conversation will be relatively unimportant where there is overwhelming evidence against an offender, so the less said the better. It is re-emphasised that the penalty that is paid for non-conformity with the Judges' Rules in the asking of questions is that neither questions nor answers can be given in the course of evidence.

Before dealing with the Rules in detail it may be appropriate to outline the fundamentals which must be observed in the course or form of questioning.

PRINCIPLES AND PRACTICE OF OBTAINING STATEMENTS AND QUESTIONING SUSPECTS

It must be appreciated that real ability in ascertaining the truth from a confused witness or an obdurate suspect cannot be acquired by memorising a checklist of points to be observed. Much will depend upon the personality of the questioner and his correct assessment of the person to whom he is speaking. Technique is only acquired after years of practice in talking objectively to a diversity of people to an extent which security officers are rarely likely to experience. This can become an intuitive skill rarely appreciated by any layman or even perhaps the questioner himself.

Principles to be observed

1 The object is to learn the truth, not to induce a pattern of deceit nor simply to obtain answers satisfactory to the questioner.
2 Under no circumstances whatsoever should any implied threats or promises be used to obtain answers.
3 If a person has a possible explanation for his actions, he must be given every opportunity to state it. If, having been given the opportunity, he fails to take advantage of it, a court may be subsequently entitled to comment on this and he would encounter little sympathy for complaints of inconvenience arising from his failure to do so.
4 Questions must be asked in a language and phraseology which is clearly understood by the person being spoken to.
5 Ambiguous questions must not be asked. If the answers are ambiguous, they should be clarified by further questions so that there is no misunderstanding on either side.

6 Questioning must be done methodically to cover all the information that is required. A person who is not telling the truth can put up a much better performance if he has intervals in which to collect his thoughts—inexplicable hesitancy will have a reason behind it.

7 Under no circumstances must the person who is asking the questions lose his temper, irrespective of provocation. Aggressiveness, abuse, ridicule, complaint, unwarranted argument, and 'talking down to' are all means employed by an 'experienced' suspect to discomfort a questioner; a reluctant witness may employ some of these in a lesser degree.

Inside the limitations imposed by these principles, there is scope for the questioner to display his virtuosity in playing on the personality of the individual to whom he is speaking. Formal approach, sympathy, incredulity, and even a degree of flattery may be needed to persuade the witness to divulge what he knows.

There are several guidelines which can be given, but these do not replace practice or make a person into the 'good mixer' which is necessary to establish confidence.

The conditions under which the interview is taking place are of importance. It is a waste of time being in a place where people are constantly passing to and fro, diverting attention and distracting by noise and conversation. Equally, if the continuity of the conversation is continually broken by telephone calls, queries to either party, or any other form of interruption, this is fatal to any complex discussion, particularly if dealing with a suspect. In the further interest of the privacy of the person being spoken to, which is desirable, a quiet, separate room is preferable, with refreshment available if needed during a long session.

In dealing with employees there may be established company rules which allow them the presence of a representative, should this be practicable in the circumstances. If dishonesty is the subject, the probability is that an employee will prefer to keep it between himself and the interviewer; if the implication is disciplinary it is more likely that the meeting will be pre-arranged and the attendance of a shop steward or other 'friend' asked for. It is worth remembering and pointing out, that if prosecution follows the 'friend' may find himself in what may be the invidious position of being required by the police to give evidence.

When taking statements of any kind, about the best way of producing a thoroughly bad one, and perhaps causing a witness to close up completely, is to pick up a note book or sheet of paper, then sit down and start writing immediately. The first step is to get a clear picture of what has happened before committing anything to writing, this will make the subsequent screed more readable, sequential, coherent and comprehensive. It will also give an opportunity to draw out the witness on any points on which he is reticent. Having achieved this end and, perhaps, having jotted down a few rough words for sequence, an attitude of 'let us' or 'should we' get this down in writing will make the witness take a keener interest in what is being recorded and be more personally involved in ensuring absolute accuracy, since he himself has had time in the preceding discussion to marshal his thoughts.

Do not hurry statements as you may omit matter that should be included. Although everyone at some time or another has done this, it reflects adversely on the capability of the note-taker and should be avoided if at all possible. Keeping to a chronological sequence helps to prevent omissions.

In the case of a suspect, anything that is said or written by him should be in his language, not that of the interviewer, and remember that the questions are equally important evidentially as the answers.

Technique

The amount of information that can be obtained from a witness or a suspect is dependent upon the rapport established by the questioner. He has to recognise their kind of emotional reaction to the situation, mentally put himself in their position, and then modify his own personality and approach accordingly to establish the desired relationship.

Correct recognition of the witness/suspect type is necessary. The talkative is easy to handle but needs gently coaxing back into the areas from which information is required. The boastful need nudging in the same direction but at the same time their vanity can be played upon so that in the process of self-aggrandisement they may throw out information they really do not wish to give. A nervous witness is a difficult type and the symptoms may be mistaken for a guilty desire to conceal facts, or as sheer stubbornness—this type needs to be dealt with sympathetically. The taciturn have to coaxed and persuaded to talk, perhaps even by questions unconnected with the main line of inquiry. The aggressive type has to be cooled down by a total lack of reaction to his attitude and by unruffled patience. The self-pitying need a good listener, but one who takes every opportunity to channel the conversation onto the subjects he is interested in. The casually evasive may react to a direct and formal approach to bring them down to earth. The outright and blatant liar can sometimes be embarrassed into truth by unconcealed and somewhat hilarious disbelief.

Of course these are generalisations and there may be occasions where a person does not clearly conform to any group, possibly due to the embarrassment of the situation in which he is found. With such circumstances a dispassionate approach as if this was an everyday matter is advisable.

The cardinal characteristics to acquire may be listed as:

1 Be tactful—do not cause unnecessary friction or resentment.
2 Be a good listener—try to create an atmosphere of interest to encourage the witness to talk confidently.
3 Be patient—above all avoid any display of boredom.
4 Do not interrupt unnecessarily, except to keep the conversation to the point.
5 Do not show annoyance or make either verbal or physical threats.
6 Do not use ridicule or abuse.
7 Do not make promises or offer inducements to obtain answers.
8 Do not be diverted from your objective by any antics of the person being spoken to.
9 Persist until you are satisfied you have found out as much of the truth as you are likely to get.
10 Do as you would be done by if the positions were reversed.

Remember that if any of your assumptions have been wrong, any annoyance felt by the person to whom you have been speaking will be mitigated by a pleasant and courteous approach.

JUDGES' RULES

The first of these were formulated and approved by the Judges of the King's Bench Division in 1912. They were added to and revised in 1918 and restated in January 1964,

The Rules are for guidance on the following main points:

1 Procedure in dealing with suspects and prisoners.
2 When they can be interrogated.
3 When interrogation should cease.
4 When to caution.
5 Form of caution.
6 Taking statements.

A number of principles are laid down in Appendix *A* to the Rules in addition to the 'foundation stone' quoted above. It is clearly stated that the Rules do not affect the following principles:

1 Citizens have a duty to help a police officer to discover and apprehend offenders.
2 Police officers, other than by arrest, cannot compel any person against his will to go to or remain in any police station.
3 Every person at any stage of investigation should be able to communicate and consult privately with a solicitor. This is so even though he is in custody, provided that in such cases no unreasonable delay is caused to the processes of investigation or the administration of justice by his doing so.
4 When a police officer, who is making inquiries about an offence, has enough evidence to prefer against any person for the offence, he should without delay cause that person to be charged or informed that he may be prosecuted for the offence.

Rule 1. Any person may be questioned

This stipulates that a police officer inquiring into an offence may question anyone from whom he thinks useful information could be obtained, irrespective of whether the person to be questioned is in custody or not, so long as he has not been charged or told he would be prosecuted in connection with the offence being inquired into.

Rule 2. When to caution

When a police officer has evidence giving reasonable grounds for suspecting that a person has committed an offence he must 'caution' that person before proceeding further with questions.

The caution shall be in the following terms:

> You are not obliged to say anything unless you wish to do so but what you say will be put into writing and given in evidence.

When, after being cautioned, a person is being questioned or elects to make a statement, the time of commencement and termination of such questioning and statement shall be recorded, together with the place and persons present.

Rule 3. Charging with an offence or informing persons they may be prosecuted

This applies only to police officers. It deals with the procedure when a person is charged or informed that he may be prosecuted for an offence and outlines the official caution that must be administered in these circumstances, which is:

> Do you wish to say anything? You are not obliged to say anything unless you wish to do so, but whatever you say will be taken down in writing and may be given in evidence.

The only questions that may be put to an accused person after the administration of this caution and in connection with that particular offence are those essential for the purpose of preventing or minimising harm or loss to some other person, to the public, or to clear up ambiguities. The procedure that should be followed for such further questioning is also laid down in this Rule.

Rule 4. Written statements after caution

It is possible that written statements may now be taken after caution by chief security officers investigating matters where the firm intends to exercise a discretion on whether or not to prosecute if an offence is disclosed. This discretion has been used unofficially for a number of years despite the moribund misprision of felony, and its validity is now accorded legal recognition by the Criminal Law Act 1967, section 5, subsection 1, with certain provisos to ensure it is not abused. Few occasions will arise in a chief security officer's service when he will need knowledge of this procedure, but it should be a sobering thought that if he did there would be matters of importance where his ignorance would be thoroughly spotlighted and could have long-term effects. They are therefore outlined below in full. All written statements after caution shall be taken in the following manner:

If a person says that he wants to make a statement he shall be told that it is intended to make a written record of what he says. *He shall always* be asked whether he wishes to write down himself what he wants to say. If he says that he cannot write, or would like someone to write it for him, a police officer may offer to write the statement for him. If he accepts the offer, the police officer shall, before starting, ask the person making the statement to sign, or make his mark, to the following:

> I, —— ——, wish to make a statement. I want someone to write down what I say. I have been told that I need not say anything unless I wish to do so and that whatever I say may be given in evidence.

Any person writing his own statement shall be allowed to do so without any prompting, as distinct from indicating to him what matters are material.

The person making the statement, if he is going to write it himself, shall be asked to write out and sign before writing want he wants to say, the following:

> I make this statement of my own free will. I have been told that I need not say anything unless I wish to do so and that whatever I say may be given in evidence.

Whenever a police officer writes a statement, he shall take down the exact words spoken by the person making the statement, without putting any questions other than such as may be needed to make the statement coherent, intelligible and relevant to the material matters; he shall not prompt him.

When the writing of a statement by a police officer is finished the person making it shall be asked to read it and to make any corrections, alterations or additions he wishes. When he has finished reading it he shall be asked to write and sign or make his mark on the following certificate at the end of the statement:

> I have read the above statement and I have been told that I can correct, alter or add anything I wish. This statement is true. I have made it of my own free will.

If the person who has made a statement refuses to read it or to write the above-mentioned certificate at the end of it or to sign it, the senior police officer present shall record on the statement itself, and in the presence of the person making it, what has happened. If the person making the statement cannot read, or refuses to read it, the officer who has taken it down shall read it over to him and ask him whether he would like to correct, alter or add anything and to put his signature or make his mark at the end. The police officer shall certify on the statement itself what he has done. (See 'Exhibits' in Chapter 12.)

Rule 5. Statement of persons jointly charged or told they may be prosecuted

Again, except where a firm intends to carry out its own private prosecution after an internal inquiry, it is difficult to visualise where this Rule would have an application in respect of security officers' work. It is, however, quoted for information.

If at any time after a person has been charged with, or has been informed that he may be prosecuted for an offence, a police officer wishes to bring to the notice of that person any written statement made by another person, who in respect of the same offence has also been charged or informed that he may be prosecuted, he shall hand to that person a true copy of such written statement, but nothing shall be said or done to invite any reply or comment. If that person says that he would like to make a statement in reply, or starts to say something, he shall at once be cautioned or further cautioned as prescribed in Rules 3 and 4.

Rule 6. Investigators other than police officers

This has already been stated (page 127) and is the authority which makes it advisable for security officers to have knowledge of the Judges' Rules.

ADMINISTRATIVE DIRECTIONS

These are almost wholly applicable to police officers. The main points which a security officer may wish to know are:

1 If, during the course of making a statement or being questioned, there are intervals or refreshments are given, this must be noted. Under no circumstances must alcoholic drink be given.

2 Care should be taken to avoid any suggestion that a person's answers can only be used in evidence *against* him, as this may prevent an innocent person making a statement which might help to clear him of the charge.

3 Where foreigners are questioned, their statements should, if it is necessary, be taken down by an interpreter of the language in which they are made.

CONCLUSION

The Rules are more than liberal in ensuring that an accused person is advised to be careful in what he says and against making an admission. Interrogating officers must be fair to the person being questioned and scrupulously avoid any method which could be regarded as being in any way underhand or oppressive.

ADDENDUM

After eight years' work the report of the Criminal Law Revision Committee on the law governing evidence in criminal trials was published on 27 June 1972. The Committee's draft Criminal Evidence Bill containing the recommended changes has now to be considered by Parliament.

The recommendations are largely concerned with procedural changes having little association with the work of a security officer with one very important exception—the recommendation that the police caution, now contained in Rule 2, be abolished. In its place the accused would be given a written notice advising him to mention any fact on which he intends to rely in his defence and warning him that if he holds this back until the trial it may have a bad effect on his case.

The changes carry with them the need to revise the Judges' Rules, which have been described in detail on previous pages, and the Committee recommends that they should be replaced by Home Office directives to the police after being approved by the judges.

The extent to which the recommended changes will be approved by Parliament and given the force of law will not be known for some time after this edition is published. It is recommended therefore that close attention be given to new enactments of Parliament to ensure that in circumstances to which the new law refers the action taken is correct.

12

Elementary rules of evidence and court procedure

It is desirable that anyone concerned in the investigation of an offence against the criminal law, or in preparation of a prosecution, should know what would be acceptable to a court as evidence and what would be rejected. Where an inquiry concerns an offence against company rules the same principles should be adopted so that disciplinary action is not taken without justification; it must afterwards be able to stand critical examination by a third party.

EVIDENCE

Law of evidence

This determines (*a*) what facts may be proved in order to ascertain the innocence or guilt of the accused person, and (*b*) how and by whom those facts may be proved.

The word 'evidence' means the facts, testimony, and documents which may be legally adduced in order to ascertain the fact under inquiry.

Direct evidence

Evidence is *direct* when it immediately establishes the very fact sought to be proved and is normally what a witness saw or heard connecting the accused directly with the offence.

Circumstantial evidence

Evidence is circumstantial when it establishes other facts so relevant to or connected with the fact to be proved that they support an inference or presumption of its existence. For example, if *A* was seen in a cloakroom by *B* to put his hand into the pocket of a coat which was not his and take out some money, later proved to have been stolen, what *B* could say would be *direct* evidence. If, however, *B* saw *A* standing beside the coat from which money was later reported stolen and *A* was unable to account for his possession of a similar amount to that stolen and denied his presence in the cloakroom at the relevant time, what *B* could say would be *circumstantial* evidence which in association with the other evidence of finding the money in *A*'s possession, and what he had to say, would go towards proving his guilt.

Hearsay evidence

Where a witness cannot give direct evidence of a fact. For example, *B*, who saw *A* steal the money, tells *C* what he saw; *C*'s evidence of what he was told would be hearsay and would not be allowed to be given in court. The evidence can only be given by *B*.

Another example is where a document has to be produced to prove a fact in a prosecution. It is not acceptable for it to be done by someone who did not prepare it. For example, the fraudulent addition of figures on a consignment note (see Chapter 24) would require evidence from the actual clerk who wrote the original figures and who would say that since he did so they have been altered by the addition of other figures thus inflating them.

Opinion

As a general rule a witness must give evidence of facts within his knowledge and recollection and not of his opinions. It is for a court to draw opinions from what he has to say. The exception is where specially qualified experts are called to give their opinion on, for example, medicine, art or foreign law. Fingerprint officers giving evidence must satisfy the court as to their experience and expertise before they are allowed to be heard.

Character

In general terms evidence that an accused has been previously convicted of a criminal offence is not admissible in evidence until after he has been found guilty. It is then helpful to the court to know of his character in deciding on the most appropriate punishment. However, if a person of bad character gives evidence of his good character the prosecution may, in rebuttal, give evidence of his bad character *before* the court arrives at its decision.

GIVING EVIDENCE

A witness may not read his evidence from a document. He may, however, refresh his memory by reference to any memorandum if made at or about the time of the fact to which it refers and either made by himself or seen by him while the facts were fresh in his mind and then recognised to be correct. The opposite party may see any such memorandum and cross-examine him upon such parts as are referred to. (See page 144.)

WITNESS SUMMONSES

Where a witness required by either the prosecution or defence is not willing to attend court to give evidence, an application is made to the court for a summons. This is served personally on the witness and calls on him to attend the court issuing the summons at a specified time, pointing out that intentional failure to do so might lead to an issue of a warrant of arrest under which he will be detained until the next hearing of the court.

If a person required to give evidence attends court but refuses to give evidence this amounts to 'contempt of court', which can be followed by his committal to prison until he is prepared to give evidence. These circumstances rarely happen but the power to deal with recalcitrant persons is there if required.

The most common instances where witness summonses are used are where bank managers are required to give evidence of the accounts of their customers.

EXHIBITS

These are tangible objects produced in court which go towards proving the guilt, and sometimes the innocence, of the accused person. They are evidence in themselves but they have to be produced in court by witnesses who speak of how and when they were found and explain their connection with the accused.

An exhibit can be a weapon alleged to have caused an injury which is the subject of a charge, for example, a blood-stained knife in the case of murder or causing grievous bodily harm. Other exhibits could be hair of the injured person found on a weapon in the possession of the accused; safe ballast found in turn-ups of trousers worn by a safe-breaker at the time of his arrest; the written statement alleged to have been made and signed by the accused; or, the most straightforward of them all, the property concerned in the charge of theft subsequently recovered in whole or in part.

STATEMENTS OF WITNESSES

In Chapter 6 reference is made to obtaining statements or proofs of evidence from witnesses who may be called upon to give evidence in the prosecution of the offender. The statements may contain hearsay which, as it has been explained, would not be admissible in a court. This sometimes is necessary to explain the actions of a witness in relation to other persons to make a complete story of what happened.

Statements in lieu of oral evidence

Since the passing of the Criminal Justice Act 1967, a revolutionary change has taken place in the administration of the courts. Whereas until then only a witness in person could be heard, evidence can now be adduced through the reading of a written statement from a witness, subject to certain conditions. This requires the editing of written statements to remove hearsay or other passages which would not be admissible if the witness were giving evidence orally.

Section 2 of the Act says that in committal proceedings (that is the hearing in a magistrates' court of an offence which, if a *prima facie* case is made out, will be committed for trial at the Crown Court) a written statement of any person shall, if the following conditions are satisfied, be admissible to the like extent as oral evidence:

1 The statement purports to be signed by the person who made it.
2 The statement contains a declaration by that person to the effect that it is true to the best of his knowledge and belief and that he made the statement knowing that, if it were tendered in evidence, he would be liable to prosecution if he wilfully stated in it anything which he knew to be false or did not believe to be true.
3 Before the statement is tendered in evidence, a copy of the statement is given, by or on behalf of the party proposing to tender it, to each of the other parties to the proceedings.
4 None of the other parties, before the statement is tendered in evidence at the committal proceedings, objects to the statement being so tendered under this section.

The following provisions have to be complied with:

1 If the statement is made by a person under the age of twenty-one it shall give his age.

2 If it is made by a person who cannot read, it shall be read to him before he signs it and shall be accompanied by a declaration by the person who so read the statement to the effect that it was so read.
3 If it refers to any other document as an exhibit, the copy given to the other party shall be accompanied by a copy of that document or by such information as may be necessary to enable the party to whom it is given to inspect that document or a copy thereof.

Although a written statement may be admissible, on a court's own motion or on the application of any party to the proceedings, the person making it may be required to attend the court and give evidence.

Section 9 says that, in any criminal proceedings in a magistrates' court where it is competent to proceed to a conviction or dismissal, a written statement of any person shall, if conditions which are basically the same as required by section 2 are satisfied, be admissible in evidence to a like extent as oral evidence.

There are certain administrative requirements which are not likely to affect others than the police or court officials.

If a witness wilfully included in his statement anything he knew to be false, or did not believe to be true, he would be liable to prosecution under section 89 of the Act and on conviction would be liable to a term of imprisonment not exceeding two years, or a fine, or both. (See 'Perjury' on page 146.)

Refreshing memory from statements

Home Office Circular No. 82/1969 to Chief Officers of Police says '. . . the Secretary of State is able to commend for adoption a revised practice which has the approval of the Lord Chief Justice and the judges of the Queen's Bench Division. This is that, notwithstanding criminal proceedings may be pending or contemplated, the Chief Officer should normally provide a person on request with a copy of his statement to police . . .'

This procedure was confirmed in R. *v* Richardson (Court of Appeal) [1970] 2 W.L.R. April 1971 where witnesses were allowed to see their written statements to police made sometime earlier before they were called to give evidence in accordance with them. Richardson appealed against conviction on the grounds, *inter alia,* that as the prosecution witnesses had been allowed to refresh their memories through this means their evidence was inadmissible.

The Court of Appeal held, dismissing the appeal, that there was no general rule that prosecution witnesses might not refresh their memories before trial from statements which they had made non-contemporaneously but near the time of the offence.

This is an important privilege and where a witness of fact is uncertain through passage of time what he said when his statement was taken he should ask to be allowed to refresh his memory before giving evidence.

PROCEDURE IN COURT

Examination-in-chief

After taking the customary oath or making affirmation the witness gives his evidence, in which he will be assisted by the lawyer representing the side for which he is appearing. This is called examination-in-chief.

Cross-examination

The lawyer appearing for the other side then has the right to ask the witness questions in an attempt to negative or diminish his evidence by casting doubts on his memory, hearing, or accuracy, etc. This is called cross-examination. After any cross-examination the lawyer representing the witness's side then has the opportunity to *re-examine* in order to explain or remove ambiguity in any answers given by him in cross-examination.

Leading questions

Leading questions are not as a rule permitted. They are questions so framed as to suggest to the witness what answers are required. For example: 'Did you see *A* take some money from the coats in the cloakroom?' will be disallowed. The proper question would be: 'What did you see *A* do?'

Perjury

This offence is committed by a witness who gives, under oath or affirmation, evidence which he knows to be false or does not know to be true and which is material to the proceedings.

CASE LAW—COURT DECISIONS

If either party in a prosecution disagrees with the decision of the court, they may usually appeal against it to the next higher court. If a new point of law based on special facts has been decided by the court, and this is not reversed by a successful appeal, it forms a precedent which will be followed in similar cases in the future. Case law is made up of such decisions. In legal reports, criminal cases are referred to as Regina (or Rex) versus (abbreviated to 'R. *v*') the name of the accused person followed by the year of the decision, for example, R. *v* Bloggs (1963).

There are a great number of cases interpreting the old Larceny Acts and legal arguments will undoubtedly arise from the decisions of the courts in respect of offences under the new Theft Act.

SUSPENDED SENTENCES

The opportunity is taken here to mention the power of courts under section 39 of the Criminal Justice Act 1967 to suspend a sentence of imprisonment. This requires the accused to be of good behaviour for a specified time on pain of suffering the period of the suspended sentence if within that time he is convicted of a later offence. This is a different procedure from releasing a convicted person on probation.

13

Scottish law

There are certain fundamental differences in the pattern of the Scottish and English legal systems, and many of the enactments which apply to England and Wales are not extended to Scotland. It is not intended to discuss court practice in detail, other than to mention that private prosecutions in Scotland are to all intents and purposes unknown and the police act as intermediaries in all cases.

For security purposes there are certain matters that must be taken notice of since they affect the security approach to incidents. This is particularly true of questioning and arrest.

CRIMINAL LAW

In Scotland this consists of a mixture of common law and enacted law with an emphasis on the former which does not exist south of the border. Enacted law includes statutes, orders, regulations, and by-laws; much of statute law applies to both countries and this is true of the whole of the Road Traffic Act.

Despite the differences there is no abrupt demarkation line and the same general rules apply in Scotland as under the English law. Every person who breaks the law within the jurisdiction of the courts is liable to punishment; ignorance of the law does not justify the commission of any crime or offence; and the person is deemed innocent until proven guilty. The old presumption however continues in Scotland that the age at which a child can be guilty of an offence is eight years whereas in England and Wales that age is now ten and likely to be raised to fourteen in the foreseeable future.

Theft

The English Theft Act 1968 is not applicable in Scotland where stealing remains a common law offence. It is defined as taking and appropriating property without the consent of the lawful owner or other lawful authority. The 'taking' must be with the felonious intent of depriving the owner of his property and appropriating it to the thief's use. It is not theft to take property under a claim of right made in good faith or under the reasonable belief that the owner had granted his permission. 'Taking' need not be for the sake of gain, the essential ingredient is the intent to deprive the owner of his property in the knowledge that the act of doing so is unlawful. From this it will be seen that, to all intents and purposes, 'theft' in Scotland contains the same basic elements which are defined in the English Theft Act.

Embezzlement

In Scottish law this is not an offence which can only be committed by servants and employees. It is the felonious appropriation of property which has been entrusted to the accused with certain powers of managerial control.

The accused must have either:

1 Limited ownership of the property or
2 Actual possession of the property with a liability to account for it to the owner.

Limited ownership can be acquired by loan or hire and in these circumstances it is necessary to differentiate between embezzlement and theft. If the loan or hire is for a limited period and for a specifically defined purpose, or for a limited period only and the appropriation takes place after the date fixed by the owner for the return of the property then the offence is theft.

There is little object in probing the finer points of difference since the person charged with embezzlement may be convicted of theft or receiving and a person charged with theft may be convicted of embezzlement.

Fraud

In this instance recent English law would appear to have borrowed from its Scottish equivalent. The term 'fraud' is inclusive of all offences which have the ingredient of fraudulent deception.

Factors which must be present are:

1 Falsehood—false representations by word of mouth, writing, or false conduct.
2 Fraud—an intention to deceive and to defraud.
3 A wilful imposition and the cheat being successful to the extent of gaining a benefit or advantage, or of prejudicing or tending to prejudice the interests of another.

Housebreaking

There is little basic difference between the Scottish interpretation of housebreaking and that which was applicable in England prior to the Theft Act, except that the term 'house' applies not only to a dwelling, etc., but also to any other roofed building whether finished or unfinished, or to any part of a building used as a separate dwelling which is secured against entry by intruders—this is a more omnibus definition than any used in England.

Opening lockfast places

This is an aggravation of theft and is a term met with in charges but is not in itself a substantive charge. 'Lockfast' places include rooms, cupboards, drawers, safes, desks, cash boxes, showcases and other receptacles, the contents of which are protected by lock and key. Where 'places' of this type are attacked in the commission of a theft the act is regarded as an aggravation of the theft or, if the theft is unsuccessful it may be regarded as demonstrative of the intent to commit the crime.

It is immaterial as to how the security of a lockfast place has been overcome and the usage of the term is shown in an instance where a house is broken into and the safe is forced—the crime would be theft by housebreaking and opening a lockfast place.

Malicious mischief

There are no Criminal Damage Acts in Scotland where malicious damage is a common law offence. It consists of the wilful wanton and malicious destruction of, or damage to, the property of another person. There must be malice either actual or inferred on the part of the perpetrator, it is not essential that there should be a deliberate intent to injure another person. If the damage is done by a person with a deliberate disregard or indifference to the property or rights of the other, the offence is complete. If the person responsible claims a private or public right, even where he is acting under misapprehension, malice may not be inferred.

Under certain Acts, specific offences of malicious mischief are in effect created, the main ones of these are related to Post Office property and the use of explosives.

Fire raising

This is the Scottish equivalent of arson and to all intents and purposes the constituents that are legally required for proof are those which are applicable to that offence in the English courts.

Forging and uttering

Though there are offences of forging under various statutes, particularly those relating to the forgery of bank notes, licences, passports, certificates, etc., it is also a common law offence. It consists in the making and publishing of writing feloniously intended to represent or pass for the genuine writing of another person. It is not essential that there should be an imitation of the handwriting. Forgery by itself is not a criminal act, the crime must be completed by putting the document to use and purporting it to be genuine. The uttering must be with a fraudulent intent and it is not material that the false uttering is immediately detected.

POWERS OF ARREST OF PRIVATE PERSONS

Under Scottish law these are identical with those formerly accorded by the English common law—now transferred verbatim into the Criminal Law Act 1967 (see page 87).

ACCESSORIES AND ABETTORS

No difference is made between commission and accession—a guilty person may be convicted as a principal either in aiding and abetting or by being an accessory.

An 'accessory' is defined as a person who aids the perpetrator with advice or assistance before or at the time of commission of a crime, or acts in concert by watching out while it is committed. An 'abettor' is a person who incites, instigates, encourages or counsels another to commit a crime or offence.

Evidence

In Scotland the evidence of a single witness, no matter how much he is entitled to be believed, is not deemed sufficient to prove a charge against an accused person, or even to positively establish any essential and material fact such as a visual identification of the accused person. The single witness's evidence must be corroborated in some other form, either by that of a second witness, or by evidence of facts and circumstances which in themselves provide the necessary corroboration. These principles apply not only to the proving of common law offences but also to statutory offences unless the statute under which the offence is being prosecuted makes provision to the contrary. Somewhat illogically, however, it is accepted that in the absence of evidence to the contrary, facts which are not basically of essential importance may be accepted from the evidence of a single witness.

Corroboration can be in a variety of ways, in the form of a subsequent admission to a police officer that what is alleged is true, or the facts and circumstances themselves may be conclusive of the veracity of the single witness. However, it has been held that even a confession of guilt—short of a formal plea of 'guilty' by an accused—is not enough to secure a conviction; there must be evidence to incriminate him other than that which has been given purely by himself.

STATEMENTS AND QUESTIONING

The Scottish courts are more reluctant than their English counterparts to admit self-incriminating statements but the question of whether the evidence is admissible is again primarily for the presiding judge.

The rules which are applicable to questioning under Scottish jurisdiction do not include any section which says they should be observed by *other* persons who have a responsibility for the investigation of crimes. So far as the police are concerned, the form of caution is the same as in England but it has been said that when the police reach the stage at which suspicion has centred upon a person to the extent that he is likely to be charged with the crime, further interrogation becomes dangerous and if carried too far, to the point of extracting a confession by cross-examination, the evidence will almost certainly be excluded.

The lesson for security officers therefore is that they should not administer a caution at any stage when speaking to suspects but they should stop questioning when they are satisfied that there is sufficient evidence existent against the person to whom they are speaking to justify the laying of a charge against him. It seems unlikely that any latitude will be extended to security officers if they trespass beyond the limits to which the police are allowed.

The Scottish law contains a verdict of 'non-proven', which is satisfactory neither to prosecutor nor accused and does nothing to clarify the facts for any civil claims that may be subsequently made. Consideration is being given to amendments to the legal system and it is possible that at least the necessity of corroboration may be relaxed.

PART THREE

SECURITY IN OFFICES AND SHOPS

14

Cash on premises

This chapter will deal with the recommended precautions respecting the holding of cash to prevent loss from theft, fraud, and mistake.

Cash in transit will be dealt with separately in Chapter 15.

REMITTANCES RECEIVED

General

These are usually received through postal deliveries. The security of the mail is dealt with in some detail in Chapter 16 but some reference to the procedures is necessary here. The opening of the post should be attended to by at least two persons, one of whom preferably is of supervisory grade. The cashier should take no direct action in the operation.

All cheques and other orders for payment must be compared with any accompanying document, usually the invoice, immediately after the enclosing envelope is opened. The document should be marked in some manner to show the amount stated has indeed been received. Any discrepancy should be noted. The date, the name of the payee, that the written amount and the figures agree, and that the cheque has been signed must be verified. Acceptable cheques should then be stamped with the company's name and separated from the accompanying documents.

The amounts on the cheques and on the documents should be separately totalled, with an adding machine if available, and the respective totals must agree. The cheques with the add list should then be sent to the cashier and the documents with their add list sent to the sales-ledger section to be dealt with according to their separate.requirements.

Receipts

Official receipts provide a limited control over remittances received. A comparison between the receipt counterfoils or duplicates and the cash-book entries should be made from time to time without notice by a person independent of the cashier and all receipts should be accounted for. These should be serially numbered. Spoilt receipts must be retained to assist that reconciliation.

Great care must be exercised over the security of books of unused receipts. These are sometimes stored in stationery cupboards accessible to anyone. They must be issued in strict numerical sequence and a record kept of who they are issued to.

OTHER CASH RECEIVED

Canteen cash

Tills with rolls which record separate cash takings and an aggregate total of receipts are preferred. The till operator must not hold a key whereby the accumulated total of the receipts can be read. When a till is to be emptied of money by someone other than the operator the money should be totalled and a note of the amount initialled by the operator in agreement. A comparison should then be made with the registered total and any discrepancies recorded and countersigned by the operator. On the till takings being passed to any other person, for example, the cashier, the amount being handed over should be signed for.

Vending machines

These are now in common use and provide a mechanical service in lieu of personal service which has become expensive even when available. The machines often issue a selection of drinks or meals at different prices. The meter in the machine records the gross number of purchases but in most instances does not register the numbers under the respective prices. Therefore the collection of money from the machines must be under good control. A recommended procedure is as follows:

1. There should be no access to the cash-box by the person replenishing the ingredients. To this end it should be secured to the machine by a lock and be of a type where money cannot be extracted without the box itself being unlocked by a key.
2. Two persons should be present at the machine when cash is collected. Cash-boxes removed from the machine should be replaced by others. The counting of the contents of the removed boxes should be under supervision, the keys being held by someone in supervisory authority in good security conditions.
3. At the time of collecting the cash-box the meter reading should be taken. This may be only one grand total but it will assist to provide a comparison with the cash received.
4. A record should be made of the amount of ingredients used by each machine, or of articles complete in themselves with which it has been stocked, so that a reasonable assessment can be made of what cash return should be expected and compared with what has been collected.

Staff sales

The removal from the premises of purchases of company products, etc., by employees is dealt with in Chapter 4. When cash payments are received a form of receipt should be issued to confirm this. This can be either a cash-till register slip or an authority to purchase order on which is shown the amount of the payment required which is stamped 'paid' when that has been done. It will be necessary from time to time to compare the amounts on the duplicate of the order with the cash-book entries. Whilst in many instances the intrinsic value of the property may be nothing, by putting a nominal value on it a much tighter control is exercised.

Sales of scrap materials

Where sales of scrap materials take place to dealers it is not unusual for payments to be made in cash rather than by cheque. A procedure similar to that in the foregoing paragraph can be adopted to ensure the payment of the cost decided by an authorised person before the material is allowed to leave the premises.

Although not strictly in context, the opportunity is taken to recommend that when scrap or waste material is being loaded it should always be done under supervision to prevent thefts, for example, non-ferrous metal being included with less-valuable metal, sound jute or paper bags being removed with those sold as requiring repair or of scrap value, or company property of other descriptions being stolen under cover of such material.

See Appendix 22.

Van sales

This is the description given to selling direct to customers according to their requirements from a van. The salesman at the time completes two copies of the sales invoice and receives cash in payment. The customer is handed the original and the duplicate is used in the stock-taking of the van and for administrative purposes. From time to time it is recommended that the customers' original copies be compared with the duplicate invoices.

By various means, including overloading the van by collusion with the loader or under-delivery to customers, the salesman can get possession of additional stock to sell, for which he has not to account. One way this can be done is to give the customer a correct invoice but show on the duplicate a different and lesser amount.

PAYMENTS

These should always be by cheque or credit transfer. Settlement of accounts by cash should be discouraged. Requests for cash from within the firm should provide particulars of the payee, the nature of the payment, the amount, the account to which to be allocated, and the signature of the authorising person. The form of request and any supporting documents should be cancelled at the time of payment with a 'paid' stamp which includes the date of payment.

A list of persons authorised to approve requests for payment and to sign cheques, and to what maximum amounts, should be prepared and approved by a senior official of the concern. A copy, with specimen signatures, should be held by the cashier.

Cheque books

These must be kept under the standard of security which would be accorded cash. Spoilt cheques must be cancelled and retained. Cheques should not be signed unless accompanied by a proper document preferably prepared by an independent person. Blank cheques should never be signed.

The security which should be provided is described on page 188.

Fradulent invoices

The printing of fraudulent invoices showing the alleged supply of goods or services is far from uncommon. The control necessary to prevent the payment of such an account is the responsibility of the accountants of the concern.

One example was when someone holding a supervisory position in a public corporation succeeded in defrauding his employers by the use of false invoices. He created and registered a firm under the Registration of Business Names Act showing an accommodation address as the place of business. He then had invoices printed with the name and address of the firm. He filled these in with details of advertising materials purporting to have been supplied to his employers and sent them off by post. After they had been delivered he obtained possession of them by some subterfuge through his knowledge of the systems used and forged the initials that a genuine invoice would require to bear before payment was made. He re-introduced the documents to the office routine system whereby they again came into his possession but this time correctly. It was his job to prepare the requisition for cheques in payment of accounts, which he did with the false invoices. The cheques were completed by another person and then posted to his accommodation address. He later collected them and paid them into a bank account opened in the name of the non-existent firm. He was caught, not through the operation of any sophisticated control procedure, but the suspicions of a very junior member of his department.

BANKING CHEQUES

There can be a tendency to relax the security precautions associated with the depositing of money when taking cheques to a bank. The visits to the bank should be done at irregular times. Thieves are not to know that no money is to be taken or collected from the bank and are likely to make attacks.

An example of this was where it was usual with a company to deposit daily the cheques received. To suit other transport requirements the times of the visits to the bank were usually about the same. One day on the return from the bank the car used was rammed by a stolen vehicle, the occupants smashed the windscreen and after striking the company's employees with pick handles snatched the empty bag and escaped in a second stolen car.

CASH OFFICES

These should not be on the ground floor and never in an isolated building. The higher the floor which is most convenient to the administration the better and they should be located as far away from stairs or lifts as possible.

Before arriving at a decision, one very important question has to be answered. Will the floor take the weight of the safe which it is intended to use? It must be as close as possible to the point of use and not, for example, in the basement or on the ground floor, with the cash office on another floor. That would mean carrying money to and from the safe at the opening or closing of business. Where a new building is being erected it might be necessary to arrange for an additional girder to be fitted. The best position for a safe is where it can be seen from the outside of the premises. If it is placed inside a cupboard or cabinet, security will be improved.

A safe should either be heavy or secured to the floor by strong bolts or another device. Modern safes are sold with provision for their fixing to the floor which must be suitable for the purpose. The floors of modern buildings can be of hollow construction and therefore fixing the safe to the wall is an alternative arrangement.

The location of the cash office within another office so that it is necessary to pass through one office to get to it is good security. Another good deterrent to attack is to have the cash office between other offices which are constantly in use and windows fitted in the dividing walls or partitions. Should an emergency arise in the cash office this is likely to be seen. A sliding panel or service hatch in the window might have organisational advantages.

The cash office should not be overlooked from other buildings but if this is unavoidable the windows should be made opaque as necessary. They should not look on to a flat roof or a fire escape. The door should be strong with a slam type lock. There are likely to be times when money in some quantity will be exposed in the office when the door must be locked. It should be an order to all staff, irrespective of grade, that entry to the office at such times is not permitted. To provide for the special occasion when this is unavoidable the door should have a spy lens or inspection window through which a caller can be identified before the door is unlocked.

The office should have a counter beyond which no unauthorised person is allowed to pass, with a grille through which business is conducted. Where particularly large amounts are handled and the office is unavoidably not in as secure a location as would be preferred, the installation of an armoured glass screen will provide additional protection. Cash drawers with slam locks should be fitted in the counter which should be wide enough to preclude anyone reaching the money in the drawers from the customers' side. It is not good security if the cashier has to go to the safe or cash box to reach the money. The cashier should sit at his desk so that he is facing the door. When holding a large sum of money cannot be avoided and more than one safe of good quality is available, the money should be divided amongst them.

An alarm to be set off by foot or hand to operate in an adjacent office is recommended. The staff in that office must have instructions in writing what to do on hearing the alarm, such as calling the security staff. It must be made quite clear that they, and the cashier, should not get involved with any intruders in an emergency. The alarm must be tested daily.

Cash on the premises

Cash in hand must be balanced every day by the cashier who should sign or initial a note to that effect. There should be no delay in investigating shortages and in serious cases all relevant documents, receipt counterfoils, and books must be removed from the possession of interested parties. Holding large sums of money overnight and particularly at weekends must be avoided. This can be achieved by calculating the money required from the bank in conjunction with the takings from a canteen or vending machines, where applicable, in such a way that it will be at a minimum on Friday night. Where salesmen return to base with their cash takings at a time when the office staff has left, the use of internal 'night banking' safes is recommended. These provide a drawer into which the money can be placed and receive the protection of the safe without the possession of the safe key. The quality of the safes should be reviewed frequently, having in mind the amount of money they will be required to hold or whether their primary function is to protect records, the loss of which would have serious consequences, from the effects of fire. Depending on the amount at risk consideration should be given to fitting a burglar alarm. The various types of safe are discussed in Chapter 33.

SAFE KEYS AND CODES

Safes with key locks are usually sold with a minimum of two keys. The person responsible for the custody of the money in a safe should always personally hold the key and not leave it on the premises at any time. It is very bad security to put the safe key in a filing cabinet and then put the key to that in a drawer, lock it and take the drawer key home. A filing cabinet drawer is not sufficient protection for a safe key.

The duplicate key should not be in the possession of anyone else regardless of position unless there are very special administrative reasons; if they do not involve money, the money in the safe must be put under separate protection such as in a locked cash-box or in a locked drawer of which only the cashier has a key. If there is good reason for this separate holding, the duplicate must be kept in first-class security. If not, the second and any other copies should be deposited at the bank where the arrangements should be that it will be surrendered only on written application signed by two cheque signatories or by two of a list of members of the firm whose specimen signatures are held.

To even the wear on the keys they should be changed over regularly, say annually. An occasion is known where a key to a safe was lost and when the second key was withdrawn from the bank it had been there so long that it would not open the safe lock which had become worn over the years.

When a safe has a combination lock the code should be known only to the cashier with a copy in the bank. Advantage should be taken of the easy facility to alter the code when the responsibility for the safe changes or at times of holiday or sickness of the cashier. The copy of the code at the bank will correspondingly require amending.

Extra security is obtained if there are two locks to a safe which can both be key or code operated or one of each. This requires the attendance of two persons in possession of the key(s) or code(s) to open the safe. A refinement is the addition of a time lock which is set to operate, in conjunction with the key or code, at a time when the safe will be required to be open. The cutting of extra keys should be prohibited without the permission of an authorised person. (See 'Safes withdrawn' below.)

Keyholders

Holders of safe keys or codes should beware of spurious calls, either by telephone or personally, from persons masquerading as police or security officers who say their business premises have been broken into and their presence is required for some reason. If in any doubt the local police station should be telephoned for verification of the request before leaving for the premises.

Holders of keys on their way home after business should not comply with a request to lend keys to anyone in apparent difficulties owing to the alleged inability, for example, to open the door of a car or car boot because the keys have been lost. Instances have occurred when impressions in wax have been taken of vital keys to jewellers' shops and a few days later the premises to which they belong were broken into with the use of a false key.

Safes withdrawn from use

When a safe is withdrawn from use and put into storage it must be ensured that all keys known to exist have been recovered. If one or more cannot be traced the safe must be marked in a distinguishing manner for the information of all that it must not be used to store money or confidential documents until the lock has been changed.

To assist in the recovery of keys, when additional ones have been cut with authority, the inventory or list of assets will be noted with that fact. Should a theft occur from a safe and it is clear that a key was used, the lock must be changed before it is used again. The possibility of the safe being opened irregularly will be much reduced if a combination lock is used, but if that does happen it is simple to change the code.

WAGES

The recommended procedure in the collection of wages money from the bank is contained in Chapter 15. When the money, whether brought by a company's own staff or by a security services company, arrives at the premises concerned the recommended security precautions apply. The additional precautions relating to employing the services of such a company are referred to in Chapter 31.

After the vehicle carrying the money enters through any road gates these should be locked until it is known the money has been safely delivered. There have been a number of instances of money being stolen on the premises so security precautions must not be relaxed. One example of this concerns a block of offices where the cashier's office was on an upper floor. Before the arrival of the armoured car a number of men dressed in stolen overalls of a well-known window and office cleaning company obtained access to that floor and began to clean the windows. As the security officers carrying the wages money approached, the 'window cleaners' attacked them, stole the money and escaped.

A similar instance occurred at a hospital where the robbers dressed themselves in white coats and stood chatting in a corner. They were taken for medical students. Another case was at London Airport where a lift is used to take money in bulk to upper floors. Men dressed in dark business-suits and bowler hats and carrying umbrellas entered the lift at a higher floor and contrived its descent to the ground-floor level to coincide with the arrival of a large sum of money protected by security staff. They emerged from the lift taking everyone by surprise, attacked the guards, and escaped with the money in a waiting car. In similar circumstances it should be arranged for the lift to be held on the ground floor for the arrival of the money.

Another way money was stolen shortly after delivery by a security services company was when a knock was heard on the door of the room concerned. On the inquiry from inside asking who was there, a voice replied saying that it was one of the security men who had just delivered the money and he had left something behind. The man behind the door had no means of identifying the caller. When it was opened, three men rushed in, stole the exposed money, and escaped.

From recent information supplied by two of the largest cash transporting companies, out of fifteen attacks by robbers six took place on the premises at which delivery was made. The security requirements for the office where the cash is received and made up into wage packets are the same as for a secure cash office (see pages 162-5). The use of blinds over windows at relevant times should not be overlooked. The office certainly must not be a one-storey isolated building which was the case in two recent large-scale thefts of wages. It must be locked and no person permitted to enter or leave until the make-up is complete and the packets under a reasonable degree of security. Visits to the lavatory by staff concerned must be made before the session commences. Where there is a delay between the completion of the make-up and the pay-out of the wages they must be locked in a safe.

Pay-out

When packets are removed to pay-out points the clerks concerned must be escorted by able-bodied men, usually members of the company security staff. If they are to be taken over a factory road, this should be by vehicle and the gates to the outside of the premises should be locked. Pay-out points should be as few as possible, preferably under good security and near the exit. This reduces the possibility of wage packets being lost on the premises. For the same reason pay-outs are preferred at the end rather than the middle of shifts. The paying-out must be under good security wherever it is and within reach of a telephone for use in an emergency. A third person, who can be a supervisor or a security officer, should be in attendance and on the outside of the station if it is a separate part of a building with a service hatch. (See 'Straw or Dead Men', page 295.)

Unpaid wages

Where it is necessary to pass unpaid wages in bulk to someone who will be responsible for paying them out later they should be counted and signed for. After a pay-out, remaining packets sometimes have to be passed to another person who will pay a later shift or return to the cash office. On each transfer the packets must be counted and signed for.

When wage packets are held between pay-outs in a safe, which is usually when the office staff has left, it is advisable to devise a system whereby two persons hold keys to locks which require opening to reach them. The principle can be illustrated by a safe being placed in a steel cabinet which is locked. One key would be held by, for example, the pay-out clerk or security officer on duty and the other by the shift manager. This prevents the theft of the packets when a single keyholder is attacked by intruders or when false statements are made of having been overcome by such persons who took the key and stole the money.

Sometimes a weekly paid employee is unable to collect his wages because of sickness and it is the practice to accept a letter of authority from him to pay his wages to the person presenting the letter. To provide against the forgery or theft of such a letter the holder should be required to take it to a department competent to verify the signature or to question the holder if there is any doubt before the wages are given to him. The confirmation could be by a stamp or signature known to the wages clerk.

Checklist

The recommendations in this chapter are summarised in Appendix 23.

15

Cash in transit

This chapter concerns the security which should be given when depositing money at, and collecting from, a bank, either by a company's own staff or by a security services company. The recommendations also refer to the use of a bank night safe.

Knowledge of the precautions taken must be restricted to as few persons as is necessary to carry them out. If an employee who is familiar with the procedures leaves the company, in circumstances where his integrity is shown to be unsatisfactory, changes from the procedures should be made but not so as to weaken overall security.

BY COMPANY STAFF

On foot

1 The use of a bag for carrying money to and from a bank presents the opportunity for a thief to snatch it when the carrier would probably be unprepared to resist. This risk can be removed by carrying the notes dispersed in the pockets of the carrier and any escort, with coins in a bag. Specially designed cash-carrying waistcoats can be purchased.

2 If a bag is used it must be strongly constructed, particularly the handle, and carried by an able-bodied man. It should never be carried by a woman or a juvenile. The chaining of a bag to the wrist or waist is not recommended because if the bag is snatched away it could cause serious injury.
3 Leather bags which incorporate audible alarms, emit smoke, or dye banknotes when snatched from the carrier are available.
4 The escort should also be a fit man who should walk a pace or two behind the bag carrier. If he holds a stout walking stick it would not be out of place.
5 The bag carrier and escort should walk so as to face oncoming traffic, keeping away from the kerb. Busy streets, and *not* unfrequented side streets, should be used.
6 Recently engaged employees should not be used for either of the duties described until they have satisfied their employers regarding their integrity.

By taxi or hired car

1 If a taxi is customarily used, one which is stationary near the exit of the bank should not be taken. One method used by thieves is to steal a taxi and wait to be 'hired' by the carrier and escort of money with obvious results.
2 If it is the custom to use a hired car to be taken to the bank and if, when it arrives, the driver is unknown, telephone the hire company before leaving the premises and, if possible, speak to a known person to check the driver's identity.

By company vehicle

1 Fittings should be provided whereby the cash bag can be secured in the vehicle. This can be an eye bolt fixed to the frame through which a strong chain can be passed and then through the handle of the bag and secured with a snap padlock. No key to the lock should be carried on the vehicle or by anyone. This would be available at the company premises. Special steel frames can be purchased for fixing in car boots to accommodate a steel box carrying money.

2 The doors must be capable of being locked from the inside.

3 An audible alarm should be fitted to the vehicle to be operated immediately an attempt is made to interfere with its progress. The alarm should be tested immediately before the journey to the bank. The police should be informed when an alarm is fitted. A police whistle is a recommended alternative. Where a company has a radio-telephone system the cash-carrying vehicle should be equipped with a set.

At the bank

The collector should be accompanied into the bank by the escort. When ready to leave the escort should go first and, when satisfied there is no danger, signal to the cash carrier to leave. If the escort is sufficient in numbers, another should remain outside the bank to warn those inside if he has suspicions for any reason. The driver should remain in his vehicle which he should keep locked.

During the journey

1 The routes and the times taken should vary as much as possible. This variation, however, does not preclude thieves from waiting at the bank and following the vehicle carrying the cash.

2 Drivers should be on the look-out for faked accidents, bogus police officers, and unexpected traffic diversions. When carrying money, the doors or windows of a vehicle should not be opened in order to speak to anyone, including persons in police uniform. In the latter circumstances an offer should be made to drive to the nearest police station instead of leaving the vehicle.

3 A watch should be kept for vehicles following the cash car. If in doubt slow down. If the following vehicle also slows down instead of overtaking when this would be safe, the cash vehicle should be driven to the next convenient premises where protection against attack can be obtained. If fitted with a radio set, the details of the occurrence and registration

number of the vehicle should be radioed to the company for the police to be informed.

4 Motor vehicles stationary in sight of the bank and containing men should be viewed with suspicion.

5 Consideration should be given to using a second vehicle to follow the cash-carrying vehicle especially when particularly large amounts of money are collected.

Night safes

These are particularly useful for depositing shop takings so that they are not left overnight in possibly insecure premises. Closing times of retail premises can be established quite easily so the time of the deposits can be prepared for by intending thieves. Consequently it is imperative that these times must vary as much as possible, which might mean relatively small amounts of money being left on the premises, but this is a justifiable risk in comparison with the larger one.

Persons making deposits must not wear overalls or other clothing which identifies them as coming from particular premises. The point of the greatest potential danger is the pavement outside the bank. While the money carrier opens the safe, the escort should have his back to it and be looking about the vicinity to warn him of any suspicions he may have and to protect him against attack. When a car is used, the driver must draw up as close to the bank as possible. He will then leave the vehicle, open the safe, and if all is clear signal to the intending depositor who will then leave the car and put the money-pouch in the safe. If more than one pouch is used they should not be carried in a bag as it is easier to snatch.

RESPONSIBILITIES OF EMPLOYERS FOR SAFETY OF EMPLOYEES

There are court decisions in two claims for damages by employees against their employers in respect of injuries received while in charge of cash which underline the duty of employers not to expose employees to unnecessary risk including risk of injury by criminals. They will be described in some detail.

Houghton v *Hackney Borough Council,* [*1961*] *3 KIR 615*

The plaintiff was a rent collector employed by the defendant local authority. Most of the places at which the plaintiff collected rent were specially constructed with a barrier or grille behind which the plaintiff sat and took the rent. However, the particular collection place in question was not specially constructed and had no barrier or grille; in fact it was part of an ordinary room and separated from the rest of the room by chairs. The defendant authority had consulted the local police concerning safety precautions against criminals and secured the presence in the immediate vicinity of the collection place of one of the authority's porters. One day, while the porter was away for a short time attending to his other duties, thieves rushed into the collection place, attacked the plaintiff and stole his takings. The court dismissed the plaintiff's claim for damages and held that (1) the defendant authority owed the plaintiff a duty to protect him against unnecessary risks including injury by criminals but (2) that on the facts that duty had not been broken.

Williams v *Grimshaw,* [*1967*] *3 KIR 610*

The plaintiff had been, since 1962, a stewardess of a cricket and sports club. She was responsible for the cash takings and it was her habit to take them home every night. Her husband, or if he were not present, another member of the club, would escort her. In April 1964, while walking home escorted at night, she suffered injury as a result of an abortive robbery. The evidence showed that, since the plaintiff's appointment, there had been five offences of breaking into the club, three of them in the autumn and early winter of 1963. On her claim for damages against members of the club, it was held, (1) that the club were under a duty not to expose the plaintiff to unnecessary risk, including risk of injury by criminals: but (2) on the facts, no breach of that duty had been established and, therefore the plaintiff's claim failed.

BY SECURITY SERVICES COMPANY

The administrative precautions which should be taken before and after contracting for the transport of cash are described in Chapter 31.

Where money is being collected for deposit at a bank, the container in which it is carried should be sealed and locked if possible. No keys to the lock should be carried in transit. The serial number of the seal should be written on the receipt by the customer's representative before the security officer conveying it to the bank signs it. At the bank the respective numbers would be compared before the consignment is opened with a key which has been previously provided.

After collection from a bank, money can be delivered either in an unlocked container, when it will have to be counted on delivery, or in a sealed container. Before breaking the seal the number should be compared with that on the copy of the receipt given to the bank. These checks are especially necessary as instances have happened of money being extracted from sealed containers in transit and different seals substituted in circumstances where the approved security procedures have not been carried out in full. Money received should always be counted by two persons together at the same time.

GENERAL

If any of the foregoing observations and recommendations are thought to be unnecessarily restrictive it must be remembered that the theft of money on its way to and from a bank represents probably the quickest way of acquiring for a minimum of effort what is usually a large and unidentifiable amount of money.

Wages office

For recommendations on security precautions see 'Wages' in Chapter 14.

Checklist

The recommendations in this Chapter are summarised in Appendix 24.

16

Security in offices

The overall security of a building in one tenancy should be the responsibility of someone with the necessary authority to see that the required standards are observed (see Chapter 2).

THE BUILDING

In large buildings with a number of floors it is recommended that someone, who can be called a security steward, be appointed for each floor, department or section, whichever is most practical and convenient. If there is more than one tenant, each should appoint someone to undertake similar responsibilities and to cooperate with his opposite numbers among the other tenants.

The security of the buildings and the property therein can be prejudiced from inside and outside. The degree of supervision which can be exercised over, for example, the entrances depends on whether the premises are wholly occupied by one tenant and also on their structure.

Fire doors

Fire doors on the ground floor and on other floors giving access to fire escapes should be fitted with breakable bolts which have been approved by the Fire Offices Committee. (Glass bolts known as 'Redlam bolts' are an example.) These permit immediate exit in an emergency and provide good protection from interference with the doors from the outside.

Entrances

As offices of commercial concerns occupy larger and larger buildings, with more and more floors, so the number of thefts in them from walk-in thieves has increased. They enter the buildings with every show of confidence, dressed in a manner similar to the regular users and consequently are extremely difficult to detect. Premises with multiple tenancies are particularly vulnerable to losses through the attention of that type of criminal. There must be as small a number of entrances as possible. Where the principal use of a door is as an exit, it should be spring-loaded and fitted with a slam-type lock so that it shuts after use in order to prevent unauthorised entry.

CALLERS TO THE PREMISES

Receptionists

Where the structural features of the entrance are suitable for a receptionist to be in attendance to deal with callers she should sit with a clear uninterrupted view of the entrance doors, the stairs and the lifts. Her view must not be obstructed by, for example, frosted glass with a sliding panel that is opened only when dealing with a caller. If the receptionist is also the telephone operator she must be given instructions in writing on the action she is to take on hearing the fire alarm, or the alarm from the cashier's office, or on being told to call the police or the fire brigade.

Where there is a receptionist a notice should be displayed requesting callers to make their inquiries of her. The caller's name and the name of the person he wishes to see should be recorded. The person the caller wishes to see should be telephoned for instructions, after which, if he is to be received, the caller should be escorted to his destination. Unless he is known, he should not be allowed to find his own way there alone. The subterfuge of thieves calling at premises to see a member of the staff, whose name has been obtained by any one of a number of ways for the purpose of obtaining entry to the premises, is well known but can be frustrated if a personal escort is usual for such callers.

An additional control, which records information that can be of later use, is for the receptionist to complete an entry pass which the caller takes with him to the person he wishes to see. After the completion of the interview the member of the staff concerned signs the pass which the caller surrenders as he leaves the premises.

Window cleaners, maintenance men, telephone engineers

These persons, including delivery men and others in similar capacities who are not known to the receptionist, must not be allowed to enter the premises until permission has been obtained from a person authorised to give it. How thieves, purporting to be window cleaners, stole wages was described in Chapter 14.

COMMISSIONAIRES' DUTIES

In buildings where uniformed commissionaires are posted at the entrances to deal with callers, the foregoing recommendations also apply. They should have their duties fully described in the form of standing orders, covering all circumstances that may be reasonably expected. (See Appendix 3 for an example of such orders in relation to factory security.) The instructions should include: action to be taken on receiving postal deliveries, which is dealt with later in this chapter; the action to be taken on hearing the fire alarm (see Chapter 26); similarly in respect of the alarm in the cashier's office being activated (see Chapter 14), which should include locking all doors and excluding callers; the duties to be performed when patrolling the premises and the frequency of patrols, with or without a 'watchman's clock' (see Chapter 4). Patrolling duties should include turning off lights and electrical equipment not required, shutting windows, inspecting lavatories and large cupboards which could conceal anyone, admitting cleaners, and finally locking up the premises. If a commissionaire or security officer remains alone on the premises at night or during the weekend it is recommended that, for his protection against the effects of intruders or of accident or illness, he should be incorporated in a mutual aid scheme with men in similar employment in other premises—this scheme is described in some detail in Chapter 4. Alternatively he should report by telephone at specific times to the operations room of a security services company (see Chapter 31).

THEFTS IN OFFICES

Chapter 14 deals at some length with the care of cash on business premises and this will not be repeated here. However, it is essential that it is impressed on departmental heads or their equivalents that they have a personal responsibility for the security of their employers' property. They also have a moral responsibility for seeing that employees look after their own property.

Office equipment

Small items of office equipment such as calculating machines, dictating machines and typewriters, which are particularly attractive to thieves because they are usually easy to dispose of, should be locked away when premises are vacated. Old safes which may be in use for the storage of old ledgers, whose real value is questionable, can be used for this purpose. The manufacturers' serial numbers of all machines must be recorded. The presence of this type of equipment should be verified from time to time, supplementary to the annual inventory check. This is especially necessary with respect to machines held in reserve because they can be stolen but not missed for some time.

Personal property

Thefts of personal property, so often accompanied by suspicions which in most cases cannot be disproved or confirmed, can have a demoralising effect on all employees, and productivity suffers. Lockable drawers or cupboards should be made available for the individual use of employees so that personal property such as handbags, slide rules, attractive ash trays, or drawing instruments can be protected against theft when not in use.

A ruling is recommended that large sums of personal money should be kept by the cashier, or whoever else in his absence has the responsibility for the company's money, in specified security protection, which would not include desk drawers or cupboards. A further requirement should be that any loss of money due to failure to conform to the security rules will not be the responsibility of the company. This may appear harsh but knowledge that the rule will be invoked where necessary will have a deterrent effect on carelessness.

Cloakrooms

These should not be situated near to entrances or exits or close to offices where callers (for example, applicants for employment) might be required to wait until attended to or lift entrances. Cloakrooms in those positions are particularly vulnerable to the theft of property from them by intruders, or even legitimate callers.

It is not unusual to find a belief amongst employees that their employers are responsible for refunding losses sustained by them whilst at work. There is no such obligation and this should be made clear by posting disclaimers of responsibility in cloakrooms and other places where employees normally leave their personal property. A suitable wording is:

> The management cannot in any circumstances accept responsibility for loss of or damage to any property or vehicle of employees while on these premises. Such property can only be permitted to remain at the owner's risk.

Notices are also recommended above or near wash-basins reminding users not to leave personal property behind, particularly items of jewellery.

Coloured posters depicting circumstances in which property can be stolen can be obtained from the local police and crime prevention officers. These can be displayed from time to time with frequent changes to stimulate and maintain interest.

AIDS TO SECURITY

A new type of compound is being produced by Camrex Special Coating Services Limited at Sunderland, for identifying the theft of articles, such as shop merchandise, money, wallets, cash boxes and packing cases, that have been treated with it. It is in the form of a paste in a handy squeeze bottle and although invisible can subsequently be detected on the fingers of the thief when exposed to an ultra-violet lamp unit, which produces a fluorescent effect. Examination can normally be carried out up to twenty-four hours after contact as normal washing will not remove all the traces.

The same company has also produced an invisible marking liquid which can be used to identify the ownership of valuable articles of a common nature such as jewellery, coins, currency notes, radios, cameras, typewriters, dictating machines and television sets. The marking can take any form the owner wishes and under the ultra-violet light will fluoresce. A battery-operated light unit can be obtained from the company.

SECURITY STEWARDS' DUTIES

One of the responsibilities of security stewards is to be on the alert, as they perform their primary duties, to notice weaknesses in security, for example, handbags left on or beside desks in unattended offices. The owners' attention should be drawn to the failure to observe elementary precautions, the consequences of which could be a theft and strained relations amongst employees, because of possibly unwarranted suspicions. The duties could be combined with those of the floor or departmental fire stewards, referred to in Chapter 26.

CLEANERS

Premises are particularly susceptible to thieves when only cleaners are in them, before or after usual business hours. The cleaners must be instructed to lock themselves in and not admit anyone who is not known to them.

CONDITIONS OF EMPLOYMENT

Like production unit employees, all office personnel should be required to agree to sign conditions of employment that include an agreement to submit to personal search by an authorised person, should this be required in the event of pilfering. The law on searching is given in Chapter 7, with the wording of some possible clauses. It should be pointed out here that searching female employees must be done only by another female.

CONTRACTORS' EMPLOYEES

When contracts are being arranged for work (and this includes cleaning and maintenance) to be done on premises, especially when they are unattended, it should be agreed in writing with the contractors and any subcontractors that their employees will be subject to the same searching clause as personnel regularly employed on the premises.

LOCKING UP PREMISES

In the absence of a commissionaire or security officer to undertake the duties preparatory to locking up, which have been described, someone has to be delegated those responsibilities. This is very often the keyholder who will be anxious to avoid being called from his home should any breaches of security be discovered after he has left. Fitting one door with a spring and a slam-type lock has advantages; staff who remain on the premises after the usual hours and cleaners can leave by that means without having to be supplied with keys.

Keyholders

It is customary to register with the local police the name, address and telephone number of the employee who is the keyholder of the premises. This is so that should something happen at the premises which requires attention, such as unusual lights left on, doors left open, or perhaps persons arrested for committing offences there, the keyholder can be requested to attend to take any necessary action. This also can be to assist the fire brigade in connection with a suspected outbreak of fire. It is important to report to the police any temporary change in a keyholder because of holidays or sickness of the registered keyholder.

The inconvenience of having to attend premises at all hours, and of ensuring the correct information is registered with the police, can be removed if the keys are deposited with the operations room of a security service company. They become the registered keyholder. (See Chapter 31 on the services of security companies.) Keyholders to premises, particularly those who also hold the key to the safe, are warned in Chapter 14 about attending premises without satisfying themselves that the request to do so is genuine. Should keyholders move to other premises it is recommended that the locks concerned should be changed. This is particularly advisable where stores or retail premises are concerned. This is a positive requirement with a well-known supermarket company.

KEY REGISTER

Filing cabinets, cupboards and drawers of desks usually are supplied with two keys to each lock. One of these, with only an identifying number, should be kept by an authorised person in a key repository which is given first-class security protection. The numbers of the keys should be on a confidential list. Cutting extra keys should be prohibited without proper authority. If a cabinet is intended to be used to keep confidential information in, but two keys cannot be produced, it should not be used unless the lock is changed. The locks of steel filing cabinets often have the serial number of the key marked on them. Unless special care in ordering is exercised, more than one cabinet will be found to have identical locks.

When a member of the concern, who has responsibilities which include the security of vital information which could be of use to competitors, leaves for any reason, it is recommended that the locks of cabinets or safes which protect the information be changed. Where especially critical information is concerned, the use of key locks is not advisable. Combination locks are preferred because the security of the contents of a cabinet or safe is much improved by the facility of changing the codes to operate them whenever required. Copies of the codes, it really goes without saying, require strict security protection and preferably should not be written down in a personal diary.

SECURITY OF POSTAL DELIVERIES

Whenever the amount justifies it, arrangements should be made with the Post Office for mail to be delivered in bulk direct to the addressee. Precautions must be taken to see that the incoming mail is properly protected against loss from the time it is delivered until it is accepted by the concern's postal department, or whoever else is responsible for it. If the delivery is before the arrival of persons authorised to deal with it, provision must be made for its protection until then. (See the reference to the security of letter-boxes below.)

A record must be kept of the delivery of all registered and recorded-delivery letters and parcels and the person authorised to deal with them must sign for them in a permanent form of record. Where the mail is small it should be handed unopened to a designated person who will open it and arrange its distribution to the proper places. Where the circumstances justify a postal department, the mail should be opened there under the supervision of an authorised person. Where employment in that department is only part time, work there should be on a system under which the duty is rotated among suitable staff.

Remittances

All remittances received must be compared with any accompanying documents and any discrepancies in the figures shown noted on them. The cheques, postal orders and money orders should be separated from the supporting documents and forwarded directly to the cashier or other person responsible for their treatment. The accompanying documents are then distributed to the relevant departments for their action. This is developed in Chapter 14.

Letter-boxes

Quite frequently letter-boxes of business concerns using the same building are found at the common entrance with separate receptacles or cages behind the door to receive the letters. These are a source of income for letter-box thieves who take advantage of them should the door be left open before the boxes are cleared. They should be locked between postal deliveries.

COUPONS AND GIFT SCHEMES

In these schemes, whole or portions of labels, wrappers or cartons are sent by the purchaser of the goods for some free gift or a special offer which requires the forwarding of money with coupons. As the coupons represent some value a strict record must be kept of them and this starts with their printing. If the printer is not a member of the instructing firm he must be required to certify the number of coupons printed and indemnify the company against loss of printing blocks, plates, and coupons in his custody. The security of the destruction of imperfect coupons must be satisfactory.

After the coupons are received from the printer the quantity supplied to the department responsible for issuing them must be recorded. Where the coupon is part of a package there will be some control on their issue from production records. The coupons, whether loose or otherwise, must be kept in good security conditions and frequently counted by a person independent of their custodian. Where they are of high value, such as those issued to traders as a form of bonus according to their sales, it is preferable that a small stock be kept available for immediate use with the balance under good security such as that which would be associated with cash. At intervals the total of the coupons redeemed should be compared with the sale of the relevant products. An excess of coupons will reveal a leakage requiring investigation. A final redemption date should be printed on all coupons.

When envelopes containing coupons are received by post, either with or without money, two persons at least should be present to deal with them. One should open them and check that what has been claimed to have been sent in money and/or coupons has in fact been received; any discrepancies should be noted on the envelope or claim form. The coupons should be detached and listed with the names of the senders and retained. Any remittances—money, cheques, postal orders—should be passed with any accompanying claim form or envelope to another person who will endorse the document with the amount which has been detached. The total of the remittances should later be compared with the total of the amounts shown on the documents—an adding machine being used for preference. The list of coupons received should also be checked by a second person. The remittances and the coupons, with their respective lists, should then be passed on to the cashier who, after satisfying himself of their correctness, should sign the lists in receipt. The claim forms or letters from the customers should then be passed to the department concerned with sending the gifts or special offers to the customers.

It often happens that coupons are received with cash but the sender has omitted to include a name and address. The cash should not be held unbanked awaiting a letter from the sender complaining about not having received the special offer that has been paid for, but dealt with in the same way as remittances from known senders. The envelope concerned should be noted at the time of opening that the sender's name and address were not included and with the remittance received. It should be put aside to assist in identifying the sender from any subsequent correspondence.

Destruction of coupons

After the count, all redeemed coupons must be cancelled with some device which will prevent their being used again. When they are to be destroyed this should be done by a designated person and certificates of the totals, which have been signed by two responsible persons, should be provided.

INSURANCE CARDS AND INCOME TAX FORMS

If these are lost in a fire, considerable time will be spent in obtaining their replacement and therefore it is recommended that they should be kept in fire-proof containers as close as possible to the desks where they are worked on. In the event of a threat of fire they then can be placed under protection immediately. Special cabinets have been designed for this purpose.

CHEQUES

Pre-signed cheques

These are printed by a customer's bank and held there in bulk to be drawn on as required in the usual way. As a security protection against misuse they are overprinted 'account payee only'. With that restriction they are valid orders for payment bearing the facsimile(s) of the signature(s) of the payer with no amount or payee's name entered on them.

The cheques held by the customer should be kept in a suitable container which has two locks, either key or combination operated, the means of operating them being held by two persons. Combination locks are preferred because they do not require a key which passes from person to person when the responsibility for its custody changes, for example, during the holidays or illness of the usual keyholder. To provide for those circumstances it is better security if the lock is a combination one which can be changed as required. The stock of cheques should be counted daily by a person who is independent of the person concerned with the preparation of the cheques for payment.

Requests to the bank for pre-signed cheques must be signed by two authorised persons.

Cheque-signing machines

These machines require two keys to operate separate locks, *A* and *B*. One, *A*, secures the signature block and the second, *B*, prevents the operation of the machine with the block inserted. The keys must be held by two persons. In readiness for use the holder of the key for *A* will insert the signature block and lock it in. The holder of the key for *B* will then use his key when the machine is ready for use.

There is a meter fitted which records the total number of cheques which have been signed. Before the machine is put away at conclusion of use the figure shown will be noted by the holders of both keys and this will be compared with the list of cheques which have been signed. Notice will be taken of any spoiled cheques and of their disposal. When the machine is next required for use the previously noted meter figures will be compared with those then shown to note any differences. Duplicate keys should be deposited in the bank under the usual strict control over their withdrawal.

SECURITY SUPERVISION

It should be part of the duties of the person responsible for security in buildings to visit them occasionally outside the usual office hours to check that security procedures have been carried out. This inspection should extend to examining documents exposed on desks and in open drawers and cabinets to see whether they are of a confidential nature. If any are found they should be removed and a note left where found saying they have been taken away and from where they may be collected. Similar action should be taken when attractive office equipment has not been stored as instructed.

Cash security

Cash security is dealt with under that title in Chapter 14.

Checklist

The recommendations in this chapter are summarised in Appendix 25.

17

Explosives and other stores and stocks

The importance attached to the safe storage and usage of explosives is shown by the scope of the Explosives Acts of 1875 and 1923 and the mass of Statutory Rules and Orders and Statutory Instruments based upon them. The safety aspects are obvious: there are few more instantaneous and devastating ways of creating havoc than by the ill-timed setting off of an explosive charge. Reduction of incidents of this type was no doubt the prime initial objective of legislation but security has become increasingly important.

SECURITY OF EXPLOSIVES

Old-fashioned safes offer only a token resistance to thieves with adequate cutting gear or the ability to get at the ill-protected backs. As improvements came in design and construction, thieves soon found that gelignite and detonators took all the hard work out of their operations and speeded them up. Thus their first need was to acquire these substances in adequate amounts. It is a sad reflection upon users that there never seems to have been any difficulty for thieves in this respect. It does not always necessitate breaking into explosives stores—checks on usage by employees are rarely so exact as to provide any knowledge of misappropriation. It is appreciated that where there is heavy consumption, as in coal mining and quarrying, opportunities will always exist for those so minded to steal the materials they are using.

It is obviously not within the scope of this chapter to deal with legal requirements in detail; for those who need minute knowledge, the *Guide to the Explosives Act 1875* (fourth edition), issued by HM Stationery Office will provide this. A number of summaries are also available from the same source (under Code 34/99). Summary Number 4 deals with 'Stores'; Number 6 with 'Registered Premises'; Number 7 with 'Sales.' In addition, the Nobel Division of ICI has a publication, *Explosives, the Sale, Stores and Conveyance by Road,* which sets out the essential requirements quite clearly and indicates Statutory Instruments on which they are based.

The sale and purchase of explosives have less importance to security requirements than the storage; suffice it to say that all users of explosives, including gunpowder, all types of detonators, capped fuses, and safety fuses must have a police certificate or licence to cover their purchases. The only exceptions are government departments and licensees of magazines—the latter are controlled by the Secretary of State who has power to refuse a licence. It should be noted that where the storage of explosives is concerned, a local authority has no power to withhold a store licence, provided the conditions laid down for its construction, proposed site and the quantities held, are complied with; the discretion vested in the police to refuse a certificate is, therefore, the means of control to prevent undesirables having control over explosives. A standard form of certificate is used which certifies the holder 'is a fit person to keep during the continuance of this certificate at his store licensed for mixed explosives (or at his registered premises; or for private use) . . . the following explosives . . .' A factor which any chief constable will take into account in considering whether a person is a 'fit person' is the manner in which he keeps the explosives in accordance with regulations or otherwise.

Practical Security in Commerce and Industry

STORAGE OF EXPLOSIVES

Page 193

Since the preparation of this Edition the Home Office has issued a circular to Police, No. 113/1972, the Appendix of which makes recommendations respecting the 'STORAGE OF EXPLOSIVES: SECURITY ARRANGEMENTS'.

These deal with standards of resistance; strengthening and anchorage of steel stores; fitting of two *steel* bolted multi-lever mortice deadlocks in welded-on steel pockets to each door; making reinforced concrete 'skins' for vulnerable buildings; ventilation and alarm systems. Padlocks are *not* advised on grounds of susceptibility to attack by explosives, use of their fittings to lever open the doors and possible disuse of the mortice locks.

The recommendations are applicable to all licensed stores.

Police generally are the inspecting authority and their advice should be sought because few stores are likely to meet the new standards. This could lead to the refusal of a licence or of a renewal to store explosives.

STORAGE OF EXPLOSIVES

General information

When mixed explosive is stored, gunpowder and explosive are reckoned weight for weight though in the lower amounts, if gunpowder alone is kept, provision is allowed for a higher quantity under the same conditions. For all types of storage, other than private use, the explosive equivalent weight of detonators should be calculated at 2.25 lb (1 kg) per 1000. Cordtex is in general use and when storing it, its weight as an explosive should be reckoned at 16 lb per 1000 ft (23.81 kg per 1000m).

The police certificate runs annually and there are distinct advantages in making the renewal of this concurrent with that of a store licence or registration of premises (see below).

At the time of writing there are special restrictions on all aspects of purchasing, storing or using explosives in Northern Ireland. The Explosives Act 1875 is normally applicable there with minor differences in respect of sale and conveyance. It is likely to be advisable for some time to consult local police offices there for up-to-the-minute guidance.

Private

Small quantities up to a maximum of 10 pounds (4.5 kg) of explosives and 100 detonators can be stored for private use without legal restrictions on the method of storage. It is, however, recommended that the detonators be kept separately and both should be contained in strong boxes which are transportable (for the event of fire), appropriately labelled, and fitted with substantial locks. This labelling, which is necessary for safety reasons, could attract the notice of a thief so when the boxes are not in use, they are best kept out of sight and in locked premises.

While in 'private' storage 10 pounds of explosives may be replaced by thirty pounds (13.6 kg) of gunpowder.

Immediate use

If a user requires explosives for immediate purposes in larger quantities than can be acquired under 'private use', a Chief Officer of Police may issue an 'immediate use' certificate to authorise this, provided they are not intended for resale. These documents only cover one transaction and must be signed by the supplier. It is recognised that the explosives will not always be used immediately but that there will be a necessity to store on many occasions. HM Inspectors of Explosives advise that if explosives so obtained are not entirely used on the day of receipt, they should be stowed with minimum hazard and maximum security and the police, or local explosives officer, informed and given an indication of the safety/security steps taken.

This type of certificate has been found extremely handy when, because of administrative delays, the renewal of licences has been delayed and a supplier has had no authority to release the needed explosives. The police will normally be prepared to issue the necessary certificates to enable work to continue.

In registered premises

There are two types of registered premises:

Mode B caters for up to 15 pounds (6.8 kg) of mixed explosives. These may be kept in any type of building, provided that they are in a substantial container, which is used exclusively for the explosives, under lock and key to exclude unauthorised persons. Detonators must again be kept separately; a fire-proof safe would be acceptable for gunpowder only (up to 50 pounds, 22.7 kg, if not mixed).

Mode A increases the amount up to a maximum of 60 pounds (27.2 kg) of explosives (up to 200 pounds, 90.7 kg, if not mixed). Dwelling houses are not acceptable and the building used must be substantially constructed of brick, stone, iron or concrete (fire-proof safe again for gunpowder); provision is made for keeping in excavations. In all cases the building or excavation must be at a safe distance from highways or places where members of the public pass or work—never less than 15 yards (13.7 m) is considered to comply with this requirement.

Obligations are laid, as follows, in respect of the buildings for mode *A* and the receptacles for mode *B*:

1 The shelves and fittings must be lined to prevent exposure of iron or steel.
2 Precautions must be taken to observe scrupulous cleanliness to eliminate grit.
3 Water must be excluded where this may have a dangerous effect on the explosive in store.
4 All tools, locks, keys, etc., must be of a non-spark-producing metal.
5 All articles of a highly flammable nature must be kept well away.

Multiple registered premises

In certain industrial concerns, where there is substantial usage of explosives, as in collieries, it is often impossible to keep to the safety distances required for licensed magazines or stores. Special provision is made so that, apart from the main central store, a number of individual points, each regarded as registered premises under mode *A* are established and are subject to the same conditions of registration and inspection.

Explosives stores

These are divided into five divisions, referred to alphabetically as *A, B, C, D* and *E*, by capacities ranging from 60 to 4000 pounds (27 to 1814 kg), in excess of which magazine licences are required. Special limitations are placed upon the siting of these stores, according to their division and the proximity of 'protected works', which are themselves divided into two main classes. To illustrate the application of these regulations, in the case of a division *E* store, 2000 to 4000 pounds (907 to 1814 kg), this could not be sited within 352 feet (107 m) of a pier or jetty (Class 1), or within 704 feet (214 m) of a hospital or theatre (Class 2).

Under no circumstances may detonators be kept with other explosives in the main store; they must be in a separate annex or licensed building. Where a few hundred detonators only are concerned, special arrangements may be allowed, however, providing there is no simultaneous access to both detonators and explosives inside the store. All stores containing more than 1000 pounds (454 kg) of explosives must be fitted with an efficient lightning conductor.

General safety rules

There are some general rules for stores which must be complied with. The more important are:

1. The permitted amount of explosive must not be exceeded.
2. The store must only be used for keeping explosives and tools and receptacles in that connection.
3. All tools, etc., shall be of some soft metal or wood to obviate sparks.
4. The interior must be lined to prevent the presence of exposed iron or steel.
5. The interior must be kept clean and free from any kind of grit.
6. The explosives must be taken out and the store thoroughly washed out before any repairs are made to it.
7. Provision must be made, by use of suitable shoes and pocketless clothes and searching or other methods, to prevent means of causing fire—matches, steel, or grit—being taken into the store.
8. There must be no smoking in the store.
9. A person under the age of 18 shall not be employed in or enter a store, except in the presence and under the supervision of some person of the age of 21 years or above; a person under the age of 16 shall not be employed in the store at all.
10. A copy of the general rules shall be affixed inside the store, together with an extract from the licence showing the amount of explosive that may be lawfully kept.

Not mentioned in regulations, but nevertheless a practice that should be followed, is that of having a pre-arranged drill to be carried out in the event of an explosion. This should include designation of individuals with responsibility for specific actions and names and telephone numbers of those services and persons who must be informed. If possible, a list of those persons who might be expected to be working in the danger area at any time should be available.

Where any accident occurs by explosion or fire in connection with an explosives factory, magazine or store, notification must immediately be sent to HM Inspectors of Explosives; similar action should be taken in respect of registered premises when the incident involves any form of personal injury. In serious cases, notification in the first instance should be by the quickest available means and the debris should not be touched except to recover bodies or treat injuries.

Workshops

Any workshop used for making up charges must be sited at least twenty-five yards (23 m) from the store or 'protected works'. It must not contain more than 100 pounds (45 kg) of explosives, and the local authority must be notified of it. To all intents and purposes its upkeep and usage should be similar to that of a store.

SECURITY OF EXPLOSIVES AND EXPLOSIVE STORES

Explosives and detonators can fetch a good price amongst the criminal fraternity and therefore there is temptation for employees to misappropriate them and to outsiders to break into stores. An accurate recording and usage documentation is essential and where any form of search clause exists it should be regularly implemented—whether it is solely that intended to stop the introduction of matches or prohibited items liable to start fires, or specifically aimed at finding stolen property.

If a daily balance of what should be in the store can be maintained by logging all incoming explosives and detonators, those passed out for use, and those returned unused, a quick check on the contents should reveal deficiencies and provide accurate information to the police in the event of a theft. A strict control must be kept on keys, which should be signed in and out daily.

Where explosives stores or unregistered premises are within a works or factory perimeter, the actual physical protection offers no more difficulty than any other important point. Alarm systems can be easily fitted and protection afforded against unreliable employees by making an enclosure with a padlocked gate and illuminating the whole area.

Stores of heavy steel construction with a welded-on detonator annex may now be purchased and installed with minimum effort —it seems that this may be the fashion rather than to construct brick or concrete types in future. They have a matchwood lining secured with copper nails and many unfortunately are only fitted with heavy brass deadlocks. Brass keys are used with these and can cause a difficulty by breaking in the lock if roughly handled; in these circumstances the services of a locksmith should be obtained, and no amateurish efforts permitted. Although no steel or iron is allowable inside the store, future stores of this type will be required to have two approved steel deadlocks fitted. There would also appear to be no reason why substantial hasps should not be welded on the outer side of the door and the jamb, and a high-quality, close-shackle padlock fitted. The stores should be firmly fixed to the ground by bricking or concreting in; it can be lagged with sandbags for protection in the event of accident.

Quarry stores are particularly vulnerable, they are invariably well away from habitation and rarely visited after working hours. The heavy metal prefabricated type, described above, is preferable, sited in open view and illuminated if possible. If a compound of the Lochrin Palisading type can be erected, so much the better. Bushes and other obstructions in the near vicinity of the store should be cleared.

If a light is to be left over the store the police should be formally notified so that its absence will at once arouse suspicion. If steps are taken to make it absolutely obligatory for thieves to use considerable force to get into a store they will not try it. They have everything to lose if anything goes wrong.

Security powers in respect of explosives stores and magazines

Anyone who enters without permission, or otherwise trespasses, on any explosive factory, magazine, or store, or land immediately adjoining and occupied therewith, can be removed by the occupier or his agents or servants, as well as by the police, and may be fined up to £5.

Anyone, other than the employee, doing any act which tends to cause a fire or explosion can be fined up to £50. (A wide power of arrest is given to a variety of persons for this offence under the Explosives Act 1875 section 78.) Notices warning all persons of their liabilities shall be posted by occupiers, but their absence will not exempt from the penalties.

CONVEYANCE OF EXPLOSIVES ON ROADS

To date, there do not appear to have been any instances of stealing explosives in transit in this country and the danger, which would attend any violent attempt to do so, limits the probability down to one of collusion with the driver. The following general regulations, imposed by a series of Statutory Instruments, are similar for both diesel and petrol-driven vehicles:

1. Two men must always be in attendance.
2. The cab must be separated from the body by a fire-resisting screen to within 12 inches (30 cm) of the ground, the whole of the exhaust pipe being in front of the screen, which must have a clear space of at least 6 inches (15 cm) between it and the body.
3. The petrol tank must be in front of the screen.
4. Adequate means of fire extinction must be carried.
5. There must be a quick-action cut-off to the petrol supply, near the carburettor, but not so close that it could be involved in a fire therein.
6. The vehicle interior must be lined with non-inflammable wood or asbestos, with a sheet metal exterior.
7. Electric lamps only should be carried.
8. Normally, the only opening in the body should be a door at the back.
9. The cab area should be rendered non-inflammable.
10. The engine must not be run either loading or unloading.

11 Speed limits must be rigidly observed.
12 Where possible, stops should not be made in built-up areas.
13 Smoking in the vehicle must be absolutely barred.

GENERAL STORES

Construction

They should be solidly built with as few windows as necessary. The door should be strong with a first-class mortise lock with a mortise deadlock or multi-levered padlock and a strong hasp. The window should be protected on the inside by iron bars correctly set in the stonework, not screwed into woodwork, or by some form of steel shuttering for erection when the store is not in use. There should be a service counter behind which only the stores staff should be allowed. This is particularly advisable where canteen stores are concerned.

Stock level

It is, of course, good practice to keep stores at a level from which the normal requirements can be supplied but not in excess, which is uneconomic. Therefore the stores which are held must be protected against improper usage and thefts for two reasons: first, the intrinsic loss, and second, if this continues unabated there can be tendency to allow for the losses by an excessive holding which, as has been said, is costly.

Engineers' stores

There should be an approved list of persons authorised to sign requisitions for stores. The storekeeper should have a copy with specimens of the authorisers' signatures. Figures should not be used on requisitions, all amounts being written in words. This makes the addition of other figures, thus irregularly increasing the amount to be drawn, more difficult. A line should be drawn immediately under the last item on the requisition to prevent the improper addition of further items.

Certain named persons may be given authority to draw repair items on request with their immediate superiors making out and authorising a requisition later. To save time and to ensure that no item is missed, goods can be entered directly on to a multi-lined requisition to be seen by the supervisor for subsequent authorisation. Requisitions which have been presented should be inspected afterwards by someone superior to the authorisors so that excessive or fraudulent withdrawals can be detected and necesssary action taken.

Specially attractive and valuable stores

Electric light bulbs, dry batteries, tyres, sparking plugs, torches, paint, paint brushes, cleaning materials, and small tools are examples of stores which are especially vulnerable to theft. They should be kept as far away from the entrance to the store as possible. Preferably they should be stored in a separate area under lock and key. This can be a compound made of strong wire-mesh within the main store. The key should not leave the possession of the storekeeper responsible for issues. To provide for special cases where issues of such nature might be required when the stores are closed a selected number can be stored outside the compound. See a later paragraph on withdrawals from stores when closed.

Special tools

Valuable tools, such as electric drills and micrometers, are usually held in the engineers' stores to be drawn for use as required. A control on the issue and return of the tools must be introduced and enforced to prevent thefts. One type of control is for an employee drawing a tool to surrender a metal disc with a number by which he can be identified. This is hung on the hook from which the tool was taken or put in the bin. On the return of the tool the disc is handed back.

A number of the tools are likely to be of a common type and make and could be confused with similar tools of outside contractors doing work on the premises. To prevent loss this way it is recommended that all serial numbers on them should be recorded and in other instances a distinctive mark be given to them.

Ladders and scaffolding

The name of the owners should be burnt into the wood. Steel scaffolding is difficult to mark but one way is to paint a series of bands of a distinctive colour on them. A security paint is available from well-known suppliers.

Withdrawals from stores when closed

Where there is no constant attention at stores the problem can arise of how items can be obtained from them in the event of an emergency happening when they are closed. If the premises are continuously attended by security officers this can be provided for by leaving the key to the stores at the security department. Supervision of the person entering the stores in such circumstances is required if he is not a foreman or of supervisory grading. This can be done by the security officer on duty accompanying the employee into the store and witnessing him take what is required from where it is kept. It is not satisfactory for the security officer to remain at the counter while the required stores are collected from the bins. A requisition or other form of record must be left for the storekeeper and a note of the occurrence made in the security log book which would always be available to the storekeeper on request. (See Chapter 4.)

Borrowing tools

See Chapter 4.

Equipment for burning and welding

This is especially attractive to thieves who require such equipment to carry out crimes and therefore it must be kept under good security. Cylinders of gas should not be kept in close proximity to the burners. There have been instances where such equipment has been used by criminals to cut open safes on the same premises. There is an additional specially dangerous risk of explosion of the gas cylinders should they be involved in fire. Therefore they should be stored in a specially marked area close to a door which will permit their immediate removal in the event of fire.

PETROL STOCKS

The issue of petrol from storage pumps is usually recorded according to the vehicle supplied and its registration number. If a mileage record is kept and the consumption is reasonable there is no cause for concern. However, losses can occur at the point of delivery if care is not used. The meter pumps, of course, provide a further check on the issue of fuel. The dipstick used to measure the contents of the tanker before delivery should be inspected to see that nothing has been cut from the end which is inserted. If this has been done the stick will show greater contents than there is. For the same reason the stick used to measure company storage tanks should be checked.

The Petroleum Spirit (Motor Vehicles etc.) Regulations 1929 (SR & O number 952) control the storage of petroleum in bulk which is not for sale and say that on the application of anyone intending to store the liquid the local authority may grant a petroleum spirit licence authorising such storage. Attached to the licence usually are conditions under which the petrol is to be kept. In Appendices 13 and 14 are shown examples of these conditions.

The Petroleum (Inflammable Liquids) Order 1968 (SI 1968/570) applies certain provisions of the Petroleum (Consolidation) Act 1928 to no fewer than 207 inflammable liquids. These provisions require notice of certain accidents in connection with inflammable liquids to be given to the Secretary of State, make provisions relating to inquests and confer certain powers on magistrates' courts and on government inspectors in respect of the inflammable liquids. The new Order also empowers the Secretary of State to make regulations as to the conveyance of inflammable liquids by road and for protecting persons and property from danger in connection with such conveyance and he has made the Inflammable Liquids (Conveyance by Road) Regulations 1968 (SI1968/927) which came into operation on 1 October 1968.

A model code of principles of construction and licensing conditions for petrol depots and major installations can be purchased from HM Stationery Office, price 10p.

The Highly Flammable Liquids and Liquefied Petroleum Gases Regulations 1972 (SI 1972/917)—These regulations were made on 20 June 1972 and will come into force on 21 June 1973 (except for part of regulation 10, requiring some workrooms to be fire resisting, which comes into force a year later).

The regulations impose restrictions, for the protection of employees who are covered by the Factories Act 1961, on the use of highly flammable liquids and liquefied petroleum gas. They cover the following topics:

Regulation	*Subject*
2	Interpretation
3	Application
4	Exemption certificates
5	Storage
6	Marking of storerooms, tanks, vessels, etc.
7	Liquefied petroleum gas—storage and marking of tanks, vessels, cylinders, etc.
8	Precautions against spills and leaks
9	Sources of ignition
10	Prevention of escape of vapours and dispersal of dangerous concentrations of vapours
11	Explosion pressure relief of fire-resisting structures
12	Means of escape in case of fire
13	Prevention and removal of solid residues
14	Smoking
15	Control of ignition and burning of highly flammable liquids
16	Power to take samples
17	Fire fighting
18	Duties of persons employed

There are two schedules in the regulations concerning the method of test by flashpoint (closed cup method) and method of test for combustility.

Oil stocks

These can be of various types, both edible and non-edible, and are usually delivered by weight. One way in which small but valuable quantities of the more expensive oil have been stolen is by the suspension inside the tank of a bucket or can. After the tanker is discharged the suspended can full of oil is lifted out. The weight is so small in relation to the total that it very often escapes notice when the net load is recorded after the empty vehicle is tared.

Canteen stores

Assessments of requisitions in respect of withdrawals from these stores may not be practicable. A control on materials used, however, is obtainable by a reconciliation of the cost of stores purchased with the proceeds of the sales, etc. This usually is the responsibility of an internal audit department, where appointed. When stores are closed, all china, cutlery and small items of equipment should be under lock and key and this left with the security officer where applicable.

Special attention is required to the security of wines, spirits, cigarettes and cigars which may be stored in fragile cupboards in dining-rooms which are reserved for the use of senior members of the staff or directors—see Chapter 4 under 'Keys'.

RADIOACTIVE MATERIAL STORES

The increasing use of radioactive materials and isotopes in industry requires protection of the handlers, other employees, and the general public against the dangerous effects of radiation to be of the proper standard. The Radioactive Substances Act 1960, contains the main provisions controlling the holding of such materials.

The International Atomic Energy Agency issued a booklet in 1958 called *Safe Handling of Radioisotopes* which gives a comprehensive list of recommendations for the storage and transporttation of radioactive materials. With the permission of the Agency these are given in Appendix 11.

LIQUID PETROLEUM GAS

There have been several fires involving cylinders of this gas and, with the permission of the Fire Protection Association, their recommendations on the handling of the containers, given in the October 1967 issue of their journal, are given in Appendix 12.

CHECKLIST

The recommendations in this chapter are summarised in Appendix 26.

18

Commercial and industrial espionage

In these days of fierce competition in commerce and industry the protection of information which, in the wrong hands, could be detrimental to one's interest is a vital part of management's responsibility.

The fundamental security principle which governs the transmission of information requiring protection is the 'need to know' principle. This simply says that information of this type must not be given by any means to any person who has not an absolute requirement to have that information for the purpose of his work.

Although the responsibility is far from being new, the possibility of espionage occurring has been emphasised in recent years through the introduction of special electronic and mechanical devices developed for the purpose. By the use of such equipment, security measures protecting confidential information, previously considered satisfactory, can be circumvented. There are many less sophisticated means by which valuable information can be obtained by interested parties and the observations and recommendations which follow are mainly directed to preventing leakages through neglect of proper precautions.

What is confidential information? A simple definition is that it is any information in the possession of the company that is not readily available to competitors and enables the company to have a competitive advantage.

ROLE OF MANAGEMENT

Security against industrial espionage poses problems which should be dealt with at board or, at least, senior management level. It is necessary first to identify the really valuable secrets which the company possesses and concentrate on their protection rather than attempt the costly task of safeguarding the enormous amount of information of lesser importance.

A detailed security survey should be conducted and reports obtained of what must be protected. 'Restricted areas' should be delineated particularly where research and development is concerned, and lists of persons authorised to enter published. The wearing of badges with photographs should be considered. The exclusion of unauthorised persons might have to be enforced by equipping doors with special locks either electrically operated by a special card, or combination locks. Once security measures are decided, regular inspections should be made to verify that these are being observed. When, in the course of an inspection, something requiring special protection is found to be left at risk, an effective way of enforcing the requirement is to remove it to a place of safety. In its place, leave a card saying that a security check was made and certain articles or documents were found insufficiently protected and may be claimed from (and here give the name of the person holding the article—in one organisation it is the managing director). To assist staff at all levels it is recommended that on suitable occasions a system of instruction in the care of confidential information should be adopted.

MEASURES FOR THE PROTECTION OF INFORMATION

A system should be introduced for control and counting of documents of a confidential nature. Particular attention should be paid to:

1 Receipt and registering of documents.
2 Storage after office hours.
3 Taking home of confidential information.
4 Internal communications.
5 Postal arrangements.
6 Duplication of records.
7 Destruction of documents.
8 Where security containers are used, a controlled issue of keys.

Most recorded industrial espionage for profit concerns operations against research establishments but it also can be used:

1 To identify key persons whose services might be sought and to appraise the inducements which should be offered.
2 To determine strengths and weaknesses of rival firms, or of firms which are being considered for takeover.
3 To discover advance information which, when published, will affect Stock Exchange share prices.
4 To discover quotations of rival companies in competitive tenders.

Some examples of information which must be protected against disclosure to unauthorised persons include:

1 That which would be of assistance to competitors, for example:
(*a*) Plans for the production and sale of new products.
(*b*) Plans for a new advertising campaign.
(*c*) Details of customers and trading terms.
(*d*) General marketing plans.
(*e*) Policy decisions.
(*f*) Proposed alliances or mergers with other concerns, or exploratory action to such ends.
(*g*) Projections in selling.
(*h*) Sources and costs of raw materials.
(*i*) Proposed redeployment or closures of centres of manufacture or distribution.
(*j*) Contractual or trade agreements.
(*k*) Details of tenders submitted.

(*l*) Research programmes and results.

2 Personnel information, for example:

(*a*) Proposed promotions, transfers, and dismissals.
(*b*) Salary structures.
(*c*) Personal files and information.
(*d*) Confidential reports.
(*e*) Rationalisation plans.

The motive for obtaining unauthorised information on those matters can range from simple curiosity, for the personal satisfaction of just knowing, to deliberate acquisition with the intentions to convert the knowledge into some material benefit or to some personal advantage. Where documents containing vital information, such as those which have been described, require to be protected from inquisitive eyes, the security precautions begin from the time the information is committed to writing, whether shorthand or longhand, type or print. There is a tendency to overlook the proper disposal of notes taken after they have been reproduced in more readable form over which a strict security blanket is drawn.

SECURITY

Principles of security against disclosure

There are certain basic principles which should be observed:

1 When confidential information is in documentary form it must be protected in such a manner that if it is copied, photographed, or stolen this quickly becomes known, and the quicker the better.
2 The decision regarding the degree of protection to be exercised is the responsibility of the person originating the document, or ordering the photograph, drawing or plan.
3 The extent and cost of the measures to be taken to prevent disclosure of information must have some relation to the consequences of disclosure.
4 The security measures everyone is required to observe must be of the kind to command respect.

To assist in making decisions on the degree of security to be accorded, specimens of classifications should be prepared and distributed.

Research and development

The desire for confidential information is not restricted to documents such as those which have been described. In research and development this is likely to be coloured by commercial advantage. This type of work costs a great deal of money and includes testing the use of new materials, the development of new techniques, and creating new formulas, but it is absolutely necessary in the competitive markets of today.

The products of research are not always of a dramatic nature and can include, for example, something which might appear to be a simple fact: that substance *X* will not mix with substance *Y* in the specialised conditions of the envisaged use. This is, however, valuable knowledge which competitors in the same field would like to have to save them expending time and money to produce the same result. As the information cost the originators of the research a lot of money and professional skill it must be protected.

An example of development work is the designing and construction of new machinery to produce improved or new products. This can involve the preparation of new packaging designs requiring the setting of printing plates and seeing specimens which need protection from prying eyes. Therefore all unauthorised persons must be excluded from the areas of work. Constant personal observation is probably impracticable and expensive, so the locking of premises and the issuing of keys to authorised personnel has to be considered. The control of keys in those circumstances is extremely difficult and they can be lost or mislaid. An alternative is to install electronic locks operated by an electric circuit incorporated in an identification card bearing a photograph of the authorised holder. The card and photograph is specially constructed to prevent interference. An additional precaution in conjunction with the use of the card is the requirement to press numbered buttons according to a code on a panel, which can be changed without difficulty to take account of changes in personnel. The locks are described in Chapter 33.

Visitors and callers

To identify visitors and callers to research establishments it is recommended they be issued with numbered lapel badges of a distinctive colour—the issue and recovery being strictly recorded. Each visitor to a restricted area should be vouched for and accompanied by an authorised person. Callers who require special attention include representatives of firms supplying equipment, who probably visit competitive firms in the course of their business.

Cleaners

By posing as an office cleaner or with the connivance of the genuine cleaners, access can be secured to rooms where confidential information is kept. The presence of such a person is unlikely to be questioned so he or she has the opportunity, and the time, to find what is wanted. This is particularly easy where cleaners are provided by cleaning contractors whose employees frequently change. At holiday times and periods of absence due to sickness the contractors are obliged to engage temporary employees. It is recommended in those circumstances that contractors should be asked to supply names and addresses and before being allowed to work they should be interviewed by whoever is responsible for security. In areas of high risk, to prevent the possibility of leakage of information, the cleaning could be done in office hours or contract cleaners could be excluded and the cleaning done by own staff. Temporary own staff engaged from agencies should be treated with similar circumspection.

Cleaning contractors recognise the special position and opportunities of their employees and when the circumstances require it, such as the cleaning of research and development areas, they will vouch for their integrity and provide them with identity cards bearing photographs and specimen signatures to assist in preventing their impersonation. Security personnel should check the identities of the cleaning staff from time to time and when patrolling note their activities for future reference.

Temporary office staff

Where the employment through agencies of temporary staff, especially secretaries for managers, becomes necessary, care must be taken that they do not have access to confidential information. They may subsequently work for another firm competing in the same field and, though innocent of any ulterior motive, disclose what they saw or heard at their previous employers.

Libraries

Confidential information on the result of work done is filed here to be available when required. The removal of papers from the library should be strictly recorded and spot checks made to see whether any have been taken without authority. Where papers are circulated to associated research establishments from a central source, it is recommended that after being seen by interested persons they should not be recorded in complete form but the essential information transferred to cards and filed in an index. This reduces the space taken on valuable shelf space and the security protection of the cards is relatively simple. If on subsequent reference more information is required, this can be obtained on request from the centre. However, one instance is on record of a young laboratory assistant removing a number of such cards from an index drawer and taking them with him to a new position he had obtained with a competitor. The necessity to check on the numbers of cards is illustrated by the fact that it was some months after he had left that the cards were missed.

The law

At the present state of the law it is not a criminal offence to 'steal' confidential information by copying it onto paper, which is not the property of the firm or by photographing it. On the other hand, if the information is written or printed on anything permanent which belongs to the firm and is taken away with the requisite intention under the law of theft, the person responsible can be convicted of stealing the material concerned.

For example, in the case of the laboratory assistant, referred to above, he was charged with stealing a number of pieces of cardboard with a nominal value of 5p. In Chapter 9 it is pointed out that to prosecute a charge of theft it is not necessary that the property concerned shall be of some intrinsic value. It simply has to be of some value to the owner. It is the practice, however, to put a nominal value on the property concerned in a charge of theft for statistical purposes.

In cases where an employee is accused of taking away trade secrets from a former employer, the courts will take into consideration whether the employee has been made aware of the secret nature of the material in question and whether the owner of the security information took reasonable precautions to preserve it in secrecy. Accordingly, all confidential information given to employees should be so marked. Although there may be a breach of contract if an employee leaves the company and takes away confidential information, and although its use may be legally enjoined, failure to install reasonable safeguards may raise the question of whether the information was in fact confidential.

Should anyone, not an employee, be found in premises in circumstances which suggest he was there to obtain confidential information, he should be detained using as much force as is necessary (see Chapter 7 on the use of force in such circumstances), and the police called. They will decide, on the evidence, what offence, if any, the person detained has committed. To assist them at arriving at a decision they will search the suspect before removal to or at the police station (see Chapter 7 on preventive arrests). This will reveal whether he has any stolen or unauthorised property in his possession or anything used in espionage such as a camera. In the latter instance a claim should be made to the police for an opportunity to check it for any unauthorised photographs it may contain.

Official Secrets Acts

It is a criminal offence under these Acts, punishable by a long period of imprisonment, to pass to an unauthorised person information concerned with work classified as secret under the Acts. See Chapter 10 for further information.

SECURITY MEASURES

Some recommendations for improving the security of confidential information in documents, which include plans, drawings, and visual aids are as follows:

1. After shorthand or other forms of written notes have been transcribed satisfactorily they should be destroyed by shredding or burning. That will be the method recommended when destruction is referred to in subsequent paragraphs.
2. If the typewriter used for the preparation of a confidential document has a once-only carbon-treated plastic ribbon, which is usually the case with electric machines, it should be sealed in a special envelope and handed to a specified person after it has been used up complete, and a new one given in return. The used ribbon must then be destroyed by an authorised person. When a ribbon which is partly used is on a typewriter, this must be kept under first-class security protection when the premises are unattended or cut off and destroyed.
3. All carbon-paper and other types of duplicating-paper used with typewriters to produce copies of confidential documents must be destroyed immediately afterwards.
4. When multiple copies of a confidential document are required, the original must be taken by a responsible person to the machine to be used where he will witness the taking of the copies. Spoiled copies will be collected and destroyed with any original which has been used in the process.
5. Duplicating machines with meters recording a progressive total of the copies taken will disclose any unauthorised use

from a reconciliation of the total of actual copies with the meter reading.

6 The number of copies of a document to be taken must be shown on a formal requisition signed by an authorised person. The number taken must not exceed those authorised and when they are being typed the paper in the machine being used should be counted occasionally by a responsible person.

7 Paper bearing confidential information put in waste-paper baskets must be destroyed under strict security conditions. The clearance of the baskets must be done at a time and in such a manner that there is a minimum of delay between the collection and destruction. In the meantime the waste paper must be under security protection to prevent interested persons prying into it.

8 Words of caution, such as 'secret' and 'confidential,' on documents and envelopes are not recommended because they have the effect of attracting the attention of the inquisitive or those with sinister interest. Their too frequent use on documents which do not justify the classification can minimise their effect, so that insufficient protection may be given to documents which do merit it.

9 As an alternative the degree of security which should be accorded to a document can be indicated by giving a prefix to any reference number it may bear, or in other instances by the use of a symbol from a code, which would be known only by the persons likely to come into authorised contact with the document. How this is done will be left to the imagination, but for example purposes the letter A will be used.

10 The number of copies which exist can be shown on the document by the figure following the reference number, or where that is absent, immediately after the security symbol. The serial number of each copy would then follow. Using the example suggested above the final reference would be A/ 1234/ 10/ 1, 2, 3, and so on to 10; or A/10/ 1, 2, 3, and so on to 10.

11 When a document is in the top security class, a signature must be required from everyone to whom copies are

delivered and a record kept so that when it is no longer required all copies can be withdrawn and destroyed.

12 For documents having a lower security classification, but which still require care in handling, a colour code can be used by affixing stickers or drawing lines across a corner.

13 Vitally confidential documents passing through the post, including any internal arrangements, should be enclosed in an envelope addressed to the intended recipient and marked 'strictly private and personal'. That envelope should then be put into another one addressed as before and otherwise unmarked. The use of the recorded mail procedure for outside post should not be overlooked.

14 Wall charts, graphs, and other forms of visual aids showing the progress of the development of, for example, a new product or a new project where it is essential to the success of the undertaking that secrecy is maintained until the time decided for general publicity, should not bear a name whereby it can be identified. A code word or number is preferred, which would be used in all correspondence, telephone conversations, etc. If it is practicable such charts should be put under lock and key when the offices are closed.

15 Drawings and tracings of new machinery, new products, or their packaging should be serially numbered and their whereabouts recorded. If they are sent outside the premises for any purpose, such as tendering for work or construction in whole or in part, the security which they are to be given must be described in writing in what is sometimes called a security clause and signed in agreement by the outside concern. The security in which the documents are held must be to the satisfaction of the instructing firm.

16 Recorded dictation must be deleted immediately after being typed—if this is not possible the tapes or discs must be destroyed.

17 When it is essential that a document should not be copied in any circumstances it is good practice to mark it clearly 'Not to be copied.'

18 Where a document, etc., has a security restriction and subsequently this is reduced or removed it must be shown by

stamp or some other mark to have been re-classified or de-classified.

These recommendations are not intended to be rigidly adhered to and can be adapted, either in whole or in part, to individual circumstances, providing the principles are observed.

Combination locks are preferred to key locks on safes, drawers, and cabinets holding confidential documents. They have the further advantage that their codes can be changed frequently.

From time to time, inspections by a responsible person should be made of all offices, to discover the standard of security practised. These inspections should include drawing offices, where confidential matter is likely to be found. Should anything be found exposed, which should be under proper cover, possession should be taken of it for disciplinary action to be taken against the offender.

MEETINGS AND CONFERENCES

When the subjects to be discussed at a meeting or conference are of a level of importance that leakage of information about them would have extremely serious consequences, the time, date, and place must be known only to those who require to know. A good security precaution to prevent the overhearing of what is said, by, say, mechanical listening devices, is to circulate a change of venue, personally to only those concerned, so as to leave the shortest practicable interval before the meeting is held.

It is preferable that such meetings take place in circumstances where the members cannot be observed through windows, particularly from outside the building. Agendas or papers on which notes have been made should not be left about the room or in the waste-paper basket but removed and destroyed with any blotting-paper which might have been used. If the result of the discussion could have financial repercussion and has to be conveyed by telephone to anyone, a pre-arranged code should be used so that if overheard it would be meaningless. If breaks occur in a meeting for refreshments and important papers are likely to be left exposed, arrangements must be made for the room to be locked and kept under observation by a trusted person.

19

Computer security

Computers have proliferated during the past few years, becoming progressively more sophisticated, more costly and more essential to the user's day-to-day activity and profitability. Their intrinsic value is substantial but interference with them may have an adverse effect on the firm that is cataclysmic beyond any other form of loss or disaster.

For this reason it is proposed to deal with their protection in detail from the planning of a new installation to consideration of staffing and policy. Much of this is really a direct responsibility of the computer manager, who may well not have had any experience in security aspects and will welcome discussion.

By comparison with the consequential loss from a shut-down, potential losses from fraudulent operations are relatively minor and the probability of their occurrence more remote, though less so than formerly, possibly due to the amount of sensational publicity which has been accorded to them. While incidents have occurred in this country, the bulk of reported frauds continue in the USA. Nevertheless systems analysts, programmers, and computer operators have increased in accordance with the law of supply and demand, which has brought in its train high wages and excellent prospects. It is too much to expect that the criminal percentage of the community will not have due representation among these. Sooner or later frauds will be perpetrated—if they are not already in being, but undetected due to the expertise of the practitioners. Fortunately, publicity has also brought precautions which make their task more difficult—these are mentioned hereafter.

STANDBY FACILITIES

A first consideration when installing a computer is a rather surprising and pessimistic one, but one which the manager has to bear in mind from the outset—it is that of ensuring alternative suitable facilities in the event of a break-down from fire, supply failure, or any other cause. In many concerns such a break-down can cause utter confusion in otherwise well-organised regimes because of the degree to which the computer has been integrated into the firm's activity.

Standby arrangements should be explored from the outset on a mutual-aid basis with owners of like machines within reasonable distances. This could have a bearing on which of two equally acceptable machines is selected for purchase in that an identical standby for one of them may be immediately to hand. There is little doubt that full cooperation will be extended, though it might be advisable to avoid potential competitors in making the arrangements !

The manual or detailed instructions kept in the computer room should contain as a minimum of information:

1 The exact specification of each standby computer contact with details of time availability.
2 The identity and means of contacting those personnel at the other installations who can authorise use of the machines.
3 A list of contacts in order of preference to minimise delays in the event of the first choice being unusable.
4 The precise location of each contact with instructions on how to reach it.
5 A checklist of the minimum software, stationery and other equipment needed together with any particular personnel to be called out.

All senior personnel should be accessible by telephone at their homes and detailed instructions should be laid down for action in the event of a failure.

HAZARDS TO BE GUARDED AGAINST

Environmental disaster—Fire, explosion and flooding can destroy not only the building fabric and the machine itself but also computer programs, information, data and records stored on magnetic media. Card and paper tape input, printed output and supplies of special paper are also at risk in this context.

Machine failure—New generations of computers should have increasingly greater degrees of reliability with their improved technology, but they are still pieces of equipment subject to mechanical failure and dependent on external power supplies which can fail.

Deliberate or accidental loss of information held on computer files—This is an area of human involvement and covers all those aspects outside environmental damage. The loss can arise from errors and acts of negligence on the part of the operating staff, faults in programs, the consequences of unauthorised persons or visitors gaining access to the computer room—or deliberate acts of sabotage.

Misappropriation of company funds and assets—The areas which effectively provide the best opportunity of using the computer resources for this purpose are those of programming and operations—they will be dealt with later.

FIRE

It would be thought that the construction and usage of modern computers made fire a minimal risk; this has been effectively disproved by several disastrous fires which have originated in the computer rooms themselves and resulted in almost total destruction. Where a central computer processes information from ancillary 'slaves' in essential component factories, even a minor outbreak might have consequences affecting the whole production schedule. Preventive precautions must therefore be designed to detect and control fire at the earliest possible stage. The means of doing so must be built into the installation during construction, and the environment in which the computer is situated must not be neglected as a source of potential danger.

Housing of computer

Although computers are usually inserted into existing buildings, the amount of work entailed may render this a false economy when all the requirements of the insurers have been complied with. It is better in every way to house it in separate purpose-designed premises. If this is not feasible, the whole computer should be enclosed in a fire-resistant compartment, protected both from horizontal and vertical spread—and from potential water damage if other parts are affected.

The isolation of the central area containing the vital computer equipment is most important. Here the devices for detection and protection must have maximum effectiveness. Segregation from the surrounding areas of data processing and ancillary offices must be by fire-resisting partitioning or solid walls. If other offices are sited above, the roof of the computer room must be fully resistant to water as well as fire. It must not be perforated by openings whether for lighting, ventilation, or any other reason. Rooms below are equally important and their occupancy should be limited to purposes which entail the least possible fire risk—there are advantages in locating the room at ground level.

These steps of segregation will meet a further essential purpose in that a reception area can be constructed to stop casual entry into the central rooms.

Contents of the central area

It is elementary that all furniture and fittings in the central area should be of metal or at least of materials which are not readily combustible. This must be remembered during the ordering of such items when the thoughts of the senior personnel, being consulted on their requirements, may be more concerned with comfort or possibly even prestige.

Tapes, cards, plans or files should be restricted to those needed for immediate use, fire-resisting cabinets should be available for their temporary storage. Any bins or containers for waste paper should likewise be of metal and regularly emptied. Smoking should be prohibited within the computer room and the ban extended to the stationery stores where it is even more important. Consideration should be given to incorporating a rest room where smoking would be allowed—this might limit abuse of the rule where it mattered.

Floor and roof voids: ventilation and wiring

Temperature and humidity have to be controlled between relatively narrow limits. Extensive fluctuations adversely affect the computer and its tapes, and could cause operating failures. There must, therefore, be a full and reliable air-conditioning system with extensive ducting which, with the huge amount of wiring, has to be housed in roof and floor voids. A false floor is preferable to cable trenches in the structural floor as, among other advantages, it allows a construction whereby computer equipment can be easily rearranged or added to. The floor itself should be prefabricated panels of fire proofed timber or metal, individual panels of which can be lifted by suction pads to give access to the services beneath; all such removable panels should be clearly marked. It is conventional for the false floor to be at the same level as the rest of the building, the void beneath it will then become a potential water trap in the event of flooding or fire in other parts of the building, causing danger from the presence of live cables. It is therefore advisable to fit a sill across the threshold of any door leading onto a false floor.

It is important to eliminate combustible materials as far as possible from these voids, materials for suspension units and brackets, insulation, sound and waterproofing panels and ducting must be considered with this in mind, ducting in fact should be of metal. A fire in these areas may not be immediately perceived and could be difficult of access for portable fire extinguishers.

Detection devices in these voids are essential and they should be coupled to an automatic fire-extinguishing system.

The extensive ducting of the air-conditioning system provides an avenue of entry for heat and smoke, particularly if the computer room shares a ventilating system with the rest of a building. Sampling of the airflow is necessary to ascertain combustion and smoke content by means of probes and detectors, fire dampers held open by fusible links or electrically operated should be built into the ducts to stop smoke or hot gases reaching areas where they could cause damage. A manual means of operation should also be incorporated.

In addition to normal detectors, heat-sensitive cables can be laid among the wiring runs to detect abnormal temperatures before they reach the stage of creating smoke or flame.

Doors, hatches and windows

In the computer room, partitions, doors, hatches and glazing should generally have a fire resistance of at least half an hour. The offices surrounding the computer room can be treated as a 'buffer' zone for fire defence purposes between it and the rest of the building by giving them a high degree of protection, though not equal to that of the central installation. Sprinklers could be used in the main areas of the building if there was no danger of water damage to the computer installation but smoke/heat detectors should be installed in the adjacent offices and hand appliances made available.

Doors in the rooms around the installation should be self-closing to function as 'smoke stops' and fitted with sills where required. Doors to the computer and plant rooms are especially important; substantial five-lever mortise locks should be fitted on these and strict control kept of keys. If these doors open internally into the rest of the building then they must be further shielded by self-closing four-hour fire-proof doors operated by fusible links. This is a normal insurance and fire protection officer requirement, which is only relaxed, and then to two-hour doors, when the rest of the building is office accommodation on negligible risk.

Where 'borrowed' lights are needed they should be of wired glass and fixed shut, any sliding or drop hatches should be self-closing and fire-resistant—wired glass if vision is required. If a viewing window is desired from the reception area into the computer room it must be expected that a solid fire-resistant, drop panel will be required as an emergency shield. This viewing window is a good idea since it permits visitors to be shown something of the computer at work while keeping them at sufficient distance to preclude damage or seeing restricted material. It also provides an opportunity for patrolling security officers to inspect the interior without entering—for the same purpose, doors in the installation offices should have clear, wired-glass panels. Having regard to the importance of the data held, this is an ideal situation for fitting special security locks under a master key system to all doors.

FIRE DETECTION AND EXTINGUISHING EQUIPMENT

Limitation of direct and consequential loss rests on the prompt detection of abnormal conditions such as may be caused by insulation failure and overheating. Detectors must be linked with a method of audible and visual warning and be capable of setting off an automatic extinguishing installation after a predetermined interval. The warning system should be linked to a repeater unit at a point which is permanently manned—either a security gate office, or a fire station in a similar manner to which a burglar alarm is linked to a police station.

The area surrounding the computer itself must have complete space protection. The usual practice is to halve the normal coverage which is allocated to a detector to double the certainty of immediate recognition of a fire source. Smoke-sampling detectors are recommended for general purposes and those of the ionisation type are particularly suitable in underfloor spaces and for airflow sampling in probe units. The computer room, underfloor and ceiling voids should be covered by separate groups of detectors so as to define the fire area immediately. The detection circuit must be so wired as to be capable of automatically shutting down the ventilating system and activating a damper unit in the duct into the computer area to stop the inward spread of smoke and heated air; it should also automatically cut off all power supplies to the computer itself. An unnecessary automatic shutdown could cause serious consequential loss—hence the need of alternative manual operation when the computer is in use. Switches or buttons for this purpose should be located near the operator's console and the exit doors. A clear visual indicator is needed to show whether the system is on 'automatic' or 'manual'.

The extinguishing agent is provided by separate banks of gas cylinders for the plant and computer rooms. These are usually housed in the plant room with interconnecting delivery piping. Carbon dioxide is commonly used but vapourising liquids of the BCF variety are tending to supplant it since they are less likely to cause the thermal shock to the computer which results from the cooling effect of a massive CO_2 release in a confined space. For testing purposes, a cut-off switch or mechanism for the bank of cylinders must be incorporated; this must have a prominent warning light or buzzer to ensure it is not left in the 'off' position making the cylinders inoperative after the test.

A predetermined delay on 'automatic' before release has a positive value in that it allows an opportunity, albeit brief, for an immediate investigation to stop the sequence if the danger is one easily contained or the functioning is accidental—this could happen with a circuit fault on changing from manual operation.

In addition to the large fixed installation, adequate hand-operated extinguishers of the CO_2 or similar type should be readily to hand in the computer room to deal with minor incidents. While water extinguishers should be available to attack carbonaceous fires in outer offices, they should not be put in the inner rooms where water could be an embarrassment. Prompt and knowledgeable action with hand appliances can avoid problems caused by the use of gas in a preventive flooding of the area.

The recommended minimum for the computer room is:

2 × 5 kg (10 lb) CO_2 extinguishers
2 × 2.5 kg (5 lb) CO_2 extinguishers
2 × asbestos cloths 1.4 m ($4\frac{1}{2}$ feet) square.

Humidity warning

The damage that can result from substantial humidity or heat changes has already been mentioned; the suppliers of the air-conditioning system must build in adequate temperature controls but it is advisable to have a separate form of warning for dangerous humidity fluctuations. This can be equipment showing conditions in the form of a graph in the computer room itself, with linkage to an indicator unit in either the gate office or some other permanently manned point to show when a state of dangerous humidity is being neared. This unit should incorporate a means of testing that the equipment and link are in order.

Indicator panels, manual buttons and switches

The main panel should be sited in an open position adjacent to the central area and in clear view to everyone. By means of coloured lights appropriately labelled—or something similar which is obvious in meaning—it should show at a glance:

1 Whether the power supply is operative.
2 The location of a fire—with separate lights for computer room, roof void, floor void or plant room (these could be extended to other parts if desired).
3 The presence of a fault in any of the sections of the installation.
4 By separate lights, whether the gas cylinders are set to operate manually, automatically or have discharged.

A push button or switch should be fitted into the panel to stop the alarms, with manual operating buttons for computer and plant rooms to be used in emergency.

Other manual buttons should be sited inside the computer room itself, preferably beside the exit doors; a distinctive audible warning buzzer should also be there. Both CO_2 and BCF are said to be non-toxic, but in fire-extinguishing concentrations the staff must get out at once. There are obvious advantages in having the buzzer of a different variety to that which functions when a fire warning is given for parts of the building outside the computer area.

Manual buttons

Apart from entailing expensive refilling of gas cylinders, inadvertent discharge will cause time loss, potential damage and the displeasure of insurers if replacement cylinders are not readily to hand. The buttons should be such that the chance of accidental firing is at a minimum; glass-fronted alarm boxes are suitable but the buttons must not be spring loaded to fire on the glass being broken. Each of these buttons should be tested at specified intervals and the test recorded.

Manual/automatic switch

While staff are engaged in the computer room, the presence of a fire condition should make its presence known to them at least as soon as the detector equipment picks it up and it would be invidious if some inadvertent action by any of them caused the system to fire while they were there. This can be obviated by coupling the alternator control which activates the computer to the release system, so that switching on the computer effects the change from automatic to manual operation and the reverse happens when the computer is closed down as staff leave.

Repeater panels

The remote panels in the security office need not be as comprehensive as the main indicator. It would suffice if the 'manual' and 'automatic' conditions were shown, together with 'fault' and 'fire'. Similarly, the humidity repeater need only consist of red and green lights. A test switch to check the circuit should be incorporated in each repeater and regularly used.

ACTION IN THE EVENT OF FIRE

All staff must be given instruction and have drill in fire prevention and the use of the hand appliances that are available for fire fighting. They must know exactly how the fixed installation functions, who to contact in the event of faults developing, and their own sequence to follow when leaving the premises, or when a fire occurs.

A notice should be displayed in the computer room on the following lines:

In the event of fire

1 Turn off master switches for machines, ventilation and ancillary plant.
2 Inform telephone operator who will call fire brigade.
3 Attack the fire with apparatus to hand (gas cylinder on electrical apparatus, asbestos blankets on waste bins etc.).

A checklist of action to be taken when leaving the premises normally should be displayed beside the departure door. An example is:

Check

1 All doors and hatches between rooms are closed.
2 All waste paper has been removed.
3 Any soldering irons or heating appliances have been unplugged from sockets.
4 All master switches have been turned off.
5 Security office notified you are leaving.

From the outset of building a computer installation, the local fire brigade should be kept in close liaison and their advice sought on any matters of difficulty; it is not enough to decide on a series of structural measures and then ask their approval, they should be consulted at the draft plan stage when their suggestions may save time, inconvenience and money later. This liaison will pay off in their intimate knowledge of the premises if they have to attend an actual outbreak in or adjacent to the installation.

SAFEGUARDING THE INSTALLATION

The Ministry of Technology has published a useful booklet, called *Computer Installations: Accommodation and Fire Precautions,* obtainable from HM Stationery Office, price 19p.

If the installation is sited in a part of premises away from immediate surveillance or in buildings unoccupied and unsupervised outside normal working hours, serious consideration should be given to incorporating an effective burglar alarm installation. The contents have little value to thieves, unless they are expert to the degree of being interested in the data content of the records, but the damage a frustrated intruder can perpetrate among that kind of equipment is frightening.

In other instances, patrols should visit and inspect the main indicator panel regularly, the first occasion soon after the staff have vacated. Lights should be left on in the computer room and approach passages to deter the unauthorised interloper.

INSURANCE COVER

Apart from structural and equipment loss from any cause, that of consequential loss might reach monumental proportions and could be induced by a simple prolonged power failure. The risk is almost impossible to assess, the extent of coverage is a matter for experts and will be related to the degree to which the firm's activities have become reliant on the computer. One thing that is certain is that the insurers will insist on all the aforementioned precautions as a minimum requirement.

The Fire Offices' Committee has issued a pamphlet, *Recommendations for the Protection of Computer Installations against Fire.*

DUPLICATION OF RECORDS

The information stored on magnetic media, such as programs, master records or data for future reference, is more valuable by far than the media. Its loss or the cost of recreation could be a serious matter. It is conventional for all master file records on magnetic tape to be kept on the 'grandfather, father and son' principle. Master files on disk are usually duplicated on tape. The hierarchical system works on the basis that 'grandfather' tape contains master records updated two processings before the current updated master file—the 'son'. The 'father' tape is that immediately before it. Both these preceding tapes, and the duplicated master record copies, should be kept in a purpose-built cabinet-type safe which in itself is thoroughly fireproof and away from any fire risks likely to affect the central area.

The compiled tape and disk programs will have source program-card packs backing them; these packs should also be separately and securely housed in a different part of the building, to be used if necessary to recreate the programs.

FRAUDULENT MANIPULATION OF COMPUTERS

The worst problem for companies is not deliberate fraudulent action but simply ordinary human error. There is no evidence to suggest that deliberate fraud is prevalent but odd instances could well lie unrecognised amongst the human errors which are regarded somewhat philosophically.

We feel that the methods whereby frauds can be carried out have already received more than adequate publicity, sufficient perhaps to influence those who are easily tempted. We do not therefore propose to deal with means whereby they are perpetrated but to concentrate on preventive measures. Fortunately, professional institutions are setting high standards of ability and behaviour for data processing personnel which should help to ensure trustworthiness.

Operators and programmers have the best opportunity to defraud and the precautions that can be taken will to some extent be dependent on the number of computer staff involved. If working is continuous with operators on the rotating shift basis, there is less chance of collusion because the job mix will change weekly and give little opportunity of an individual running one job on a permanent basis; operators thereby acting as automatic checks on each other. In writing programs, the work can be organised so that no one programmer is employed in writing a complete suite for a system.

A commonsense checklist of points in accordance with normal usage is suggested.

1. A library should be established for the safekeeping of programs and magnetic tape files. The librarian should keep an accurate record of usage and should not be a member of the computer operating staff; programmers should not have unsupervised access to the library.
2. There should be prior authorisation of all computer usage by a senior operations officer. A tight control should be exercised on computer time; the job sheet for a program should be endorsed with the estimated running time, a run of unexpectedly long duration should be queried and an analysis of computer use should be made periodically.
3. A formal standards manual should be laid down and supervisors must see its standards are maintained.
4. No operator should be allowed to work the computer alone, a second operator should sign the log book which should contain times of starting, stopping and reasons for any delays.
5. Programmers should not be allowed to operate the computer.
6. When amendments are made to a program in use, it should then be tested by an independent person. Program documentation should include a list of all changes.
7. Operators should not be involved in the preparation of any operational programs, nor should they be allowed to alter input data.
8. In the punch room, punching and verification of data should not be carried out by the same person; the work should be batched and controlled so that unauthorised batches cannot be inserted.
9. A log of all errors should be kept with a note of the remedial action and a copy of the print-out.
10. With a main application to banks, master files should be printed out periodically and checked for accuracy, any changes in information on them should be handled so far

as possible by personnel other than those handling day-to-day transactions.

THEFT: 'INDUSTRIAL ESPIONAGE'

There have been occasions where computers have featured in this context. A magnetic tape containing the entire name and address list of a company's accounts could be removed under a coat or in a briefcase: alternatively it could contain confidential records of a company's financial position or a set of computer programs. Such records could be of considerable importance to a competitor and at takeover times would be invaluable to any one interested in stock manipulation.

File security under computer conditions is likely to be more effective than under normal manual systems. However, a computer program represents a huge assembly of information which can be rapidly copied and used to advantage. Moreover junior staff in systems and programming will have access in the compiling. Thus in selecting and training staff, apart from ability, due consideration must be given to qualities of loyalty and honesty. If an employee does intend to pass on information from matter he handles daily, it is almost impossible to prevent, but steps can be taken to reduce the opportunities of deliberate damage or theft.

1 A reception area should be created and manned at the entrance of the computer area.
2 Visitors, except with high-level authorisation, should not be allowed in the computer room.
3 Unknown engineers, cleaners or others who might legitimately claim access should be asked for their credentials in reception.
4 Mutilated copies of print-out should be destroyed and not disposed of with the ordinary waste paper.
5 Duplicating paper used in print-outs must be destroyed by shredding or burning.
6 Staff who are discharged for any reason, or give notice under a cloud, should be paid off immediately in lieu of working

notice. Consideration should be given to applying this rule to all leavers.

If the last suggestion would appear to be detrimental to a firm's rights or an employee's interests, remember that a disgruntled programmer has been known to wipe off a complete program and at least one individual leaving for another position has taken all details of his employer's current research with him.

Where a firm takes advantage of time-sharing on a large commercial multi-access computer and has matter to feed which is vital to its functions it should query what controls are incorporated to ensure that other users have no access. Elaborate password systems are now commonplace and can be supplemented by other measures.

20

Commercial fraud

There is a strong body of opinion in the USA that major criminals are moving out of the sphere of violent crime into that of fraud. There would appear to be some evidence that the same may be happening in this country. Between 1970 and 1971 there was an increase of 6 per cent in overall crime, but fraud went up 16 per cent, false accounting by 35 per cent and, though over 74,000 offences of fraud were reported in the latter year, it has been suggested by a criminology research establishment that only one in twelve of those committed were notified to the police. There is obviously a variety of reasons for not doing so, one of which for businessmen is that of not wishing to be exposed to derision for being foolishly credulous. This is regrettable since criminals obviously rely upon this reluctance in planning their activities.

Under modern monetary trading and taxation procedures there are areas where there is a decidedly blurred dividing line between what is regarded as legitimate and what is punishable by law. Nowhere is this more evident than in some methods of carrying on business.

Statistics show an increasing number of small companies springing up and then failing within a matter of months, occasionally with substantial losses to suppliers or customers. Often these failures are sufficiently suspicious to be investigated by the Department of Trade and Industry but without adequate evidence to prove criminal acts. The potential magnitude is shown by the fact that in 1971 the Metropolitan and City Police Fraud Squad, which only deals with major crimes, was involved in 258 cases to a total value of £36 000 000. In 1964 in the United States one single case had a cash involvement of 175 000 000 dollars.

One of the many factors in favour of the large-scale fraudsman is the delay in getting such cases before the courts. This results in part from their very complexity in documentation and from the difficulties in finding out where missing money has gone to. One of the hallmarks of a firm that is trading fraudulently is frantic activity in the movements of its money to the confusion of all interested parties who stand to lose.

The law keeps out of business fraud inquiries as long as possible because their commencement can complicate the continuation of a firm's existence if the fact that an investigation is taking place leaks out. It follows that firms may trade on a border line of dishonesty which is not always obvious to an unsuspecting and gullible manager intent on increasing business. There should be an element of self-interest and self-preservation in any manager's attitude towards fraud. His career can be jeopardised if he is a victim to the detriment of his company or his shareholders—he should realise this and take precautions accordingly. The law has provided a number of punitive and other safeguards which it is unnecessary to deal with in detail here. Those of main interest to internal security departments are listed in the Theft Act 1968 (see Chapter 9). The offences in question include the following:

Obtaining property by deception, section 15.
Obtaining a pecuniary advantage by deception, section 16.
False accountancy, section 17.
False statements by directors, section 19.

MAJOR FRAUDS

There are a number of frauds which will primarily involve the Department of Trade and Industry or specialist investigators; these include:

Market frauds

In which speculators or unprincipled directors endeavour to manipulate share prices for personal gain by means of questionable practices.

Management frauds

Which involve the utilisation of authority by members of senior management for personal gain, contrary to the interests of shareholders, by methods which may range from borderline misapplication to actual theft.

Fraudulent trading

This is the type which is most likely to concern manufacturers or wholesalers. It is an offence against section 332(3) of the Companies Act 1948 which was applied in R. *v* William C. Leitch Brothers Limited (1932), in which the judge said:

> If a company continues to carry on business and to incur debts at a time when there is, to the knowledge of the directors, no reasonable prospect of the creditors ever receiving payment of those debts, it is in general a proper inference that the company is carrying on business with intend to defraud.

There is of course a fine but proper distinction between a company which is trading fraudulently and one which is trying desperately to survive.

The credit manager should be able to spot the symptoms of the latter. Invoices will take longer and longer to pay, necessitating letters of complaint and repeated threats of legal action. This becomes a matter of 'teeming and lading', whereby commitments are increasingly deferred pending monies coming in to pay them, and at any given time the company is quite insolvent.

Long firm fraud

This is dishonest in every aspect and one which every credit manager and supplier should be on guard against it. The object is a simple one—to obtain as large a quantity of goods as possible without paying for them and then to disappear before retribution can follow.

For this to be successful, the criminal must have capital at his disposal, the patience to carry out a lot of preparation, and sufficient business acumen, particularly in the line of goods he wishes to acquire, to deceive his prospective victims into believing they are engaged in bona fide transactions. It follows that this is a field for intelligent and organised crime.

Having established the objective, which will invariably involve goods that are easily and quickly disposable through market stalls, cut-price shops, or an outwardly genuine chain, the next step is to create an apparently legitimate business. This can be done by purchasing any suitable established one which comes on the market cheaply for any reason, such as death or pending lease expiry. The trade connections are then immediately available and can be impressed by hints of capital being poured in for expansion with the prospect of considerably increased orders.

Alternatively there has to be a start from scratch by leasing premises, going through the motions of legitimately setting up a firm, probably with an imposing name, and above all creating trade references. This can be done by opening simultaneously a number of one-room 'company offices' or even using accommodation addresses in different towns purporting to be suppliers with which the new firm has had satisfactory trading relationships. Enquiries are then placed with selected suppliers for goods and the credit terms allowed—naturally trade references are then asked for and those given are the pre-arranged ones, all of whom naturally answer in glowing terms and using impressive letter heads.

First orders may be small and promptly paid but as soon as confidence has been created, maximum ones will be placed in quick succession and the goods got away as quickly as possible until the inevitable happens and direct enquiries are made about non-payment. The organisers then decamp at speed leaving empty warehouses and little indication of their identity. Obviously this is a fraud which will come to light and a description will be available of the main negotiator. He will very rarely be the brains behind the scheme, but a 'front man', whose photograph may not be in police files. He may not even know who the organisers really are and in any case will put up a show of innocence if caught saying he is only acting on instructions and giving fictitious details of his 'employers'.

The temptation that is placed before representatives is that of the 'front man' allowing himself to be led into placing orders for slow-selling lines which the rep is elated to get rid of. Signs at premises to arouse suspicion are the absence of any apparent filing system, invoices and papers piled in trays, a secretary without knowledge of the business, whose main function is to send out letters inviting reps to call, who has to refer all enquiries to 'Mr Smith' who appears to be the only person with authority, and his invariable absence when needed in connection with complaints. At the warehouse itself, delivering drivers of suspicious mind can assist—and long-serving ones have a 'nose' for such things if they are consulted. Goods not checked off, delivery notes signed without scrutiny, an empty warehouse in dirty condition and no signs of order, carelessness in handling—in fact a general 'don't care' attitude and lack of familiarity with the job—all these will register with the intelligent driver.

There are two official sources of information which can be used to verify the background of any firm with which business is to be transacted. Where it is shown as a limited company then useful information as to its shareholders and standing can be found at the Registrar of Companies and Limited Partnerships, Companies House, 55 City Road, London EC1. If 'Limited' is not shown in a name that includes 'and Company', then lesser but useful information as to the registered owner may be obtained from the Registrar of Business Names at the same address.

Whether a firm can avoid being defrauded in this fashion depends mainly on its credit control measures and the enquiries it is prepared to make about new contacts. The services of the various trade protection associations or societies can be used but perhaps, even better, a call by an experienced representative would show the true nature of things.

Subscription or investment frauds

These involve offering shares in various speculative enterprises. Subsequently the money is put to the benefit of the organisers rather than the subscribers. This type of fraud uses the inducement of a promised quick and abnormally high profit on monies invested. Typical frauds of this kind are the inducement to invest in pig breeding schemes, where nothing like the stated number of pigs is ever purchased; sales of vending machines for cigarettes and other articles with a promise of excellent siting—rarely if ever fulfilled, and at an excessive price for the machines; and various plans involving the sending of money for placement in connection with bets. One common factor runs through all of these—it is difficult to imagine why any organiser, with such a profitable proposition at his own disposal would wish to openly invite so many others to participate. The Prevention of Fraud (Investments) Act 1958 deals with offences of this kind.

MISCELLANEOUS MINOR FRAUDS

Directory frauds

These will be aimed at persons responsible for a smaller firm's publicity. They are less likely to be attempted at larger establishments. They are a regular occurrence and impartially distributed over the country. The fraudsman introduces himself as a publisher of trade directories, telephone book covers or similar means of advertising in which those wishing to make an insertion pay pro rata for the space used. He invariably has an assortment of excellent material to show, which may have been specially prepared for him, or may be that of genuine firms but without indication of origin. Payment or part payment is requested in advance but the directory is never published or is eventually produced in cheapened form after very prolonged delays and threats of reference to the police. Many of those defrauded do not complain, and if they do there is always a series of excuses to rebut varying from personal illness and delays by printers to alleged indecision on the part of clients as to requirements.

The only certain way of avoiding losing money in this manner is to deal directly with bona fide established firms and not casual callers.

Hire purchase frauds: The Hire Purchase Act 1965 which provides for a four-day period during which a purchaser may revoke a contract after signature, has drawn the teeth of the fraudulent salesman who by persuasion, misrepresentation or intimidation has induced the buying of unwanted goods. However, legitimate commercial firms who accept payment by hire purchase as part of their normal method of trading are always at risk to the fraudulently minded customer. Modern selling pressures induce a haste to dispose of goods which militates against checking the bona fides of the customer and his ability to pay. False names, temporary addresses, false references and papers are all employed to obtain delivery when the intention is to sell for cash as soon as possible thereafter and certainly not to complete the cycle of payments.

The motor trade and the finance companies who specialise in providing money for car purchase have particular problems and many of the latter employ their own investigators to trace and reclaim vehicles on which payments have lapsed. Systematic fraud can reach high proportions where payments are continued for a period after cars have been resold but this can be limited if the services offered by firms who keep records of hire purchase defaulters are consulted. Such a service is H.P. Information Ltd, Greencoat House, Francis Street, London, SW1. There are also many organisations which will make status enquiries as well as keeping records of individuals' credit status and transactions.

Pyramid selling

Affects individuals rather than concerns; in view of the amount of publicity that it has been given it is surprising that anyone is taken in by it. Essentially the system is that positions of authority in a sales organisation are apportioned to applicants for varying payments irrespective of ability or experience. At the bottom of the scale are door-to-door salesmen who dispose of the goods under the direction of a diminishing pyramid of supervision. Those who indulge in this eventually find themselves with large quantities of unsold and unsellable merchandise and little legal prospect of recompense.

Holiday travel agency frauds

These again concern individuals, or perhaps organised parties of employees. Small new firms with glowing literature and low prices solicit advance bookings accompanied by deposits. The firm collapses and the facilities offered are then found not to be available. The answer is to deal with established firms of known integrity, or those new ones which are subsidiaries set up by large concerns normally outside the holiday market.

Distribution frauds

Most frauds in this sphere involve employees but, if a firm's procedures are lax, it may be possible for an order to be rung in and accepted apparently from an existing customer asking to collect immediately and quoting an acceptable order number from a known sequence—this of course involves either a member, or former member of a customer's staff or one's own. Collection without documents follows and the fraud does not come to light until the customer refuses the invoice.

A simple way of stopping this is to insist on proper documentation being presented even in connection with a most urgent order. Also in connection with distribution, any telephone instructions to change a delivery point, unless there is no doubt about the identity of the caller, should be questioned by ringing back. As mentioned before, if any comments are made by drivers concerning the customers, they should be taken seriously and looked into.

Internal corruption

A malaise of trading in some types of industry is that of a firm's buyer virtually demanding a personal consideration from a supplier for inducing his employer to place an order. This may be reflected in the price which has to be paid. The 'consideration' may take the form of cash, on a percentage basis, or 'presents' in the form of goods. In spheres other than purchasing, a person in authority may be prepared to grant similar favours in the form of licensing permissions, or by employing certain firms to do work for his masters, with a percentage for himself, or by specifying the materials of a particular firm to be used in connection with some project. The same malpractice can happen in a reverse sense where goods are in short supply or subject to quotas, in this instance the customer pays a 'consideration' for the privilege of preferential delivery.

Suspicions and rumours of this type of 'fiddle' in a company may be prevalent, but it is extremely difficult to get evidence to substantiate. The only possibility comes when a supplier of services or goods who has been approached for an honorarium, is too principled to comply and is not only prepared to complain, but also to cooperate in trapping the offender. This pernicious type of fraud ought to be stamped out in the interests of both morality and efficiency but under modern trading conditions it is difficult to see how this can be achieved.

There is legislation under the Prevention of Corruption Act 1906 (page 120) to deal with any instance it is desired to prosecute, but unless a firm has suffered considerable loss by its employee's actions, there is a tendency to deal with the matter internally. If a decision is then taken to refer to the police, those involved should not be questioned, otherwise the police investigations may be impeded by pre-arranged stories.

Christmas traditional gifts in recognition and appreciation of services rendered or trading relationships can raise suspicions. A dictum has been laid down by an Association of Chief Purchasing Officers which could be widely applied with advantage—it is an absolute ban on any gifts being accepted by members of their staffs other than at the Christmas season—then there is no objection, but the presents and their origin must be declared to the chief purchasing officer.

Embezzlement

The legal aspects of this, now covered under the general defini tion of theft, are dealt with elsewhere (see Chapter 9) and surprisingly enough, more often than not it involves trusted employees of long standing. This is probably due to the fact that anyone else's activities would be subject to closer scrutiny; falsification of records and the embezzlement of monies usually comes to light when the trusted servant is ill, or, against his wishes, has to take a holiday. It is then that irregularities in records or alterations and forgeries are noticed by the substitute.

Corruption

Another type of criminal offence by an employee is where 'a consideration' in the form of cash or kind is required from a customer in return for favours or is offered by him in expectation thereof. In other words an agreement to secure for a customer, who is prepared to pay, privileges which give him an advantage over others who have not paid. Some examples are when goods of a particular type are in short supply; not pressing for payment of goods supplied; agreeing to use materials of one supplier or giving business to a company to the exclusion of others. The Prevention of Corruption Act 1906 is the legislation under which prosecution can be taken against either or both parties for the offence or for an attempt (see page 120).

When it is believed that an offence has been committed and a prosecution is desired, those involved, including own staff, must not be interviewed prior to notifying police as this might interfere with their investigations.

The difficulties that arise, however, in deciding whether the presenting of gifts at, say, Christmas time in accordance with an old custom, is a corrupt gift or is a genuine appreciation of services rendered with no sinister overtone.

Forgery

Most internal frauds will involve forgery in some form or falsification of records. An alteration or erasure may, of course, be a simple correction, but it might also be a deliberate act, designed by the person making it to cover up a theft. Increases are easy to make to amounts shown on cheques and in records by adding a zero to the last figure and a 'ty' to alter, say, six to sixty. Wherever anything of this type is suspected, a thorough check is indicated and usually what has happened soon becomes evident.

Where there are no alterations but, contrary to usual practice, pencilled figures have been over-written in ink, these should be regarded as suspect. Forgery of signatures is most frequently done by taking a genuine one and either practising copying it or writing over it, then following the indentation in the paper beneath to form the forged signature. False signatures may seem evident to the eye but proof in court is almost always required by expert evidence to contradict denial of responsibility. Over-written signatures may be spotted by a minute failure to follow the indentation or by the stilted appearance of writing which seems to have been done laboriously.

THE INTERNAL AUDIT

The ideal deterrent in a large company is to have an internal audit department with wide powers of inspection without reference to departmental heads. It is important that they should not need to give courtesy notification of impending visits and should be responsible to a central high authority, preferably the finance director, so as to be unaffected by those in positions of authority who may be involved or who may be embarrassed by their findings.

It is essential that where there is an internal audit organisation there should be the closest cooperation between that and the person in charge of security. He may have received rumours of malpractice by a particular individual or in a particular department which it would be inadvisable for him to investigate directly—for that matter he may not have the facilities or knowledge of the record keeping to investigate personally. He can however direct the attention of internal audit to this and they can apply their specialist abilities in what would ostensibly be a normal check. Conversely, internal audit may throw up discrepancies for which there is no apparent reason other than sheer carelessness. These should be made known to the security department where they may have relevance to other information that had been previously received.

There is a by-product from such a liaison; auditors are primarily concerned with records of all kinds, books, documents, vouchers, invoices, cheques and records in general, and some frauds will become evident purely from discrepancies in figures and perhaps alterations which cannot be justified. After involvement in inquiries of a security nature they would probably then start looking more closely to see whether receipts and other documents that they may be examining are in themselves genuine, by comparison of signatures, watermarks, number sequence and the like. This liaison is a form of cross-pollination which must be of value to both departments.

INVESTIGATION OF FRAUDS

Whether investigation is carried out internally or whether assistance of outside investigators is to be solicited, it is absolutely essential to seize all relevant documents at the very earliest opportunity. If not 'accidents' will happen and material evidence will be destroyed or other steps taken to hamper the enquiry.

It may be that certain books must be retained in a department for use in the day-to-day work of the firm; if this is so, they should be photographed and inspected in the department itself where they would be available for reference if needed and such new entries as need to be made can be done under supervision. It is however preferable to remove everything of evidential value and in the case of a police inquiry they certainly will.

To sum up, fraud can exist at every level in an organisation. At the top it can prejudice even the continued existence of a large firm; at the bottom it can be responsible for a regular drain by way of petty theft which, if a firm's internal organisation is inefficient, may go on for a long time. Externally, a company could suffer a grievous loss through the more complex market frauds, or by over-enthusiastic acceptance of large orders from suspect or unknown customers. It has been said with some truth that the credit manager handles the most dangerous loss-risk in any company.

21

Shops and supermarkets

A well-known and respected company supplying shop detectives reports that from its experience and observations it believes shortages of goods in shops and supermarkets can be divided into 25 per cent by customers, 25 per cent by staff pilferages, 25 per cent by staff pilferages in collusion with an outsider and 25 per cent represents honest errors in receipts, dispatches and so on.

The observations and recommendations which will be made have reference to all forms of retail trading. However, as losses of goods and cash are particularly prevalent in self-service premises selling food and groceries, much of what will be said will be based on experience in that form of trading. What is essential to good security is a standard of management which will enlist the willing cooperation of the staff in applying the necessary rules and controls and exercising vigilance against all forms of theft.

STAFF

Before anyone is offered a position, his or her previous occupation, if any, should be verified—this is particularly important when engaging anyone who is to be concerned with handling money. This precaution may delay filling a vacancy, but the experience of one supermarket manager who could not wait to make minimal inquiries is worth remembering. Desperately requiring a cash-till operator he immediately engaged an applicant for the position but made no check on her answers to his questions. At the end of her first day she disappeared with the contents of her till and it was then found that she had given a false address. Inquiries of her previous employer showed that she had been dismissed for stealing.

New staff should not be sent alone to deposit money at or collect money from a bank.

Conditions of employment

A list of conditions of employment applicable to all employees should be prepared and on engagement every new employee must be required to agree to them and sign his name in confirmation of this. The following conditions should be included:

1 Should staff bring any parcels onto the premises they will be obliged to open them if required by the manager or any other authorised person.
2 All handbags and shopping bags are to be left in lockers provided.
3 Staff whose normal place of work is the shop floor are not allowed to leave that area without the permission of the senior member in that department.
4 Should the manager or other authorised person request to search their handbags, shopping bags, baskets, or motor vehicles they will agree. (See Chapter 7 under 'The Right of Search'.)
5 At the completion of their work, staff will leave only by the designated entrance.

If the firm employs security staff or engages the services of store detectives from a security company, employees should be informed that they are liable to be stopped by them after leaving the premises and asked to explain their possession of any company property which they may be carrying.

Staff purchasing conditions

The privileges which staff may enjoy in the purchase of goods should be included in the conditions of employment. A suggested procedure to control these is:

Time of purchase. All purchases must be made immediately before an employee leaves the premises to which he will not be allowed to return. If employees make purchases as they leave for a lunch-break, any handbag, basket, etc., will be subject to inspection on their return by the manager or another authorised person.

A variation of the procedure can be that employees are allowed to select the articles they require in their lunch-break. These are not paid for then but taken in a wire basket to a designated place at the rear of the premises until the close of business. Then the articles are taken through the check-out.

Dress. Shopping must be done in outdoor clothing except in circumstances described immediately above.

Baskets. Wire baskets will be used when making purchases.

Payment. This will be at a check-out point, through which *all* staff purchases will pass. All purchases and payments will be checked by the check-out supervisor, if one is appointed, or a member of the management of the shop. Sales will be for cash only, no credit being allowed.

Till slips. All sales will be supported by a till slip which will be signed by the checker of the purchases. This can be used to assist the employee to prove ownership of any goods in his possession should he be stopped outside the premises by a member of the security staff or the police.

Goods in short supply. Bread, milk, and other goods which may sell out before the close of business may be reserved for an employee by a manager. They will be placed in a bag or wrapped and marked with the employee's name and the price to be paid, signed by the manager, and placed in a special location to await collection by the employee immediately before leaving the premises as described above.

Goods reduced in price to clear. These goods, when offered to the staff to purchase, should be placed adjacent to the check-out point. These goods should not be reserved by any member of the staff.

Discount. The amount to be allowed to employees can be established from the cash-register slips signed by the persons supervising the check-out. The slips can be surrendered weekly or monthly to a designated person. Discount calculated from the amount of the purchases should be paid in cash.

THE PREMISES

Keys

The manager is responsible at all times for the opening of the premises. He must also be responsible for locking up at the close of business but on that occasion he should be accompanied by a senior male assistant. The keyholder of the premises registered with police should be the manager. (See Chapter 14 under 'Keyholders'.)

Rear entrance

Apart from the front entrances, all doors to the premises must be locked during business hours and opened only for goods to be received or to put out empties or rubbish. The keys should be held by an authorised senior employee. Fire doors, which are required to be unlocked when the premises are in use, are one of the means by which stolen property is removed from the premises. A good deterrent is the fitting of electric contacts to the door and the jambs, which, when separated by the door being opened, will activate a loud bell or buzzer.

Motor vehicles of employees parked near rear exits are worthy of attention to detect their use in thefts of company property. If they are on company property, the right to search them should be one of the conditions of employment.

Lighting

Good lighting of the exterior of a shop, particularly at the rear, is an excellent prevention of unlawful entry. Also a well-lit interior is a cheap deterrent to would-be thieves. In areas of particular risk, such as cash offices or where cigarettes or spirits are stored, double-filament bulbs are recommended to provide for the failure of one of the filaments through normal wear and tear.

Ladders

Ladders which have to be left on the premises must be chained and padlocked to some permanent fixture so as not to be readily available to an intending thief hoping to obtain access through high windows or the roof.

GOODS

Goods inward

Probably the greatest losses of goods occur at the goods reception area. These can be from dishonest delivery-men or collusion between company staff and such men, thefts through open doors, or mistakes with no sinister intent. In bulk deliveries by weight, differences from the weight charged can go undetected so that claims cannot be made if they are not carefully checked on efficient weighing apparatus. Delivery-men, who should not be admitted through the front door, should not be allowed to enter the selling area to replenish the shelves without the goods being checked in at the shelves.

If any goods are to be removed, because of staleness, for example, these similarly must be checked out by an authorised person. Where more than one delivery takes place on the same day, a delivery note must be required for each delivery which must be checked at the time. A composite delivery note covering all deliveries is not satisfactory.

The reception area must be kept clear of deliveries by their being removed as soon as possible to the particular store-room or storage area concerned. A steel mesh cage with a lockable entrance is recommended for particularly attractive or valuable goods. No unauthorised person should be allowed alone in a store-room. The keys should be retained by a responsible person.

Employees engaged on the reception of goods should be responsible persons of known integrity, well paid and not likely to be corrupted by dishonest delivery-men. They must thoroughly check all deliveries and be satisfied that all goods on the delivery note have been delivered before giving a clear signature on it. If goods are to be tallied off a vehicle, the person keeping the tally should not have the delivery note at the same time. The comparison of the tally and the goods on the note should be the responsibility of an authorised person. All overs and unders must be recorded. The fact that the delivery man has been bringing goods to the premises for a long time is not an excuse for not checking his deliveries. The use of a goods inward book has advantages.

Expensive goods on display

Expensive goods or goods of an immediately consumable nature should not be racked close to the exits to the rear of the premises or the fire-doors.

RUBBISH AND SALVAGE

Rubbish and salvage bins are favourite places to secrete company property which has been stolen and these should be examined closely quite frequently before the bins are allowed to leave the premises.

CASH

Encouraging the honesty of employees is especially necessary when their duties include taking cash for purchases. Therefore their supervision must be of a high standard so as not to present opportunities to steal which they might be unable to resist.

Where several persons use a cash-till which is not constantly attended it is more difficult to operate control procedures than if one person is responsible for it, such as in self-service retail trading. Where no till-slip is handed to the customer, the till should be placed in a position where the customer can see the amount recorded by the assistant taking the money. Placing goods or displaying material in such a way as to obscure that information is a common way of covering up under-tilling—deliberately recording less money than has been received, the balance being embezzled (see Chapter 9).

Till drill

A cashier at a self-service check-out should not be allowed to take any money or a handbag, basket, or cigarettes to her till. Overalls with only small pockets for small items of make-up and handkerchiefs should be worn. The cashier should be instructed that she is not to serve a relative or friend. The till drawer must be shut after the completion of each sale.

The float—If a float, usually a small sum made up of coins of all denominations to assist the cashier to give change to early customers, is collected each morning, this must be checked, together with the date recorder of the till, and the previous day's recording of the total takings to see this has not been changed. Any difference must be reported before any further cash is received. The cashier should have the key of the till so as to be able to lock it during meal-breaks or other absences.

Cashing-up procedure—The cashier should count the till contents and note the total. The money should be checked by another person who could be the supervisor or clerk concerned with the collection, recording, and banking of the money. The records should be signed by both persons. The till reading should then be taken by the clerk and compared with the total of the cash received. The cashier should not be able to obtain details of the till reading. Any overs or unders should be recorded and brought to the notice of the manager. Further action will depend on the amounts concerned. Repeated differences will raise the question of whether the cashier is competent for the position.

The float should not be left in the till drawers overnight or at weekends but kept in a safe. Some tills allow the cash drawer to be completely removed and a steel cover is provided which can be locked in position, so protecting the cash contents from interference. This type of till has an advantage where more than one operator uses the same till—the drawer can be substituted by another to be used by the operator taking over. The total cash takings registered on the till are recorded at the change-over times.

Instances have occurred of bank-notes being stolen from the till or when exposed at the time of cashing-up. Losses by those means, and time spent in the cashing-up, can be reduced by the clearance from the tills of the notes beyond immediate requirements by the accounts clerk or supervisor at intervals. On each of the occasions the totals of the notes should be recorded and signed by the cashier and the person removing them.

Spare, reserve or accumulated cash should be kept in a safe which should be locked and the key carried on the person responsible for its care.

Banking

Depositing money at a bank should be done at different times each day so that no pattern of behaviour is established for thieves to take advantage of. Consideration should be given to depositing some of the takings during banking hours and not leaving the total for one visit to the night safe, if available, at the close of business. If the latter facilities are available they should be taken advantage of so as to reduce the amount of money left on the premises when closed (see Chapter 15).

Safes

These should not be obscured from view by being situated at the rear of the premises but placed in a position immediately inside or adjacent to the front window so that intruders attacking them may be observed by patrolling police officers or passers-by. A further protection is to have a light shining directly over the safe, the bulb of which should have two filaments to provide against the failure of one which might cause a false alarm. The safes should be secured to the floor by being bolted or by some other means. The key should not be left on the premises. Any duplicates should be deposited in the local bank used. (See Chapter 14 on security of keys at banks.)

SECURITY STAFF

If security staff are employed by the company they should be independent of shop and distribution management and be responsible to someone with overall responsibility in the organisational structure (see Chapter 2). Their efficiency will depend on the support received from top management in the proper performance of their duties. Employees, on their engagement, should be told of the existence of such staff and their responsibilities.

Security staff should visit or keep observation on company premises to prevent and detect losses without giving prior notice to anyone. It should be made clear to all employees, irrespective of grade, that they are not immune from their attention. The policy which every trading concern must have respecting thefts by staff, customers, or anyone else will be referred to later.

Store detectives

To reduce losses of stock or cash the employment of such outside security aids is one measure which can be taken. They can be employed in their own right or to supplement a company's own security staff, who can in time become well known to shop staff and regular customers. The presence of the known security staff has a deterrent effect and thefts or laxities in control procedures occur less frequently.

The employment of experienced and trained store detectives of a security services company has many advantages including reports on the efficiency of the sales staff, their approach to customers, and the standard of discipline maintained. They can be engaged for any period, by the hour, day, or week, according to the times when the greatest losses are sustained. The detectives should be, and usually are where the larger firms are concerned, protected under an insurance policy against the consequences of wrongful arrests. Their employers take full responsibility in such cases so that no liability rests on the client employing them (see Chapter 31).

AIDS TO SECURITY

The belief that their actions may be observed by someone in an unseen position has a good deterrent effect on staff and customers in the selling area who may be tempted to steal. This can be achieved by fitting special glass partitions having the appearance from the sales side of being mirrors but through which from the other side a clear view can be obtained unseen. Another way is to install optical lenses in the ceiling through which a comprehensive view of the activities of staff and customers is possible from an upper floor.

A device recently introduced from America for use in clothing stores is an alarm beam which flashes lights and operates a buzzer when anyone passes out through a door with an article which has not been paid for. Each garment on display has a specially treated metal price tag attached which is removed by the assistant when it is sold. Stolen garments in the possession of the shoplifter, possibly being worn, will activate the device as described. One difficulty might be identifying the person carrying or wearing the stolen garment, should other customers leave the premises at the same time.

Closed-circuit television

Closed-circuit television—with cameras strategically placed about the premises to produce pictures of what is going on in the sales and reception areas on a screen in, for example, the manager's office (see Chapter 30)—is an even better deterrent than mirrors or lenses. Dummy cameras can be used effectively in conjunction with genuine ones.

POLICY RESPECTING THEFTS

Thefts by staff

The policy which is to be adopted towards staff, customers, and others found stealing company property must be decided at high level. It should be drawn up in writing and circulated to all employees. The authority to decide what action is to be taken must be given to selected persons whose names are shown.

It is recommended that, without exception, an employee found stealing is dismissed. Discretion could be used on whether the police are called from the merits of the case. What is *not* recommended is a rule to the effect that theft by an employee of company property of a value less than a specified amount will be dealt with by dismissal only, whereas if the value exceeds that amount police will be called. The choice of action is especially invidious where the value is slightly below or above the specified amount. A court which became aware of the distinction may be critical.

Details of the dismissal and/or prosecution for theft of employees of all grades should be distributed to all branches of the firm concerned and exhibited so that they are brought to the knowledge of all staff.

Before making a decision whether to prosecute for theft, in any circumstances which have been described, consideration should be given to the time of management and possibly other employees likely to be taken up by their attendance at the police station and later at the court dealing with the charge, and the value of the property involved which might have been recovered. Chapter 9 mentions that by the Criminal Law Act 1967, section 5(1), it is not now an offence to abstain from prosecuting for theft on the return of, or compensation for, the property involved.

Police assistance should be called for only when the owners of the property are prepared to support a prosecution by the attendance of their staff as required by the police or the court. The publicity of court proceedings in a local paper is one of the best deterrents to further thefts, but that must be weighed against the possible loss of customers who, whilst completely honest, may fear that through forgetfulness they may be unjustly charged with theft.

Theft by customers, juveniles, pensioners, and pregnant women

The policy decisions must extend to the action to be taken in respect of customers, with special attention given to juveniles up to a specific age, old age pensioners, pregnant women and those in the menopause who have been found stealing. It is not unusual that after the recovery of the property concerned no further action is taken on the grounds that a prosecution might not have a favourable reception by the court and the resultant publicity would be damaging to the firm's image.

SHOPLIFTERS

Where any of the mechanical aids or personal observation is used to detect shoplifters, extreme care must be used if one is discovered to ensure that he is under constant surveillance from the time the article is stolen from where it was displayed until he is stopped for questioning to detect whether he has disposed of it.

Shoplifters frequently operate as a gang in the larger stores and, to confuse security staff, pass stolen property to their confederates. If a gang is believed to be operating, and one of them is detected stealing, he should be followed from the premises, at the risk of losing touch, in an attempt to discover where he is taking the property. This might be a parked car with or without a driver to which each member goes with the property he has stolen. If this happens police assistance should be sought, giving the registered number of the vehicle used. It may be possible for them to maintain observation on it to secure the arrest of all the thieves.

Where the objective of a gang is something particularly valuable it is not unusual for one of them to deliberately behave in such a way as to divert the attention of security staff from his colleagues engaged on the main operation. This may go so far as to lead to his arrest for stealing something of small value thereby providing the required diversion.

Another technique of a different nature is for two or more men or women to visit a well-known store. One of them makes the theft of an article so obvious that it is seen by one of the staff. Whilst the 'thief' is being followed to the door the article is passed to a colleague who leaves the premises unobserved. On leaving the store the 'thief' is stopped and asked to return but no stolen property is found. The evidence of theft is that of one employee which is strenuously denied by the person detained who declares that court action will be taken for damages against the store for unlawful arrest. Rather than face such an action the claim is settled out of court. This practice is repeated in other premises and becomes a regular source of income for the persons concerned.

This type of crime can be defeated if those stores which experience it would circulate details confidentially to other firms likely to suffer similar experiences. Commercial associations provide a channel through which to exchange such information.

Rules for guidance

1 Only managers should deal with customers suspected of shoplifting. Sales staff should report their suspicions, but care should be taken in acting on them alone.
2 The incident on which action is to be taken should have been witnessed by the manager or some other senior responsible person. The suspect must not be lost sight of at any time afterwards and until being stopped.
3 The evidence of stealing is when the customer takes the article, deliberately conceals it, and leaves the premises without paying for it (see Chapter 9).
4 Wait until the suspect leaves the premises before taking any action. This is not legally necessary to prove theft but it goes to show the dishonest intention of the suspect to avoid payment (see Chapter 9).
5 Approach the suspect, tell him, or her, who you are, and ask him to return to the shop. Do not say anything about arresting him. It will be seen from Chapter 7, dealing with

the power of arrest by anyone, that this can be done if the circumstances justify it.

6 Call a senior member of the staff to be present to witness what is said and done.

7 At a place not in sight or hearing of the general public or customers tell the suspect he has taken property from the premises without paying for it.

8 Ask to inspect the contents of any shopping bag carried and compare them with the till receipt where one is issued. Do not search the person of the suspect.

9 If the comparison of goods carried and till receipt reveals clearly identifiable company goods which have not been in-included in the receipt, inform the suspect of this and note any reply.

10 When the nature of the goods permits, ask the suspect to write his name on the package. This should be counter-signed by the manager and the witness, whether the suspect has signed or not.

11 Where it is in accordance with the policy inform the suspect you are going to call the police and note any reply. Keep the suspect under constant attention until they arrive (see Chapter 7 under 'Manner of Arrest' on requests to use the lavatory by a person detained).

12 In the presence and hearing of the suspect the manager should tell the police officer who arrives what he saw, produce the unpaid-for goods if available, state their value, and repeat what he said to the suspect and what he replied.

13 The police officer will then ask the suspect whether he has anything to say and caution him that he is not required to say anything unless he wishes to do so but whatever he does say will be taken down in writing and may be given in evidence (see Chapter 11 under 'Judges' Rules').

14 On the officer being satisfied that a *prima facie* case has been made out on the evidence he will arrest the suspect and take him to the police station. In certain forces that procedure is not adopted but the full details of the accused are obtained by the officer and he is informed that the facts will be considered before deciding whether a prosecution will be instituted. If this is decided on, a summons

is later served on the accused to attend court on a certain date.

15 What had been seen by the manager and the witness(es) may be required to be recorded in a written statement to the police to assist the prosecution or in deciding the action to be taken.

16 The evidence of the police officer who arrested the accused, of what was said to him by the manager in the presence and hearing of the accused is not sufficient and the manager and witnesses may be required to attend court to give evidence. This will not be necessary if a written statement of the evidence he would give is allowed to be read to the court in his absence. See Chapter 12 which gives in detail the conditions which must be complied with before that can be done. If a plea of guilty to the charge is made, the evidence of the manager and witnesses may not be required by the court.

17 The property in the case will be handed to the police and will be returned at the conclusion of it.

Cautioning of shoplifters by shop manager

A case of particular interest to security staff and shop managers, illustrating the necessity to caution a person accused of shoplifting before asking questions, and showing to whom the responsibility to do so is intended, is R. *v* Nichols. This was dealt with by three judges in the Court of Appeal and was reported in the Times Law Report of 8 February 1967. Further legal commentary was given in the *Security Gazette* of March 1967. The facts are that a store detective stopped in the street a shoplifter named Nichols who accompanied her to the manager's office. The detective told the manager what she had seen and the manager questioned Nichols about the allegation. He replied that what the detective had said was true. Nichols was later charged with theft and convicted. He applied to the Court of Appeal for leave to appeal against this on the grounds that he was not cautioned before being questioned in the shop by the manager. The question before the court was therefore: Is there any duty in the circumstances described to administer the caution under the requirements of the Judges' Rules? (See

Chapter 11 for these.) Rule 2 applies to 'police officers' and Rule 6 to 'persons, other than police officers, charged with the duty of investigating offences' (and this could include security staff). As the store detective had not questioned the accused, Rule 6 did not apply. The court, in refusing Nichols leave to appeal, said the shop manager, who had questioned him without administering a caution, was not a special investigator—he was employed to manage shops—and was not, therefore, bound to caution Nichols before asking him for an explanation.

The decision of the Court of Appeal could be taken as an indication that the judges realise that the danger of a strict interpretation of the Rules makes it unduly difficult to acquire in circumstances such as these the evidence necessary for a successful prosecution. It is at the discretion of the judge or magistrate in the case whether to admit in evidence a statement, written or verbal, alleged to have been made by the accused after being told the conditions under which it was taken or made.

Recording information

The names, addresses and ages of all persons prosecuted for theft and the result of the court proceedings should be recorded. If there is more than one retail shop in the firm, such information from all of them should be centralised. Similar information regarding persons stopped for stealing but not prosecuted should also be recorded. By reference to these records it may be possible to show that some of the group of persons it is not the policy to prosecute are repeatedly taking advantage of their immunity and special action may have to be taken.

Information should also be recorded which might assist in connection with subsequent events involving previously known persons such as their methods of stealing, concealment, use of children, work with others, excuses made, request to use the lavatory, claims of forgetfulness to excuse their actions, etc.

PRECAUTIONARY MEASURES

Fire

The recommendations under this heading will be found in:

1 The London Building Act 1939, section 20, and associated regulations, by-laws, and codes of principles and practices issued by the Greater London Council.
2 The Building Regulations 1965, under the Public Health Act 1961.
3 The applicable clauses of the Offices, Shops and Railway Premises Act 1963.
4 *Fire Prevention and Fire Protection in Departmental Stores,* issued by the Fire Protection Association and the Retail Distributors Association.
5 'Protecting the Shopper' in the *Security Gazette,* volume 10, number 1, January 1968.

Burglar alarm systems

These will be described in Chapter 29.

Checklist

The recommendations for security in retail premises are summarised in Appendix 27. (See also 'Retail store supervision' on page 427.)

PART FOUR

SECURITY OF BUILDINGS AND SITES

22

Protection of buildings and factories

It is wellnigh impossible to make a building, still less a complete site area, inaccessible to determined thieves. Only an optimist with no practical experience worth speaking of would assert otherwise. Nevertheless, by far the greater number of illegal entries are made by criminals who are able to do so simply because no thought has been given to the steps which would deter them.

To carry protection to extremes would be to make individual buildings into virtual 'Banks of England'. This could reduce the amenities to the staff to a point which would not be acceptable. The cost of such a degree of protection has also to be taken into account; in the case of older buildings this could involve structural work which would be prohibitive in its expense.

The happy medium would be to achieve a condition where prospective thieves would consider the difficulties and risk too great for the possible gain. Where particularly valuable property is being guarded, obviously structural security combined with electronic alarms is the best solution; unfortunately circumstances where the latter cannot be used will be encountered—liability to false alarm may be a deciding factor—and a close survey to see what can be done by purely physical means will then have to be made.

Apart from merely getting into premises, the thief has to get his proceeds out. This can be the more tricky and dangerous part of his operation, particularly when heavy or bulky goods are to be removed. A second factor is that he wants to attract as little attention as possible. Not only does he wish to acquire a profit from his endeavours, he also wants to strictly limit the chances of arrest. In other words, he wants to get in and out as quietly, safely and quickly as he can, with a minimum of effort for the removal of property. He neither wants to be seen nor heard and will be looking for a clear and easy escape route in case of being disturbed. If he has to carry heavy and awkward burdens he will want it to be for a minimum distance before he can again be under cover or safely inside a vehicle.

If it is accepted that the criminal cannot be stopped from entering, it follows that the objectives of the defence are to make him as uncomfortable and apprehensive as possible when entering and leaving and to make the attractive goods on the premises as inaccessible to him as possible. This latter point is especially applicable to money and small valuable objects which should always be kept in the protection of a suitable modern safe.

ARCHITECTS AND NEW BUILDING

Altering any building to increase its security is a costly matter and may be disruptive to the work going on inside. In new building work, this can be obviated by applying security principles at the design stage. These will not excessively hinder the architects' plans nor radically affect the final cost of the building.

Architects cannot be expected to be fully appreciative of the peculiar risks involved in every firm's business, nor of the importance that it attaches to the security of its property and products. However, once these matters have been fully discussed with them, their expert experience in materials and construction will be applied to meeting what may be to them an interesting challenge. Several draft plans are prepared before the final one is accepted by architects, local authorities, and the owners of the property; it is too late to consult security and fire specialists when this stage has been reached. There will then be a natural reluctance on the part of the other bodies to make amendments and it follows that the earlier consultation can take place the better for all concerned. It has been mentioned elsewhere that there may be conflict between security and fire prevention requirements, this can be resolved and accommodated early in the planning stage if a full exchange of points of view takes place.

What is applicable to shops and offices is also applicable to factories and to perimeter fencing—these should all be discussed at the design stage and the precise type of protection that is required should be established, having regard to the usage of the premises and the surroundings.

In evaluating what can be done to defend premises, either new or old, a systematic examination should be made of the possibilities of applying security principles to each aspect in turn.

One police authority at least—Bristol—has recognised the importance and mutual advantages of creating a liaison with the building industry and architects. An officer has been given the opportunity to become thoroughly conversant with building and architectural principles and graft them onto his own crime prevention knowledge—they have not been found incompatible. This has resulted in vastly increased cooperation with lectures, visits and demonstrations from the police side and the submission of plans for critical comment to an almost embarrassing extent by the other. The long-term benefits accruing from this could be considerable. Where a security specialist is employed in industry or commerce, he is not exploiting his value to his employers as he should unless he becomes similarly involved with their projects.

THE PERIMETER

Perimeter walls

Building brick walls purely for security purposes is a very expensive business and would rarely be contemplated on that ground alone if any alternative existed. An ordinary wall has little protective potential other than its height—it would not stop anyone getting in, but could be a formidable obstacle to getting anything sizeable or weighty out.

Nothing less than 8 feet (2.4 m) in height is worth considering as a security protection, and walls of any height should have further reinforcement by setting spikes or barbed wire on the tops. Broken glass in concrete is only a deterrent to the unassisted climber and it is easy to negative its value by throwing sacking over it. An additional advantage of coiled barbed wire, particularly if arranged vertically, is that disturbance of it will give an indication to patrolling policemen or security officer that someone has climbed over the wall and may be inside the premises. This can easily and cheaply be applied to new and existing walls by setting mild steel bars in concrete or bolting them onto the wall to carry the wire. This must not be at a height where it could cause injury to an innocent passerby.

In the older types of factory property the tendency was to build high walls which were excellent in themselves but remarkably prone to having obscure doors and gates set into them in the darkest corners, inevitably fastened by large rusty padlocks with hinge shackles, simply appealing for the insertion of a steel bar. Modern construction much prefers a perimeter fence and where an unbroken wall does exist, it will be for convenience and part of a building facing inwards.

Perimeter fencing

Perimeter fencing is a much cheaper means of protection than brickwork or even concrete panels. The conventional type is that of wire mesh on concrete posts 6 to 8 feet (1.8 to 2.4 m) in height, with angled arms carrying several strands of barbed wire. The green or black plastic-covered wire mesh lasts longer and looks better but can be opened up something like a zipper and then easily folded back to allow entry.

It has to be appreciated that purely as protection against thieves, this form of mesh has only a nominal value—it is a delaying factor and minor deterrent only. It can easily be cut with wire cutters and bent up at the base to allow crawling beneath. It may keep out marauding children and it demarcates the premises, but it does require additional measures if any confidence is to be placed in it. There are two ways in which the base can be protected: by setting in concrete, or by threading barbed wire through the bottom mesh.

Much more formidable, but much more expensive, is palisading of the 'Lochrin' type, although even this could be forced apart with a steel bar or cut with large bolt croppers. This is made from mild steel 3 to 4 inches (7 to 10 cm) wide pressed into thin pieces normally with a maximum height of 10 feet (3 m) although any desired height could probably be produced. These are bolted onto a metal framework and the tops can be forked in several alternative ways to deter persons intending to climb over. The bolt heads should be burred over when installing this fencing, otherwise portions can be unbolted and removed. The manufacturers now produce this palisading in sections which can be transported between sites and are not unboltable.

All types of perimeter fence need adequate lighting along the full length. If this is not feasible because of distance from the buildings, the mesh type becomes purely a boundary demarcation and has no security value. The value of lighting is obvious: to approach and force a hole in full view of anyone who is about would give a most unpleasant feeling to the person concerned, who could be in the position of feeling that he was being seen but could not see. Assuming, as will often be the case, that a perimeter road follows the fence for convenience of production, additional lighting will also have a safety and handling compensation to offset against the expense.

Bushes or piled materials should be kept away from the fencing because they could afford concealment to the intruder to cut his way in and the aperture will not be easily detected by a patrolling security officer. Moreover wear and tear on the fence will necessitate earlier replacement.

Perimeter gates

From the point of view of supervision, and economy also, it is advisable that the entrnces to any perimeter should be kept so the minimum to suit the needs of production. This will reduce the number of security officers who are needed and will ensure that all personnel and vehicles leaving are subjected to scrutiny. It must be anticipated that some union objection would be raised to the closing of existing gates where these are sited to the convenience of personnel and rearrangement of clocking points would be necessitated. By their structure gates are often easier to climb than the adjoining fence. They should be of the same height and style and hung close to the ground to prevent crawling under. Barbed wire on the top would prevent easy climbing up and over and this wiring could possibly be extended into the mesh or palisading of the gate itself.

Wooden and metal gates both need top protection as they are equally easy to climb from inside. An aperture to allow scrutiny of the interior area by patrolling policemen is a good idea—there is a sort of psychological urge which draws the eye where these exist. Where lightweight gates are in use the suspension must be arranged so that they cannot be lifted off bodily.

To carry out a major theft requires the assistance of transport as near as possible to the desired loading point so, from the security point of view, the further such vehicles can be kept away the better. Access for these will invariably be through the gates and it is therefore essential that the shackles and padlocks fastening them should be of good quality. A habit of thieves is to force off an existing padlock, open the gates, drive a vehicle through, and replace the padlock with one of their own so that patrolling police cannot see that the premises have been entered. This enables the thieves to load their vehicle at will and in peace and drive out, immediately locking the gate behind them so that they may have several hours of freedom before the theft is discovered. Thus a good-quality close-shackled padlock should be fitted, and the gate area especially well lit.

Traffic barriers

Where the main requirement is the regular stopping of vehicles for checking or like purpose, barriers may be substituted for or used in conjunction with gates. They can now be of two types

—the more usual with pivoted arms has a nominal stopping capability but the maximum security barrier is a recent innovation which should effectively stop the passage of anything less than a tank!

The former is in common use; it is balanced against a counterweight for easy operation but the factor rarely appreciated by users is that a safety device is usually incorporated to allow the arm(s) to shear off rather than damage the pedestal in the case of accidental collision—and will do precisely the same if someone decides to drive straight through it. It may be manually operated, or electrically by pushbutton from a gate office—which is decidedly preferable on grounds of efficiency and modernity. Collapsible 'skirts' may be fitted to the arm and both should be clearly painted in red and white; double-sided stop signs should be affixed to them so that there can be no question either of their purposes or of being visible to drivers.

Security barriers

These function entirely differently; a hinged steel frame sunk into the ground supports a heavy hinged flap, heavy-duty air cylinders cause the flap to rise and present a vertical face to oncoming traffic in a maximum time of eight seconds. A third cylinder forces in locking bolts when the barrier is fully up. Safeguards are incorporated so that in the event of tampering or cutting the piping between the installation and the operating lever valve, the barrier will rise and lock automatically. There is little wear and tear in ordinary use and the compressor unit has sufficient reserve capacity to carry out one or two operations in the event of a power failure.

These are expensive but have an obvious application for roads and entrances with a high security risk.

EXTERNAL DOORS

No matter how formidable these may appear, their strength will be no more and no less than the means of securing them. Solid wood, steel, or steel-faced doors offer the best protection. Extra strength to wide doors can be given by a swinging padlock bar across the full width of the door or fitting a substantial drop bar. Attention should also be paid to the jambs and surrounds of the door—the value of a substantial door is strictly limited if the jamb and surround are flimsy and would be forced away from the brickwork by putting pressure on the door itself.

The best locks for general use are the five-lever mortise deadlocks, and to carry these the door and jamb must be thick enough to accommodate them. Padlocks should be of the close-shackled variety to prevent them being forced off with a steel bar, they should also be used in conjunction with hardened steel locking bars—to make an expensive close-shackled lock dependent on a poor hasp and staples is a waste of money.

Where the exterior of the premises is subject to observation by patrolling police, there is no reason why the padlock should not be on the outside. In other circumstances, it would be best to have it on the inside where it could be difficult to get at. It should not be forgotten that a hacksaw blade can often be inserted between double doors to attack the bolt of the lock or the padlock bar. Mortise locks are now available with a hardened steel roller incorporated in the bolt; when these rollers are reached by hacksaw blade they revolve freely giving no bite to the saw edge. If cross-garnet or Scotch tie hinges are fitted they should be bolted through the door and the bolt ends riveted over.

A weak door can be strengthened by facing it with a steel plate; this should be bolted through the framework from the outside and the nuts riveted over on the inside—there is no object in using screws for securing the plate to the outer door. Light doors consisting of timber frame and hardboard covering should never be used externally. Glass panels are not desirable, but there is no objection to the slit-type panel with wired glass. With existing glass-panelled doors, protection can be afforded by using grilles behind the doors—again providing the screw or bolt heads are burred over. Where two locks are fitted to any door—for example, a Yale lock for routine daytime use and a mortise lock for night protection—they should be sited as far apart as possible to spread the resistance to violence over the widest area of the door. Further strengthening can be added by nailing diagonal braces across the door.

Double doors

Where appearance is not an important consideration a substantial padlock and bar is the best method of securing double doors. The padlock should be close-shackled. The standing door of the pair should be bolted top and bottom preferably with riveted bolt fittings in the door end. A point to check on leaving premises is that the standing door is firmly fixed, otherwise even the best lock of the mortise type is no use.

Fire-exit doors

These can be a main source of conflict between fire and security requirements. They afford an immediate escape route for criminals and the crash-bar type can be opened from the outside by boring a hole in the woodwork and inserting an instrument, or occasionally by simply repeatedly shaking—if old and worn. This can be prevented by using the spring-loaded bolts of the Redlam Panic-bolt type which will not prevent a criminal leaving but cannot be got at from the outside. They have a further use in so far as an employee wanting to leave the premises before the end of his duty cannot do so without breaking the glass, thereby releasing the spring-loaded bolt. Some fire authorities may object to doors fitted with the normal crash bar being locked at night by means of a chain, padlock, and hasp when no staff are in the building—this should be raised with them before it is put into practice.

There is a new trick to watch for with this type of bar: an employee attaches a length of thin wire to it when the door is open and closes the door on the wire; later he can pull the wire from the outside to release the bar.

Roller-shutter doors

These can either be manually or electrically operated. Manually, the operating chain can be secured against movement by means of a steel pin and close-shackled padlock. Electrical varieties necessitate a cut-off of some kind of the electrical supply, which can easily be arranged. A further precaution consists of setting a ring into the concrete floor and welding a hasp to the bottom edge of the roller-shutter door, joining the two, with the door closed, by means of a close-shackled padlock. When this is done, care must be taken on opening up the premises: if an electrically operated roller-shutter door is switched to 'open' with a padlock in place, the result can be a U-shaped door!

Concertina-type or Bolton doors

These are normally fitted with a specially designed type of claw lock, which in itself offers limited protection. Padlocks can again be used by welding on suitable lugs to the inside edges. A point to remember in connection with these doors is that they are often used with steel-framed sheet-walled buildings. It has been found on occasions that the attachment of the door edges to the H-type steel girder forming the door jamb, has been via a small number of bolts which it has been possible to get at and unscrew from the outside, thereafter simply pushing the door open from the edge. This will also avoid the normal electronic alarm fitting and care must be taken that where these bolts are accessible they are burred over to prevent them being turned.

All-glass doors

These are usually only encountered in shops and entrances to large office blocks. Whilst they would appear to be a weakness, they are of specially thick armour-plate glass and there have been few instances of entry being forced through them. The fear of attracting attention by the tremendous crash that would occur if anyone smashed a hole through them seems a very real deterrent. Where these are used, appearance is obviously a first consideration, and the best thing to do with them is to include them in a burglar-alarm system, with either a ray across them or suitable contact.

WINDOWS

What can be done with windows depends on the importance that is attached to the appearance of the building. On the outside of a works where security is paramount a steel mesh can be bolted into the wall covering the windows or bars can be fitted. In both instances bolt-heads should be hammered over. If bars are fitted spacing should not be more than 5 inches (125 mm) apart and steelwork used of minimum diameter of five-eighths of an inch (16 mm). The same type of bar can be fitted to fanlights where it should be in a cradle shape to allow opening for ventilation.

New construction

Consideration could be given to using glass bricks or roof lights instead of windows where possible. The modern tendency in office blocks is to seek a maximum of light; in these circumstances, if sheer vandalism is not a danger, it is advisable to go the whole hog and use panes as large as possible and preferably of strengthened glass. A potential intruder is much more likely to be tempted to break a small pane, barely adequate to admit himself or to admit his hand to open a pivoted window, than to smash a large hole under circumstances which must attract attention.

PROTECTIVE GLASS

Armour plate glass

The location and type of building may well militate against the use of grilles, bars, or shutters for the protection of windows, there is no gainsaying that these detract from appearance. Added security against both damage and theft can be gained at a comparatively moderate expense and without too much effect on the amenities by using a type of strengthened glass. All glass manufacturers produce some specially toughened kinds of the armour-plate type and others of laminated construction. These will withstand blows of varying strength according to their construction. Breaking such glass involves effort and creates noise, it also increases the time and difficulty in making a hole big enough to get through.

Wired glass

For windows in the factory type of building and the smaller office building, particularly those which it is desired should be translucent rather than transparent, the Georgian-wired type of glass can be used. Both transparent and translucent varieties are ordinary glass, approximately a quarter of an inch (6 mm) thick, with an electrically welded steel wire mesh reinforcement. The mesh usually ranges from half an inch (13 mm) square to approximately seven-eighths of an inch (22 mm).

This wired glass has an added advantage in respect of fire: whilst it will crack under heat, it will remain in position even when badly damaged, thus reducing draughts and retarding the spread of flames. For the same reasons it has a safety value too, since there is little chance of injury from falling glass when a window is broken.

Locks

Whatever else may be done with windows, the means of opening should be made as difficult as possible for someone trying to do so from the outside. All the better-known lock firms supply key-operated bolts, window locks, or window stops for every kind of window; ample literature is available on application to cover all contingencies. One kind of stop, fitted with a chain, allows ventilation whilst retaining the window secure against illegal entry; mortise bolts operating on a rack and pinion principle can be used for wooden casement windows and special locking handles can be fitted to metal windows. As in other cases, illumination is of deterrent value for window protection.

Skylights

These are in a class by themselves; they are difficult to cover and the only real protection is by means of substantial bars beneath them. It is little satisfaction to know that when these are seen by an intending thief he can very often remove tiles equally easily and climb through, forcing a way through the ceiling. Consolation can be found in that his egress might prove complicated and his carrying capacity be limited—he might also fall!

Bars

Where the bars used are more than 1 yard (91 cm) in length, cross-ties, also of mild steel, should be welded to them to give strength. Cross-ties and bars should have ragged ends firmly cemented into the masonry to a depth of at least 3 inches (75 mm). Where bars have to be fitted to wooden window-frames substantial crutch-head screws should be used for securing them or the slot should be drilled out.

GRILLES AND SHUTTERS

As an alternative to bars and mesh, grilles and shutters offer protection without being permanently unsightly. There are many types of grilles on the market—collapsible, expanded metal, welded wire, etc. These must be adequately secured if they have to be fitted outside but should be fitted internally if possible. Collapsible grilles are manufactured for individual windows and tend to be expensive—these are most often used for the protection of small valuable items on display to the public, as in jewellers, camera shops and, more recently, the protection of licensed bars where there is a sizeable display of spirits almost within reach.

Shutters of any type provide excellent security for windows when fitted internally. Steel and wood are the materials which come most readily to mind and these are quite effective when fixed firmly in position by a drop bar across the back—this can be fitted with a padlock if deemed necessary. One point that should be borne in mind is that they equally effectively conceal the inside of the premises from the eye of a patrolling police officer, so a peep-hole is desirable to allow inspection of the interior, particularly if this contains a safe—it is a common practice to illuminate safes so that if there is any interference it can be readily seen.

CELLARS

It is by no means unusual for cellars to be completely forgotten when the security of premises is being checked. The cellar-flaps, more often than not, are held in position purely and simply by their own weight. Gratings in footpaths giving access to unguarded cellar windows are apt to be treated with the same degree of neglect.

Cellar-flaps and gratings

Both these should be held by substantial chains and padlocks to prevent their being lifted. Cellar-flaps can be further strengthened by means of bolts padlocked in position on the inside edges; flaps are normally of quite substantial construction but if strengthening is needed it can be given by putting cross-ties across the back of them. Where gratings are concerned, unless their removal for cleaning purposes is essential, they should be cemented into their surrounds.

Cellar-head door

This should be kept properly secured with mortise bolts on the inside and the fixing of a steel plate if necessary to provide a higher degree of physical protection. In fact, where cellars are not being used, it may be better to seal off entirely at the head of the cellar steps by bricking in.

FALLPIPES

New buildings

Access to a vulnerable roof and unguarded skylights can be prevented by making fallpipes so that they cannot be climbed. In new building work it may be possible to encase them in the shell of the building or to shield them with plastic or asbestos sheeting. Plastic fallpipes and light-weight brackets are no good at all to a climber as they will break away from the wall easily;

moreover they are exceedingly competitive in price with more traditional materials and need no maintenance by way of painting.

Existing pipes

There is a variety of ways of protecting existing pipes:

1 Wrap with barbed wire starting about 3 yards (2.7 m) from thc ground.
2 Cement into the wall over the pipe a semi-circular set of downward pointing spikes, again about 3 yards from the ground.
3 Cover a section of the pipe, well above the ground, with one of the new non-drying paints which are decidedly effective.

If none of these suggestions are practicable, ensure that any windows that are accessible from the fallpipe are adequately guarded against entry.

CONCLUSION

To sum up briefly: it is economic to consider security requirements in the initial stages of building; to implement them after construction is costly and inconvenient. Ordinary doors, windows, and locks can be substituted by similar ones with a security value with limited additional expense if included in the original specification. Bars and mesh can be built in during construction as can strong points for ring bolts and locking bars.

23

Protection of building sites

The building industry is one which is notorious for a floating population of workers; it is also one where property is dispersed over wide areas on occasions, under circumstances where it is difficult to give anything like the degree of supervision that would be desired. The workforce has a high percentage of semi-skilled and unskilled labourers; the conditions can be most unpleasant and it is not surprising that the percentage of these casual employees who have passed through police hands for dishonesty is as high, if not higher, than in any other industry. This makes security measures not only essential but most complicated to make effective.

NEED FOR SECURITY

This need is pinpointed by the fact that it is customary to take into account probable losses due to theft in calculating costs—and the percentage that is allowed in some areas is high! It would be assumed by this acknowledged wastage of capital that, if reasonable steps could be suggested to managements, adequate finance would be made available to implement the recommendations. Too many firms hold to the attitude that they are in business to build houses, not to catch thieves—and their profits suffer accordingly. Once a suspect labour force gets the impression that a firm is indifferent to its losses a free-for-all ensues.

Insurance premiums for property on building sites are so prohibitive that many contractors philosophically accept that their losses will be the less expensive—the eventual customer is the sufferer in the higher prices which are charged. It is easy to blame theft for all that goes astray on sites, but this is by no means true and it is conjectural whether a greater amount is stolen than is wasted due to sheer carelessness, damage, and bad book-keeping. Good security measures are not solely aimed at theft—they should be directed to any sphere where the employer's assets are being needlessly dissipated. This conception will not be popular in some quarters as the building industry has many long-established 'perks' where interference can cause resentment—and not always at the lowest level of employee.

Plain wanton damage, by either children or youths, is a major factor on some sites; this is particularly true where there are limited recreational facilities in adjoining estates and the new building site becomes the accepted 'sports' area. On a compact site this can be restricted by fencing, but on a larger housing development the only real solution is preventive patrolling and for this the use of trained dogs is invaluable—their presence is worth that of several security officers.

SITE SECURITY: COMPACT SITES

This type of site occurs, for example, in the construction of a block of flats or offices. A chief security officer or security adviser should be consulted in the planning of the site at the outset; not only could he make suggestions to improve the security of the final building but he should specify what is advisable in the way of fencing or hoardings, entrances, positioning of storage huts and areas inside, lighting, parking space, materials delivery procedures, and security manpower coverage during erection.

Fencing

Consistent with the needs of the builders, entrances and exits should be as limited in number as possible; this at least ensures a better chance of checking vehicles on and off site. Where a perimeter fence or hoarding can be used, it should be. In congested town centres hoardings may be an essential safety requirement for passers-by in any case. Construction sites exist for a limited duration and whatever is used in this manner would need to be transportable to the next job to add to the economic viability of the firm. Wooden hoardings are probably the most practicable solution, set with the planks vertically to make climbing difficult both internally and externally and to a height of 10 feet (3 m) or higher. Gaps should be left to allow a patrolling policeman or security officer to have a good view of the interior but not wide enough to permit squeezing through. Sufficient lighting should be laid on inside the compound to make it obvious to an intending intruder that he would be able to be seen from outside.

All valuable materials should be put in storage in site huts or buildings under lock and key—and this does not just mean a first-class padlock put through two loops of wire or on a tinny hasp and staple screwed into the wood with the screw-heads left intact. Where screws have to be used the heads should be burred or drilled out or a bolt put through the staple and the threads tapped over the nut inside. With mortise locks precautions of like nature should be taken to prevent a kick on the door breaking away half of a thin jamb and leaving the lock intact. There is a limit to what can be done with wooden site huts but these should be as strongly constructed as practicable and always kept locked when vacated.

Chain link fencing is the logical alternative to hoardings; this is usually set on scaffold poles to facilitate subsequent removal—or even on cranked concrete posts set into the ground less substantially than on a permanent site. This has the disadvantage of being easily cut by thieves and liable to damage by careless drivers. Metal palings of the Lochrin type or concrete panels are further alternatives, the former is probably best of all but is also more expensive than either of the more commonly used varieties though the new concrete panels are easily erected, portable and durable.

Gates

These should be fixed so that they cannot be lifted off and they should be locked with substantial chains and close-shackle hardened-steel padlocks—it is a favourite procedure for thieves to cut off the padlock and chain, substitute their own, and lock themselves in to accumulate what they want at leisure. As in more permanent structures, the gates should be of the same height as the adjoining fencing and should not be easy to climb (see Chapter 22, under 'Perimeter gates').

Alarms

In addition to attractive materials like copper tube, boilers, gas fires, and electrical equipment, which should be locked out of sight, other items which damage easily should be put where they cannot be damaged by stone throwing—a chipped lavatory suite is as much a loss to a firm as a stolen one and one ricocheting half brick could wreck several beyond further use. Where the value justifies it temporary alarm systems can be installed by the burglar-alarm firms, either for operating bells or the 999 warning to the police. It is not beyond a firm's electricians to rig up some form of similar audible deterrent for nominal cost; alarms are not expected by the average building-site thief.

Security officers, guard dogs and security firms

The question of live protection will to some degree depend on the locality—in a town centre with police patrols constantly passing, sheer physical protection of the nature outlined might be considered adequate. Additions can be either to use full-time protection in the form of the firm's own security officers, to tether a guard dog inside the compound, or to enlist the services of a commercial security firm to give periodic check visits.

If a dog is used it must be given as much latitude on a running lead as possible. It is not unknown for dogs to have been poisoned and indeed they have been caught with a noose round their necks and choked. The dog is really entitled to visits to check that it is safe; besides, if left to itself, it could be a nuisance to those living or passing nearby. If a commercial firm is to be used some check must be kept on their performance; the quality of these varies considerably and if a security adviser or chief security officer is employed it should be his responsibility first to obtain competitive quotations and then to ensure that his employers are getting the reliable and efficient service they are paying for. It could be found more economic to use one's own labour but in this instance also some check must be made by managerial personnel to see that the job is actually being done as required.

A further alternative, worthy of consideration, rises from the fact that some of the supervisory personnel may be living away from their home towns during the course of the contract and a comfortable caravan on the site and some financial remuneration might induce a reliable man to 'live on', perhaps with a dog provided; this would, with telephonic communication to the police readily to hand, meet most contingencies.

At all times of day private vehicles should be kept in a segregated parking area and commercial vehicles should be got off site as soon as possible, once they have delivered.

It is a waste of time and money to employ watchmen who are physically and mentally inert. They are traditionally associated with coke fires, cabins, and sleepy old-age pensioners and they are dangerous in so far as an unthinking management believes it is doing all that is necessary when in fact the man's well-being may be in hazard and his protective potential is nil. Notices warning 'guard dogs' should be freely displayed with those offering rewards for persons apprehending or reporting thieves (see Chapters 31 and 32).

SITE SECURITY: DISPERSED SITES

Except that compounds, similar to those on a confined site, can be established, these offer a more complex problem, if only because construction needs compel a virtual decentralisation of much material liable to be stolen. For economic and storage reasons scheduled deliveries are made to the site areas where the items are to be used, rather than to a compound, and this may involve them being left in uncompleted houses or, in the case of timber window-frames, stacked outside in the open. One manner in which this could be counteracted, without radical change to the building sequence, would be to complete out-buildings of the garage type first and then utilise them for lock-up storage. This would have the further advantage of reducing damage by vandalism and weathering; inadequate storage is the curse of building operations generally.

All that has been said concerning compounds on a restricted site is equally applicable to those on a housing estate; construction, lighting, and general precautions are the same. Wire-mesh fencing is much more likely to be used and, if this is so, materials must not be stacked against or near it on either side, as that would make climbing over easier and could conceal a hole from the eye of an observer or shield the actions of a thief. With adequate lighting being thrown on the fence and buildings which might contain a guard it would be risky to cut a way through.

Protection

It is even more important to protect materials from damage on housing estates—stone throwing is more prevalent and the children have more latitude. The decrepit watchman, too, is even more of a risk—children are ruthless and may create damage to annoy and ridicule the unfortunate individual, who is physically incapable of catching them. There is more scope for the employment of the able-bodied man who will be needed to patrol away from the compound; this will relieve the tedium of long hours of comparative inaction. If right persons are selected, as in all other fields of security, management should be able to incorporate profitable tasks for them which are ancillary to, but do not impede, their main one of security. With several isolated compounds there is more scope for the use of dogs both on patrol and static—if the latter, they should be periodically visited. If no one is employed on a site after working hours, specifically to look after the constructors' property, no amount of mere structural protection will stop stealing. It is unrealistic to think that the police, with their multiple responsibilities, will be able to pay more than cursory attention where the road conditions may be such that they would be unable to gain quick access with their vehicles.

Tidiness aids security

An untidy estate will give an impression of lack of interest on the part of the contractors and there is a much greater temptation to take that which is scattered about and apparently unwanted than that which is reasonably and neatly stacked. Even commonplace items like bricks are much less liable to be commandeered if they are stacked and covered with polythene to guard against weather, than those in a shapeless heap. Similarly, when timber is stacked off the ground in racking and sheeted over, there can be no excuse that it was thought to have been thrown away, unwanted. If window-frames and door-casings are stacked against the side of a house and not for immediate use they should have a wooden batten nailed inside to the top and bottom units to prevent them being thrown about or stolen; if unprimed and likely to stand for any time they, too, should be sheeted over against the weather. The enforcement of tidiness and of general security precautions for materials should be a matter of concern for the site agent or surveyor, and if he disregards elementary steps this should be brought to the notice of senior management, for it is neglect of his job.

Where materials and tools have to be left in unfinished houses they must be put out of sight and the doors made as secure as possible. If work can be so scheduled that heating pipes, boilers, fires, and so on, are virtually put in as they are delivered, so much the better. Things of value should not be left readily accessible—tradesmen take a degree of care over their tools that they do not exercise in respect of their employer's property.

It is even more important on dispersed than on compact sites for 'guard dog' notices to be displayed everywhere, together with warnings that all persons causing damage or removing property will be prosecuted and offers of rewards for the detection of offenders.

Finally, where unusually large amounts of metals are held at any time, the attention of the local police should be drawn to their whereabouts and nature. These constitute one of the major risks; copper and lead are easily defaceable and disposable and can be costly to replace—it is worth spending time and money to protect them.

PRECAUTIONS AGAINST THEFTS BY EMPLOYEES

Casual labour

In an industry where much casual labour is employed and men are thereby more likely to have records for theft (estimates place this as high as 20 per cent) it is absolutely imperative that no impression should be allowed to develop that management has a *laissez-faire* attitude towards both the theft and the handling of its property. Itinerant workers cannot be expected to have feelings of loyalty; those who are fundamentally dishonest will only restrain their natural instincts if they know that watch is being kept and prosecution will inevitably follow.

There is a great reluctance in the building industry to prosecute thefts unless they are of magnitude; this is particularly true where their own employees are involved. It is a bad attitude where petty theft is commonplace and simple dismissal is no punishment to the casual labourer who will automatically gravitate to the next building site—how often are references taken for this kind of employment? Prosecution will mean inconvenience in making statements to the police and perhaps attending court, but this will be more than repaid in the deterrent effect of a clear-cut policy on other similarly minded individuals.

Security policy

Where subcontractors are used, insistence should be placed upon their conforming to a common policy; trouble could be caused by different treatment being accorded to those committing identical offences. The same should apply to safeguarding property in their care, particularly where this belongs to the main contractor; any reports of theft from them should be carefully looked into to establish whether they are indeed genuine or are a cover for deficiencies arising from other reasons.

'Labour only' and 'piece work' are accepted components of the industry's manpower and wage structure; this does not help in checking who is actually employed on a site and when they come and go—it is quite impossible for a security officer to survey more than a percentage of workers when their work is finished. This gives ample opportunity for the removal of materials without being stopped and checked. The best protection during the day is that afforded by a reliable team of foremen and chargehands in constant contact with the labour force. There is an argument in favour of a periodic change round of foremen between sites, where the firm's size permits, to prevent familiarity or undue influence warping the foreman's judgement of what can or cannot be tolerated. The practice of a firm supplying transport to and from sites has an advantage over and above convenience and labour relations—it takes a certain amount of determination to carry stolen property in a lorry or bus load of fellow employees and foremen.

Tools

Whether issued by the firm or owned by skilled tradesmen tools are a persistent target for theft. By and large, despite the proportion of the unreliable, most workers respect other workers' tools on the same job; one tradesman rarely steals another's—perhaps because he appreciates the importance attached to them as virtually the person's means of livelihood and plying his trade. Where there are losses these should be inquired into fully and every effort made to trace the missing property; this is the ideal way to get the confidence of those on the site who are likely to be helpful and provide future cooperation. If a theft occurs during the course of a day as many employees as possible should be checked before they leave the site, the assistance of union officials, if these are present, will be freely given.

The safekeeping of tools is of importance. There is an onus on the employer to provide a locked place where they can be kept —then at the owner's risk, except in case of fire where the liability reverts to the employer. This is a trade where tools are widely insured by their users and, when looking into thefts, it should be borne in mind that the complaint might be ill founded or exaggerated. A record of thefts proves valuable for reference purposes and could provide indication of previous suspect claims.

Spot checks on persons and vehicles leaving always have a salutary effect beyond those who are actually involved, and spasmodic visits to the site by those whose prime interest is security will prove a cautionary factor and be commented on.

CONTROL OF VEHICLES

Employees' cars

As in other branches of industry, employees' cars offer the best opportunity for the removal of property in bulk and the safest way to avoid the notice of security or managerial personnel. Under no circumstances should private cars, even those of senior staff, be allowed to be parked in compounded areas. There is no need for them to be there and, apart from any considerations of theft, they cause congestion and obstruction and their presence inevitably leads others to seek the same privilege—with the result that the refusal leads to grumbling and ill will. Cars can be a thorough nuisance on the roadways of the site itself; it is essential that heavy vehicles dropping materials shall do so at the nearest convenient point to where they will be used. This may be prevented by parked cars and damage to temporary kerbs can be extensive when lorries turn over them to get through to where they need to be. If possible a special hard-core stretch should be laid down near to the site office for all employees to put their cars; if this is kept in reasonable condition there will be no excuse for leaving them elsewhere and anyone who does so can be regarded with suspicion.

Delivery vehicles

The checking of materials onto a site is the responsibility of the site agent's clerk or checker, storeman, where one is employed, or foreman/chargehand where sand, bricks, cement, and so on, are dropped out on a site. Whilst the latter are hardly of sufficient value to justify anyone coming out of hours to steal them in quantity, there is a ready market for them to be delivered by the carrier to places where small private building is taking place. This can easily be done by dropping off part of a load destined for a firm if it is known there is little or no checking or supervision of incoming materials.

No building site has its own weighbridge and loading notes have to be accepted for most bulk materials—a count could be feasible for the more expensive tiles, bricks, or bags of cement. The right should be reserved, and notified to suppliers, of occasional spot checking of loads by taking over a public weighbridge before and after delivery. When this has been done once or twice the word will spread and have the necessary effect.

Unless they are supervised, delivery drivers can clock in on a site, be directed to where they should drop their loads, and then only drop part if no one is about, or they can bribe someone to sign their consignment note that the full quantity has been received. Where there is any suspicion that this is taking place a watch should be kept on the driver in order to catch both him and the accomplice. It is a good principle to have a central reporting point from which drivers can be directed; they can also be clocked in at the time and asked to report back after delivery. Most suppliers would gladly accept this, for it gives them a check on the performance of their drivers; it also gives the site contractors the opportunity to make sure that they are not taking part of the load out again and, by comparison of times, fix a possible source of loss if items are found missing after the driver has left.

There must be a satisfactory system of recording receipt of materials. A responsible person must sign for them and accept the corresponding consignment note to pass to the site agent or his clerk. With both subcontractors and delivery firms, anything in the nature of security requirements—checking-in and parking—should be made in writing so there is no excuse of lack of knowledge and objections can be discussed before they develop into incidents.

CONTROL OF PROCEDURES AND PREVENTION OF DAMAGE

Prevention of damage has already been mentioned, with emphasis on tidiness on the site and stacking of bulk materials, but there are sources of wastage by rank carelessness during use, such as the following:

1 New sewer pipes left where they will be crushed.
2 Tie-wires in large quantities dropped in mud and used in hard core.
3 Floor-boards damaged by plaster being mixed on them and soaked with water.
4 Cartons of nails broken open and left lying in uncompleted buildings.
5 Window- and door-frames left where they will be damaged by passing transport.
6 Cement bags left on damp surfaces, part-used bags left outside.
7 Sand tipped on soft ground with inevitable high proportion of wastage.
8 Tools, fittings, wiring, electrical gear, left in partly completed and insecure units.
9 Lavatories, wash-basins, and baths carelessly handled and damaged.

All these add up to an inaccurate assumption of theft when estimates of the quantities used are reviewed.

A reasonable system of control should be instigated so that damaged fittings may be utilised on sites in workmen's lavatories, canteens, site huts, etc. It is ridiculous to install new ones in these and either scrap or store damaged ones. All surplus stock should be returned to a central store and not retained where no one is really interested in it. Book-keeping should be tightened up to prevent over-ordering and to provide constant knowledge in a large firm of 'what is where' at any given time.

Tools issued to a site are rarely all returned at the completion of a contract; each contract seems to necessitate a fresh issue. This is a matter for the site agent—consideration might be given to providing him with an incentive in this direction by giving him a percentage bonus for all tools which are transferable to a new site.

Scaffolding poles are lent out, stolen, lost in mud to the extent that in 1967 it was estimated some £2 000 000 worth was no longer accountable. There seems little importance attached to collecting these when their immediate use is over. They should be stacked for convenience and consideration given to marking to identify—different coloured paint splashes in a sequence would do this with minimum time and cost. If something of this nature is not done any length or clip is indistinguishable from some other firm's property.

Timber can be stored under cover to prevent weathering, or it should be stacked and sheeted over. Employees should not have the privilege of taking short lengths, unless they are totally unusable.

Cement huts or garages should be used for cement storage and a watch should be kept on firms delivering ready-mixed concrete. The accuracy of the reputed load can only be estimated but an experienced foreman will soon have a good idea whether he is being defrauded—this is another instance where watching may have to be done from supplier to delivery point to substantiate suspicions. After working hours, a few bricks should be rotated in concrete mixers and the displaced hardening concrete swilled out; the practice of taking off starting handles should then be implemented. On a carelessly run site, mixers not standing on a wood base and regularly moved may become set in concrete which may necessitate breaking the wheels to move them.

Metals, in the form of unused ends and scrap, are the greatest 'perk' for employees in the industry. Some plumbers regard the disposal of any waste as their prerogative, which can amount to a major loss, with the present prices of copper and lead. A mandate to dispose of short lengths will inevitably lead to the deliberate creation of short lengths and scrap. This must be absolutely stamped out and, irrespective of protestations of 'we have always had it,' exemplary prosecutions should follow. It would be interesting to know what savings would accrue if all short ends and scrap were saved for further utilisation or sale.

The same is applicable to surplus electrical fittings, conduit, and wire. If there is no return-to-stock procedure, this in itself is an incentive to over-order.

Perks of all kinds, including work done for senior staff, must be rigidly controlled—if they lay themselves open to criticism, how can they enforce their authority in cases of theft on those whose services they have used? As in more static industries, nothing should be taken off the site by way of tools on loan or materials on purchase or otherwise without a written permission from someone in authority. Damaged items of value should not simply be scrapped on the authority of anyone on the site, but should be inspected and 'written off' by managerial dispensation from the head office.

'STRAW' OR 'DEAD' MEN

The possibility that the record of the labour employed on a site may be fraudulent should not be overlooked and independent controls must be instituted to prevent this. The building and civil engineering industry is particularly vulnerable to fraud through wages being claimed and paid for fictitious employees or the continued inclusion on payrolls of employees who once worked on the site but who have left for other employment. These are known as 'straw' or 'dead' men. The type of fraud has also occurred in other circumstances where the employment of seasonal labour is usual either on full or part time. One form of control is the rotation of supervisory staff, particularly site foremen where this is possible, but it brings with it other problems.

An unannounced 'head count' by an independent person from time to time is recommended. Before this is done all time cards and/or any other means of recording labour must be put under lock and key because experience has shown that if that is neglected they will disappear during the count and vital essential evidence will be lost.

The possession of insurance cards and PAYE forms of employees is not sufficient confirmation of their continued employment because unfortunately the control by the offices concerned with the issue of further cards can be overcome by the use of false names. This is particularly taken advantage of by labour which has its origin outside the United Kingdom.

If the paying out of wage packets is done by persons independent of the site or department and against the individual surrender of a time card or some token, and if checks are made on the quantity of work carried out, especially where overtime is claimed, the chance of a fraud of the type described through collusion between persons in supervisory positions and other employees can be eliminated.

One instance of a fraudulent claim for overtime allegedly performed at weekends came to light when it was realised that for it to have been done would have required a constant supply of steam whereas the boiler house records showed no steam was supplied or was available to the machinery which would have been concerned.

One final matter relating to personal possessions of employees: make it quite plain that no money or valuables should be left in clothing or drawers and the onus is on the loser if anything is stolen.

SUMMARY

To summarise the deterrent action which can be taken:

1. Create as substantial a compound as possible for all materials which are easily stolen.
2. Ensure that the compound perimeter will be hard to climb from either side and any breaking into be immediately observable.
3. Establish a secure store within the compound for particularly valuable items: copper tube and fittings, fire-back boilers, gas fires, immersion heaters, etc.
4. Illuminate the compound—in particular the vicinity of the store for valuable materials.
5. Where the site is insufficiently large to justify the employment of a full-time security officer or capable watchman, con-

sider using the services of a private security firm to give periodic visits.

6 On a large site, where security officers are employed, consider using guard dogs in each separate compound and for patrolling.

7 Segregate employees' cars to a given parking area which is easy to supervise.

8 Have a reporting point for incoming delivery vehicles and establish a practice that they 'clock out' at this point when leaving the site.

9 Thoroughly examine any accepted procedures whereby employees could remove any kind of materials from the site for personal retention.

10 Thoroughly check all means of storage of materials to reduce wastage due to weathering and carelessness.

11 Where materials of value have no means of identification, consider whether it is worthwhile to put some identifying mark upon them.

12 Post warning and reward notices to deter thieves and mischievous children.

13 Try to build up a security consciousness amongst all permanent staff, offer rewards for the detection of theft and for suggestions to minimise waste.

PART FIVE

SECURITY AND CONTROL OF ROAD TRANSPORT

24

Security of vehicles and drivers

Of all the offences coming under the heading of 'simple theft', that of property from unattended vehicles represents the biggest single problem to the police. Private cars are those mainly affected—this follows by virtue of their numerical preponderance, but individual incidents rarely represent loss to the user comparable with those where commercial vehicles are concerned.

COMMERCIAL TRAVELLERS

Private cars fall into two categories: those that are used for strictly private purposes and those that are used as travellers' cars, either as ordinary saloon cars or shooting-brakes. The casual thief is responsible for the greater part of the thefts from the ordinary private car user—portable wirelesses are a particular attraction—but thieves attacking travellers' cars are a specialised species.

The very nature of a commercial traveller's work entails leaving his vehicle as near to his prospective customer as possible; this means that it will have to be left more often than not at the kerb-side rather than in a park. The values of goods carried are often out of all proportion to their bulk and it is incumbent upon the traveller to take every possible precaution to protect his vehicle. It is absolutely essential that he should have a powerful and efficient alarm fitted and it is hardly less advisable that he should also have an immobiliser.

It is rare that the integrity of commercial travellers, in relation to their own goods, is ever in question, nor are their vehicles left at night other than in their own locked garages; with these exceptions, what is said in respect of commercial vehicles and their drivers is equally applicable to them. There is one special precaution that they must take: to glance round on each occasion that they stop near a customer's premises to see whether any vehicle, which he has seen at previous premises visited, is just parking or has just parked. If he does see this happen more than once he would be well advised to ring the police from the customer's office before returning to his car, giving the number of the vehicle and a description of any of its occupants. Another point is that of insurance; where high-risk goods are carried special conditions and limitations are certain to be applied and the user must ensure that these are complied with to the letter or claims may be invalidated.

INSURING COMMERCIAL VEHICLES

Commercial vehicles represent a great temptation to any criminal and far too many firms fail to appreciate the likelihood of their being attacked, regarding insurance cover as being all that is required by way of precaution. Unfortunately for them insurers are tending to take a much stricter line with thefts of and from vehicles. A recent case typifies this new approach and it is obvious that where obligations are laid down upon a firm to ensure the locking of a vehicle and the taking of other precautions the insurance company will not be prepared to pay, and will be upheld in not doing so, if there is evidence that these precautions have not been taken.

One law-suit worth bearing in mind is: Ingleton of Ilford Ltd, *v* General Accident Fire and Life Assurance Corporation Ltd, [1967] 2 Lloyd's Rep. 179. Here the policy excluded loss or damage by theft while the van was left unattended in a public place, unless securely locked. The van and contents were stolen whilst the driver was in a shop making delivery of goods; the van was unlocked and the ignition key left in the van. The defendants' insurance company were held not to be liable under their policy.

Only a relatively small proportion of all goods vehicles are actually attacked, but if thieves decide that a particular vehicle is a desirable objective its safety may well depend upon the protective factors that have been built into it, the intelligence, loyalty, and determination of its driver, and the planning of routes and procedures that are being carried out by supervision.

REQUIREMENTS OF THIEVES

When considering how to combat a particular type of theft it can be advantageous to change places mentally with an intending thief. By doing this one can evaluate the things that the thief will most need to achieve his objective and, from the assessment of those, counter measures can be determined.

First of all, there are certain characteristics of all thieves which should be accepted: on the whole they dislike physical exertion immensely; they wish to get their theft over and done with as quickly and quietly as possible; they want to keep the stolen goods in their possession for as short a time as they can, converting them into hard cash at the first opportunity; and, above all, they want to incur a negligible risk of detection. If these characteristics are accepted, it is easy to see why theft of loaded vehicles is so popular with certain members of the criminal fraternity.

The essential requirements for a thief in respect of stealing loaded vehicles are listed below:

1 The capability of removing the vehicle and/or its load.
2 The opportunity to do so.
3 Sufficient time to carry out the theft without interference or attracting attention.
4 Adequate time to ensure being a safe distance from the scene before the theft comes to light.
5 An early market for his goods with reasonable reward.
6 Negligible risk of detection.

It can be taken for granted that any organised gang will have arranged a market for the goods before the theft has even been seriously considered.

METHODS OF STEALING

Thieves have several means whereby they can achieve their objective; these are:

1 Stealing the vehicle or from its load, in the absence of the driver, whilst parked, either temporarily or over-night.
2 Taking possession of the vehicle by force by either stopping the driver by a trick or attacking him whilst his vehicle is stationary under normal circumstances.
3 Persuading the driver to join forces with them in simulating the circumstances of a theft.
4 Impersonating the driver, in his absence, to obtain possession of his loaded vehicle or, by using their own vehicle, persuading a firm to give them a load on the assumption that they are bona fide carriers.

There are many lines of diversification amongst these main avenues of attack and there is little doubt that others will occur to fertile minds with ample leisure for contemplation.

The more of these lines of attack and requirements that can be negatived, the less likely it is that an attempt will ever be made, and if the steps can be offensive as well as defensive, so much the better. The greatest deterrent of all is the prospect of being caught—especially on the job.

VEHICLE PROTECTION

Without the cooperation of the driver the fitting of any device is useless and a waste of money, but by far the majority of long-distance drivers are honest men who will utilise any equipment placed at their disposal by their employers for their personal protection and that of their vehicles and loads. With no special equipment there is still much that they can do as individuals, but they should have more than the ordinary door locks to assist them.

Internal bolts

Circumstances could arise where a driver might be attacked in his cab. This would be difficult in the first place since he may be 2 yards (2 m) or more from the ground.

If the vehicle is required intact, breaking windows to get at him is out of the question, the thief therefore has to get into the cab. A very simple precaution is the provision of 3 inch (75 mm) tower bolts inside each door, to be shot by the driver as and when he requires; these will solve this problem almost entirely. If, added to this, the driver is furnished with a push-button in the cab to activate the alarm system and his own means of possible counter-attack—for example, poking a fire extinguisher through the window to discharge at his attackers—an attempt to get at him will probably soon be abandoned.

Vehicle alarm systems and immobilisers

An increasing number of commercial companies produce a variety of both alarms and immobilisers. The electric circuitry of alarms, in particular, is not complex and, should they so desire, firms can construct their own variations that may be equally effective as the commercially made ones and have the advantage of being unconventional. It is possible to incorporate both alarm and immobiliser in one unit but the disadvantages of this offset the advantages—if a means is found of bypassing the whole unit both forms of protection will be eliminated together.

Both devices are really delaying ones and any skilled and determined criminal, with ample time at his disposal, could eventually overcome them. The presence of an immobiliser represents a challenge and an inconvenience, rarely a complete barrier; the alarm, a temporary embarrassment. It is no use at all fitting an alarm siren on a vehicle which is then left in a position where no one can possibly hear it; even if it is heard there is no guarantee that anyone will take a great deal of notice of it. The lack of attention paid to alarm bells sounding at jewellers' shops near opening times is proof of public indifference to such noises. Above all, no system is better than the operator—an uncooperative driver can nullify the effects of the best devices; an uninstructed or incompetent one can prevent them even commencing to fulfil their function.

Types of alarm systems

Modern alarms are almost invariably electrical and either use the existing horn of the vehicle or are specially fitted with a siren of a distinctive note. There are several different means of activation, the most common is that of a key-switch turned after closing the doors and windows and with a separate circuit from the electrical equipment of the car. Contacts on doors, windows, boot, and bonnet, may be included as desired. Occasionally the alarm is wired into the ignition system or connected to the mechanism to operate when an attempt is made to start the engine. The switches themselves may be one of the following types: tumbler, combination cap or insert, relay, hand-operated, or key.

Other types of activation, which depend on the actual movement of the vehicle, are the pendulum type of contact, mercury switches, and vibrator contacts. These may have an additional value in being usable for protecting the load itself if they are sufficiently sensitive to react to movement on the vehicle or interference with either doors or sheeting. With this type of contact it must be borne in mind that the natural phenomena of wind or rain may occasionally be sufficient to cause false alarms. Few of the ways of totally protecting a load itself against interference have been entirely successful and a firm with a load of consistent nature and shape should be able to evolve its own alarm device connected to the locks, ties, or sheeting.

Under present-day traffic conditions it is absolutely essential that any alarm siren should be loud and distinctive, otherwise it will attract little or no attention. A high pitched or a warbling type of sound which causes irritation to the listeners is probably the best, but whatever is fitted should be tested under the conditions in which it will be required to operate.

Types of immobilisers

These are of two main varieties; mechanical and electrical. Consensus of opinion in the past has been that the mechanical types are more reliable and more resistant to interference.

Electrical immobilisers—These work on similar principles to the alarm systems, but cut off the ignition, starter motor, or the fuel supply. Those stopping the fuel supply are particularly suitable for diesel motors where they operate a plunger to lock the fuel injection system. Operation is similar to alarms in the form of a key or combination lock mechanism, set after the driver leaves the cab. Variations have been introduced, dependent upon movement in the cab or again on an attempt to start the engine.

Mechanical immobilisers—The physical methods applied are, more often than not, in the form of locks upon the gear-shift, brakes, or parts of the engine and steering mechanism. More complex methods involve the transmission and the differential or the hydraulic braking system. There is ample scope for ingenuity in devising exclusive types—if these are of the lock and rod principle they should be in a position where bolt croppers cannot be applied to them and also be in a position where they are not obvious so the thief will have to spend valuable time locating them before determining how to counteract them. The unassisted driver quite often has simple and very effective means at his own disposal—removal of rotors, leads, or parts of the mechanism or slipping a piece of paper between the points and terminals can stop a thief, as effectively as any expensive mechanism.

Cost of equipment can vary immensely—the more complex the more secure, but also the more difficult to service and reset if accidently operated. Simplicity is a virtue for the driver—if he has to carry out a complicated and difficult procedure he will be tempted not to bother.

TRAILERS

Trailers represent a problem since the normal immobilising

devices are not functional when they are separated from their tractors as may happen at loading and unloading points. It is a practice with many firms to load the trailers over-night for picking up early next morning. Unless precautions are taken there is nothing to stop thieves backing their own tractor onto such a trailer and removing it immediately.

Chains around suspension and brake locking devices are apt to be more dangerous than effective. A pivoted steel clamp with a hacksaw-proof padlock on the trailer pin seems to be the ideal fitting and a commercial appliance is available, although this is something which can quite easily be made. It is essential to remember to use a lock on the fitting which will be impervious to hacksaw blades and sufficiently close shackled to prevent it being easily forced. Indeed, from time to time there have been spates of empty trailer thefts and there is a market for them after respraying.

If these trailer pin clamps are used they can form a line of defence for the tractor. If a driver can drop his trailer so as to leave his tractor between it and a wall with no room for manoeuvring, there is nothing a thief can do about it without forcing away the clamp.

LOAD SECURITY

Where large vans are concerned there is no difficulty at all in wiring up the rear doors to the alarm system of the vehicle. In any case, these rear doors should be fitted with a sizeable close-shackled padlock and bar to discourage an attempt in the first instance. If such a fitting is to be made it must be substantial and not the flimsy hasps and cheap locks that are very frequently encountered.

If the value of regular loads demanded it there would be no difficulty in protecting the walls of the vehicle as well with closed-circuit wiring of the type which is used with burglar alarms.

Open lorries carrying goods covered by sheeting represent a much more difficult problem. All loads must be sheeted down tightly to prevent casual petty theft during the course of deliveries, and also to conceal the nature of the load. A common practice is to steal from loads parked overnight at transport cafés by undoing the tie ropes on the sheeting, taking out property from the centre of the load, and resheeting down, so that the theft may not be discovered until the vehicle is several hundred miles away. More often than not this type of theft is almost certain to be the work of a dishonest lorry driver also parked at the café. Each driver has his own means of sheeting down and if he uses knots of a particular type or sequence, or in some way ensures that interference will be detected by him when he inspects his load before leaving, the chances of the police detecting the theft are infinitely increased.

With a little ingenuity alarms can be attached to sheeting and for that matter, as mentioned earlier, switches functioning upon movement of or upon the vehicle can be installed. The easiest and cheapest way to stop thefts from loads is to put the vehicle on a guarded or well-lit car park.

Where deliveries are made of comparatively small packages to a number of customers and the carrying vehicle has to be left in public streets it is highly advisable that two men should be used so that the vehicle is not left unattended at any time. A limited number of firms have experimented with carrying alsatian dogs in the back of their vehicles—this is effective but has caused complications in some instances with members of the public!

Tanker vehicles

These fall into a special category; the sheer bulk of their contents seems to militate against the vehicles being hi-jacked. The main source of temptation is the possibility of repeated pilferage particularly of fuels, especially petrol, and wines and spirits. Preventive action is really required therefore primarily in respect of employees.

Where fuel is concerned, a driver's obvious ploy is to ensure that he has a quantity left after making all his authorised deliveries—any time before then has danger of detection for him in some form or another. To do this he has to ensure that he has the cooperation, either intentional or unintentional, of the person who is supervising or accepting delivery—sometimes this is unbelievably slackly carried out. If the form of measure is to use a dip-stick, this is frequently left to the driver. Despite the fact that he may have been visiting a particular site for years, an occasional check should be made to ensure that he is using it correctly and not just accept his measure. Drivers often have regular runs and it is easy to develop a sense of confidence and familiarity with them.

Where wine or spirit is concerned there are three main methods:

1 Short delivery (wine left in the tank).
2 Dipping through the top of the tank and transferring liquid to a container.
3 Removing customs seals to attach specially adapted couplings and pipe to the outlet valve.

Added to these possibilities there is the risk of dilution being used to cover volume deficiencies caused by the theft. Frauds can of course also occur during loading and offloading and the temptation is perhaps greater with these goods than most.

Security seems often to be based upon a touching faith in the use of customs padlocks and seals. These do not appear to have been changed for years and a criticism that they are antiquated might be thought valid. Where they are mandatory there is no reason why they should not be supplemented by more modern types. The vehicles should be surveyed to ensure that every possible means whereby liquid can be abstracted are either protected by padlocks of quality or are fitted with seals sufficiently reliable to give an immediate indication of interference.

Seals

These do not provide additional security for the hasps, bolts or other means of fastening to which they are affixed. They are simply a means of indicating a possibility of tampering and theft and the necessity that a full check should be made.

For many years the conventional form was a wire with a lead seal which was crimped hard on to it and may or may not have carried an identifying number. With time and care these could be opened and refitted so as to avoid detection, but many new and improved varieties have come onto the market. Of these, those made of plastic seem to be advantageous both in price and performance; they have to be broken to be removed, and their nature makes it easy to carry a clear identifying number or marking so that they cannot be replaced by an identical seal. A system whereby the number of the seal is shown on a driver's consignment note to be checked at the delivery point before unloading takes place can easily be inaugurated with little additional clerical cost.

There are of course numerous other obvious applications—on cash bags, coin containers, and rope lashings on the motor vehicles. In any instance where it is necessary to recognise interference at a glance their use can be considered. A less obvious use is to carry a fire door key on a hook screwed into the wall or door in lieu of the conventional glass-fronted box, which has become absurdly expensive.

Routes and schedules

There is a limited amount that can be done in this respect. If long-distance trunking vehicles are to operate economically and to a schedule suited to the customers' needs they are virtually confined to the main route between producer and customer. Regular variations are ideal in theory only, and no route which involves passing through quiet country lanes, as opposed to main roads, should be acceptable at all. Using a trunk road with other drivers on it denies opportunity to the thief to carry out his theft in peace and quiet.

So far as variations in time are concerned, commercial considerations will be paramount and, in any case, except in the transport of wages, there is little likelihood that a regular time schedule will be adhered to.

Radio linkage

A commercial security firm, Securicor, offers a system of radio contact with vehicles through the medium of their own chain of radio stations. A radio transmitter/receiver (transceiver) is installed on loan in each vehicle operating on wavelengths allocated to Securicor throughout the country. This enables a driver who is being attacked, or who has reason to fear that he may be attacked, to call the local control who will in turn notify the police. Other amenities that are offered by this service are those of passing messages, passing weather reports, and giving information on road diversions, etc.

There are arguments for and against the value of this scheme and whether it is of use to a particular firm will have to be determined by that firm itself, bearing in mind its mode of operations and its requirements and need of the facilities offered.

Overnight parking

This is possibly the main danger area; there are two difficulties, one the lack of adequate supervised parking facilities throughout the country, secondly the difficulty in obtaining the cooperation of drivers. Fortunately the Government, through its Home Office Standing Committe on Crime Prevention, is fully aware of the problem and has allocated very large sums of money for the provision of guarded lorry parks in strategic parts of the country during the next five years.

At the moment these are limited in number outside London—the local police or road haulage associations will be pleased to advise on their location. Well-lit parks are a second best but, where very valuable loads are concerned, it may be worthwhile to consider coming to an agreement with a local firm or garage which can provide enclosed accommodation. Every large group of firms should insist that vehicles belonging to any member of the group should be accepted overnight into any of their factories which is convenient; there is surprising parochialism to overcome in attitudes on this point. Where vehicles are to be left at a depot or warehouse with alarm facilities, it is possible to arrange external points where the vehicle can be joined to the circuitry so that any movement of it will trigger off the alarm.

Using supervised parks usually requires payment by the driver for which he will re-claim. As mentioned elsewhere there is a temptation for him to park in the open and claim the allowances. This should be clearly laid down as a disciplinary matter which will be regarded seriously. In no circumstances should lorries be left in back streets near lodgings for a personal convenience of drivers who are creatures of habit when it comes to having favourite stopping places. The regular presence of their vehicles there will eventually attract the attention of interested thieves in the area who can then pick their opportunity at convenience.

Police manpower resources are not, and never will be entirely adequate for all calls, but if they are told of the presence of a high-value load in their area, they will try to periodically visit it and at least, if they see any interference taking place, they will know that action is required—even to checking the bona fides of the actual driver.

THE DRIVER

The driver is all-important—he is not just another employee. His integrity is worth any amount of the alarms and immobilisers which are useless without his cooperation. The best place for a suspect driver is in some aspect of transport where his potential to steal is strictly limited. There are numerous jobs of this type in the industry and to place temptation in such a person's way is foolish and unfair.

Selection

Regrettably, integrity has acquired almost a cash value. The higher the risk the higher the standards to be set and the higher the rates of pay that should be offered to ensure a good choice of acceptable applicants. The idea that security is a vital matter for the employer is one which should be instilled in the driver right from the outset. This can be made clear by the application form to be completed and in subsequent interviews.

The Traders' Road Transport Association has designed a special form for the employment of drivers; a variation of this is suggested for those drivers who are intended to carry valuable loads. The object is to reduce the number of unsuitable applicants who cannot meet requirements and also to provide a ready avenue for dismissal, if it is subsequently found that they have deceived the interviewer. (See Appendix 19 for specimen application form.)

This form of application makes it incumbent upon the driver to declare any criminal convictions that he has. The object is not that these should be a bar but to allow the interviewer to use his discretion in full knowledge of the circumstances; some criminal convictions are so trivial or so long ago that they can be virtually disregarded. These are interviews that should not be hurried; qualities like honesty, reliability and integrity are difficult to assess purely by interview but the effort must be made. A good and stable employment record is a first consideration; a man who has jumped from one firm to another at regular intervals for the hackneyed reason of 'advancing himself' is unlikely to have the steadiness of character required. Neither is a driver with a hard-luck story, or claim of victimisation, or that he now wants to settle down. However sympathetic one might feel, the firm's interests are paramount and such a man is not a good risk.

Long service with previous employers, a settled background with domestic responsibilities, and personal recommendation by well-regarded long-serving present employees are all advantageous points. If there is any reluctance to answer questions it can be accepted that the matters are detrimental to the person concerned. It should be borne in mind that it never occurs to an honest man that he has to convince any one that he is. A complete record of previous employers, as far back as possible, should be checked with the applicant; all gaps in the record should be queried and accounted for; references should be sought from employers—preferably the last three—and not from personal acquaintances furnished as referees. In one instance, those given were from a father, uncle, and brother-in-law and all three had long lists of criminal convictions.

Any interviewer, no matter how experienced, can misjudge a convincing liar so before any new driver is sent out with a load his references should have been verified and his driving licence, form P45 and National Insurance card inspected. The time expended on these precautions will not be wasted—if it does nothing else it will impress upon the driver that security is the basic consideration of his post. A degree of casual supervision should be continued after a man's employment has been commenced. It must be repeated that the percentage of experienced lorry drivers, especially those in long-distance haulage, who are not entirely trustworthy, is remarkably small.

It is obvious that there should be a uniformity of performance on given runs, so if a new driver displays abnormalities in the time taken and the distances covered note should be taken and a special watch kept upon him. Even so, variations in mileage and times at stopping places are more usually for personal reasons rather than more doubtful motives. Breaches of discipline militate against a firm's interest and it may be equally necessary to dispense with the driver's services on these grounds as on those of dishonesty.

Driver collusion

No one knows how many hi-jackings are faked, but they are far in excess of the genuine—these are estimates from persons in the transport industry and not those of sceptical police officers—and this is a major threat. The vast majority of drivers are strictly honest and it is a matter for conjecture how many casual approaches are made before a driver is found who is prepared to cooperate—at a price. Drivers appear to think that informing reflects upon themselves, and so hesitate to report these overtures; this is a pity, for they could be stamped out by police utilising the common law offence of incitement to commit crime. It is immaterial that the solicitation or incitement of the driver has no effect upon him—if he is asked to help in stealing his load an offence is committed even though he refuses point blank. Proceedings would not normally be begun by the police unless there was a sequence of at least two such acts which could be substantiated to form corroboration.

A dishonest driver incurs infinitely more risk than the thieves; it is unlikely that they will let him know their true identity or even see him after the incident if this can be avoided. Such payment as he receives will be trivial in comparison with what he loses by way of trust and possible liberty. Having achieved their objectives it is more than likely that the thieves will simply ignore him as his usefulness is finished. He dare not talk and he has no redress of any kind. If they are caught they will have little hesitation in putting him forward as the instigator.

If the risk justifies carrying two men on the vehicle the danger of collusion is more than halved. Careful selection is the only counter-measure that can be taken, though it should be borne in mind that a driver who is contemplating participation in such a theft is apt to be transparently uneasy and worried before it takes place.

Observations on drivers

With experienced drivers this is a most difficult thing to do. No driver of any calibre at all will fail to spot a following vehicle within a matter of miles and, if there is no justification for the suspicion, the practice will cause strained relations with the transport fleet drivers. If it is considered essential to 'tail,' several cars of differing types should be used so that there is no continuity in the following vehicle. Police cooperation should always be sought in these cases—they have wireless facilities and inter-communication between vehicles will enable observations to be kept at a distance both from front and rear. Electronic devices of the radar type have been tried with some success, but this is a matter for the police.

PETTY FRAUDS BY DRIVERS

It is somewhat surprising that a proportion of drivers who would not dream of stealing from their loads sometimes indulge in practices to obtain what they regard as 'perks,' which would however be regarded with equal severity by a court of law. The overnight allowance is a prime example and the consequential loss arising from the steps taken to justify this far outweigh the financial gain. A driver who could reach his home base within his permitted hours deliberately does not do so, but parks up near his home where he spends the night, then running in early the next morning purporting to have stopped some distance away. Perhaps £1.50 or £2.00 accrues to him but the employer loses a night in which to offload and reload and the whole transport schedule may be upset. A further security aspect is that the lorry will almost certainly be parked in streets at risk.

In a similar manner, books of parking tickets can easily be purchased and submitted as receipts for stops on established parks when the vehicle has been left in streets near lodgings, or the driver may have slept in his cab. Again, whenever one of these is submitted it implies there has been an unnecessary risk to the load.

If the duration of a journey is such that refuelling away from base is needed, the employer can cater for this in a number of ways—none of them foolproof. A cash float may be allowed, agency cards supplied, a voucher system used, or purchases allowed only from accredited garages where a credit account scheme is maintained. Connivance between a driver and a dishonestly cooperative supplier can result in less fuel being put in the tanks than will be shown on the receipts, the difference being shared between the two individuals. Even at the accredited garages, there can be the same type of conspiracy between driver and pump hand.

Trunking drivers, who are most likely to indulge in these practices, are usually well paid and the first deterrent step should be that of making it perfectly clear that the strictest disciplinary action will follow if anyone is found committing them. Drivers who are prone to overnight stopping near their base will show up repeatedly on their log sheets as being early morning arrivers into depots from certain runs. An enquiry to a driver as to why this is happening will bring an improvement at least for a period, the alternative is for visits to his alleged regular overnight stopping point and a tour of the area around his home. This is a matter which a firm may think fit to deal with internally, but it should also be borne in mind that inevitably a further offence of falsification of drivers' records is committed and drastic action is needed.

The difficulty with parking tickets lies in that they will be normally attached to expenses claims for auditors' purposes and will not be retained in the department. Where identical tickets are submitted, or those from a single numerical sequence, this may not be immediately obvious to supervision. Unfortunately this is one offence which sometimes is connived at by supervision and it is therefore one which should be mentioned to and borne in mind by the cashier's department who pay the expenses or the auditor's department who check the receipts.

Whether the fuel receipts which have been falsified show up at an early stage depends upon the record keeping of the concern. If a mileage record is kept against the fuel usage, it will soon appear that some drivers are getting less mileage from their fuel than their contemporaries. Scrutiny of receipts may show up the regularity of a particular garage with perhaps identical writing but differing signatures or even fuel purporting to be drawn from that garage when the vehicle is nowhere near that area which may happen if the driver has acquired a book of receipts.

Agency cards

Agency cards are equally subject to abuse and it is not unknown for a large user to pay accounts rendered from petrol or diesel suppliers without any cross-reference to records submitted by drivers or other query. If any of these cards is lost or stolen it is imperative to inform the issuer at once and confirm by letter. Lists of lost cards are circulated to the garages of the fuel company concerned at monthly intervals and it would not be unreasonable for a user to refuse to pay if the card was accepted after such circulation has been made. Liability does rest on the user and if the drivers have regularly tendered the card at garages operated other than by the issuing firm a relationship by custom will have been established and the user will have to pay in respect of any fraudulent withdrawals by means of a stolen card despite the fact that they are not from garages of the issuing firm.

Own tanks

If the company owning the vehicles has its own tanks of petrol or fuel oil and insufficient supervision of issues is exercised discrepancies are likely to be made up from that source by the dishonest driver. Petrol and fuel oil pumps usually have meters registering issues and it is recommended that at least weekly checks of the figures and issues are made by someone independent of the person who is primarily concerned with the intake of the liquids and their issue.

Vouchers

The system of vouchers used by some major carriers does reduce risk; they are usually in the form of a triplicate book and provide for work done or parts in addition to fuel. Two copies go to the supplier who forwards one with his account to the user who already has the driver's copy for reference. There is less opportunity for fraud but it still exists.

Even when company's pumps are used fuel may be supplied over recorded amounts or into vehicles for personal purposes. As has been mentioned this can be prevented by tight procedures and checks or by fitting the pumps with a special device only responsive to a magnetically taped token applicable to a particular vehicle. This causes the amount issued to be shown on a recording panel away from the pump or on printout in the device. It is costly to install but can show substantial savings in wages as well as cutting pilferage. Casual defalcations involving fuel will be hard to detect if done systematically but good supervision should bring them to light reasonably quickly.

PROCURING A LOAD BY FRAUD

This is rarely attempted on a large firm with well organised traffic arrangements and insistence on precise documentation. The procedure is that the prospective thief with a large lorry or articulated vehicle, either genuine or stolen for the purpose, and in any case carrying false number plates, calls at a firm on the pretext of seeking a return load for a journey, usually to London. The vehicle may have a fictitious owner's name, address, and telephone number painted on the cab, and the driver will be in possession of copy delivery notes and other 'proof' that his offer is genuine and honest.

The field of operation is usually that of provincial cities well away from London and the driver will have documents showing a London telephone number which can be contacted if the firm so desires. This number may be that of a telephone kiosk or some other place where an accomplice is waiting to answer. By the time a check on non-delivery has been made the goods are beyond redemption and identification of the culprit is remote by both time and by distance. If the thief is purporting to be employed by a well-known haulier he may even take the risk of contacting a local transport clearing house to be sent where transport facilities are needed at short notice.

That any firm should fall to a fraud of this nature seems ludicrous; nevertheless it is a regular occurrence—perhaps the convenience of a prompt dispatch at reduced rates is too tempting for the operator. The easy solution is not to use 'casuals' at all unless the carrier has been previously utilised, even those sent from a clearing house should be suspect until the driver's licences and his vehicle have been thoroughly checked and examined to ensure they are in order. A further check that could be made is to establish the identity of the customer to whom he purports to have delivered in that area and check with that person whether what has been said is true.

If it is thought that this could not happen refer to the case of Garnham, Harris & Elton Limited *v* Alfred W. Ellis (Transport) Limited, [1967] 2 AER 940. This is a judgement given on the liability of a contractor who subcontracted the carriage of a valuable load of copper to what was subsequently found to be a non-existent firm. Damages were awarded against the contractor. Another large metals firm lost a complete load valued at tens of thousands of pounds in this way in July 1972.

DRIVER IMPERSONATION

This is a rare form of theft, mainly confined to the London area. Loaded vehicles have been taken from guarded car parks by thieves posing as the legitimate drivers. Obviously, either a considerable amount of planning must go into an operation of this kind or there must collusion with a driver to obtain suitable keys for the vehicle and knowledge of times and loads.

Driver identification card—This is the apparent answer. National Car Parks Limited have circulated a request that the firms using their facilities should provide their drivers with a form of identification. A suitable card should carry: a full or three-quarter face photograph of the driver, his full name and address (and employer's address), date of birth, height, weight, and hair colour. His signature and that of someone of authority in the firm should be appended and the card stamped with an official stamp. This would not preclude this type of theft but would certainly make it considerably more difficult where guards are employed on the car park.

VEHICLE OBSERVER CORPS

The menace of lorry thefts caused the Road Haulage Association to create this body in June 1962; it is now operating in several important centres in the country and is not restricted to members of the Association.

Fundamentally, the scheme is one of immediate search on notification of the theft of a lorry and load. In each locality a permanent area control is established which furnishes member firms with details of the stolen vehicle as soon as they are known. These firms in turn supply cars and crews to patrol small pre-determined areas of an extent that can be covered in a matter of ten minutes. After an initial check they report back to their own base and then make a fresh search. With the cooperation of a large number of firms all these areas, which make up an entire district can be searched simultaneously. No action is expected from the crew finding the vehicle other than to inform the police where it is and to follow it, if seen on the move. This is a form of self-help in the transport industry which has had success, particularly in London.

INTERNAL FRAUDS INVOLVING TRANSPORT

Once a vehicle has left a firm's premises carrying goods in excess of what should have been loaded, detection of the theft becomes difficult. On many occasions it will be impossible to determine whether stock deficiency is due to bad documentation or theft from inside the premises, either by intruders, or by progressive pilferage by dishonest employees. A driver and warehouseman acting in collusion represent a dangerous combination and could function for some time without detection. Any signs of undue affluence on the part of two friends in these categories should be viewed with suspicion.

There are only two physical ways of establishing whether a loaded vehicle carries what it should: by offloading and checking against the consignment notes or, where the load is composed of items of known weights, taring and grossing the vehicle over a weighbridge and comparing the net weight with the cumulative weight of the individual parts. The use of a weighbridge will have limited application with many products and especially with mixed loads but where bulk metals or raw materials are concerned it is always of value—if only because it will have a cautionary effect on those concerned.

Offloading is not a popular practice and could have repercussions in relationships if nothing were found; it should only be done where there is virtual certainty of fraud or where it is accepted that periodically such action will be taken. Spot checks can be made on odd items to establish that they are shown on the consignment note—in most cases a driver will want to get rid of unauthorised material as soon as possible and his most likely sources of disposal will be local, so this will be on or near the top of the load and readily accessible.

DOCUMENTATION

Firms will have differing administrative requirements in this field, and no universal system can be recommended. Incidents have occurred, and no doubt will in the future, where goods in excess of an order have been deliberately labelled and consigned to a dishonest customer—this has even been known where no legitimate order of any kind has been in existence. Adequate documentation and good stock records which show up deficiencies immediately will effectively limit this type of fraud.

Some goods are inevitably lost in transit sooner or later, without indication of whether it is by theft, carelessness, or by misdirection; complete documentation will be necessary to support a claim on insurance and its absence may prejudice payment. Claims for reimbursement, particularly where outside carriers with trans-shipment points are used, may prove difficult in laying liability on a particular insurance company if the documentation has not been meticulous.

Each firm will have its own system of producing interlinked orders, job cards, advice notes, consignment notes, and invoices in sequence. The opportunity of a fraud really arises at the consignment note stage. If these can be created in a warehouse without the other notes coming into existence, it might be possible to conceal the method of goods getting out since the automatic procedural action, which would normally follow outside that department would never begin. The existence of an interlinked system where warehouse staff could only dispatch in accordance with notes sent to them would eliminate this possibility.

Normal distribution of consignment notes is from a set of three—in different colours for convenience and ready identification:

1 To be signed by the driver and retained in the warehouse.
2 To be handed over with the goods to the consignee.
3 Driver's copy for signature of the consignee on accepting delivery.

In a fraudulent arrangement where the third copy goes back to the warehouse originator and the second copy is destroyed, the whole could disappear leaving no trace of what has happened.

The third copy of this set is most important in another respect: claims of non-delivery of goods to a customer are commonplace and without this third copy, signed by the recipient, there is no basis on which to contest the claim. There may be no dishonesty on the part of the customer, whose own records may be faulty, but he will be convinced he is right in the absence of proof to the contrary. This third copy should therefore go back to someone other than the person responsible for making up the load and it should be filed and retained for a definite period, adequate to preclude possibilities of any claim being made—six months is a reasonable time.

FALSE SIGNATURES

These come to light when a customer refuses to accept an invoice for goods thought to have been supplied to him. When proof, in the form of a signed consignment note, is produced, for the first time it is found either that the signature is indecipherable and bears no resemblance to that of any of his employees or the name shown does not belong to any of them.

To complicate inquiries there will have been an inevitable delay between dispatch and the realisation that something is wrong. Several alternatives exist: the driver may have forged a signature to cover his own theft or a genuine loss from his load—if he thought that might be held against him; a member of the customer's staff might be camouflaging a theft of goods inward; a total outsider may have posed as a customer's employee and deceived the driver; the goods may have been dropped with the wrong customer by genuine mistake and the latter has not brought the fact to notice.

There is a reluctance to accept that the loss is due to anything but mistake or bad records and the supplier, if the customer is a valued one, sooner or later has to come to the commercial decision and accept the loss despite annoyance and uncertainty. There will be little chance that his insurers will accept any liability in the circumstances.

If all firms would use an official receipt stamp as well as the initials or signature of the recipient, this increasingly frequent form of fraud would soon stop; unfortunately this is most unlikely and suppliers should take such precautions as they can.

The possibility of fraudulent acceptance should be among the matters impressed on drivers. A driver should be instructed to ask the name of the signer when he cannot read what has been written; he should then print the name alongside the scribble. In the event of any complaint of non-delivery, the first step should be to forward the signed note to the customer with a request for immediate comment if not in order—not to go through several stages of correspondence before this is done. Finally, the driver should be asked whether he can throw any light on the discrepancy at the earliest opportunity. A check should of course be made with other customers.

If no logical explanation, other than theft or fraud, can be established, the customer and supplier must decide whether to inform the police; in the case of substantial loss, insurers will not be sympathetic unless this is done. The best protection again lies in a commonsense, well-instructed and reliable driver, in this instance backed up by a laid-down and speedy system of dealing with alleged shortages in deliveries.

WEIGHBRIDGE

The weighbridge, like any other mechanical device, is as good as its operator and no better. This is another post where honesty and reliability are of greater value than any other traits. If the operator is tempted to defraud his employers he has ample opportunity to do so and the chance of early detection, if he is not too greedy, is small. However, modern improvements in weighbridge construction and recording have provided safeguards which make the deliberate misreading of weights almost impossible. This is a good post for a reliable disabled man.

Types

There are two types of weighbridge in common use: the 'dial recorder' and the 'steelyard'. A high percentage still rely for accuracy upon a visual reading of the operator but these are being increasingly converted to automatically printing the weights on a card; the dial recorder, by nature of its construction, can be fitted with a device for this purpose which is less subject to interference than similar means on a steelyard. On the latter, the recording is complicated by the progressive movement of the blade, and it would be possible for the weighman to stamp a card at a false reading. With the newest device associated with dial recorders it is impossible to mark the card when there is any movement of the indicator or vibration on the platform, in fact a one and a half second standstill period is required and the possibility of accidental or deliberate error is therefore removed.

Faults

It must be remembered that there are several ways in which a weighbridge can make a false recording without there being fraud. The worst offenders are weighbridges of inadequate length where double weighing has to take place. This is now technically illegal as errors of up to 200 kilograms (about 450 pounds) are common, even when both driver and weighman are expert in positioning the vehicle at the time of both taring and grossing. The absence of flat approaches to either end of a weighing platform will further complicate matters—there is also a legal requirement that these should be level but in the case of the older weighbridges this is rarely so—and this will render double-weighing even more unreliable.

On an old weighbridge with mechanism beginning to wear it is essential there should be even distribution of the weight—if excessive point loading at any part of the weighbridge exists this can lead to an excessive deflection on the recording mechanism and a false reading. If water and dirt are allowed to accumulate on the weighbridge, or a suddent shower soaks it, an error of considerable magnitude can arise unless the operator adjusts his zero accordingly. The steelyard type is particularly prone to error if the operator lacks experience and expertise—with too quick a movement of the blade, readings will inevitably be incorrect.

Frauds and precautions

There are two main legal aids to security at weighbridges: firstly, the weighman, from his position by the dial or steelyard, must be able to see both ends of the weighbridge clearly; secondly, the weighbridge office must be so constructed that the driver can see the readings being taken whilst his vehicle is being weighed, so that he can challenge them if he thinks fit. The virtue of these requirements lies in the facts that the weighman has the capability of seeing that the vehicle is correctly positioned which gives him the opportunity of questioning anything he thinks should not be on it, whilst the driver also has a potential check on the weight of his vehicle—this is of less value for there is little chance that the weighman would be involved with anyone else except in collusion with the driver.

Where stockpiling or selling large quantities of raw materials or scrap is being undertaken a fraud in weighing a loaded vehicle can lead to a loss which is difficult to detect or occasionally even to be aware of. This is a type of dishonesty which leads to over-confidence on the part of the perpetrators and will no doubt be practised regularly; the persistent drain should then eventually attract comment. One particular set of circumstances which are recurrent are those involving the purchase of scrap. Once the vehicle has loaded at the seller's premises the weight shown over a weighbridge will be that accepted for payment. If an agreement has been entered into between the buyer and the weighbridge operator it would be easy to earn quick money by under-recording the gross weight. The prevalence of this type of offence is demonstrated by the number of occasions these facts have been repeated in courts.

It is easy to establish that a load is under-weighed once suspicion has been aroused. Either the security staff or the police can stop a vehicle and send it back over the weighbridge or to another weighbridge. It is more difficult to prove that it is a deliberate act with fraudulent intent and not an accidental error that has taken place. If there are strong grounds for suspicion this is taking place it is better to enlist the services of the police and try to ascertain where the driver is dropping the excess material that he is carrying. A defence of negligence and inefficiency is a difficult one to rebut unless there is positive proof of illegal disposal of materials. It may be possible to gain additional proof by an examination of documents over a period showing regular deficiencies in the amounts that should have been carried. Where a publicly owned, as opposed to a private, weighbridge is being used an independent person from the firm should always accompany the vehicle to the weighbridge, even if this causes inconvenience and loss of time.

The opportunity that a driver himself has of perpetrating a fraud is mainly concerned with producing too high a reading for the tare weight of his vehicle. He could, of course, remove materials from a load in transit and substitute disposable matter to be jettisoned after gross weighing but before offloading. However, the procedure would be complicated, dangerous, and unlikely to be carried out.

There are a number of ways in which he can falsify his weights, examples of which are listed below:

1 Fitting a large spare petrol tank which can be filled with water, to be emptied after tare weighing and before loading.
2 Lorries carrying sideboards and tailboards having a 'bowed' tarpaulin over the empty lorry bearing an appreciable amount of water to be tipped off before loading.
3 Carrying paving stones, old spare wheels, skids, old tarpaulins, and so on, to be discarded prior to loading and reweighing.
4 Remaining in the cab or carrying dogs or family during the tare weighing, but all getting out after having loaded.
5 Where he has tared correctly and wishes to reduce his apparent gross, fractionally overlapping the weighbridge edge with a wheel. The errors that this will show may be too sizeable to avoid notice.
6 Where a driver is conversant with a short weighbridge, by carefully positioning his vehicle during double weighing he may induce a false reading which is constantly to his advantage.
7 Where a driver has discharged his load and wishes then to inflate his tare weight he can do so by filling to capacity with fuel before reweighing.

When looking for places on a vehicle where objects can be stored to increase its apparent weight, favourite places are those used for the concealment of stolen property, such as:

1 Inside the spare wheel.
2 On top of the cab.
3 Under old tarpaulins or skids.
4 Under seats (common spot for half paving stones).
5 Wedged under or along the chassis members.

It is as well for an independent person to occasionally cast a paternal eye over what is happening at the weighbridge and ensure that the weighman is doing his job as he should.

A precaution that an honest weighman can always take, with regard to the tare weight, is that of asking the driver what his vehicle is marked as weighing and comparing what he gives verbally with what is recorded upon the dial. A further check is that of noting and comparing the tare weight printed on the side of the vehicle—though this is often inaccurate. Where regular records are kept of the same vehicle coming to the premises and it suddenly sustains a noticeable variation in weight, the reason for this should be challenged. Regrettably, in one instance this was noted after a lorry had called to collect scrap for the last time before being sold—it suddenly appeared to lose precisely 10 hundredweights (508 kg) after being constant for nearly two years. This was coincident with the police making inquiries at the weighbridge at the time of taring, the unresolved suspicion now remains that a profit of half a ton a trip had been enjoyed by the driver and his friend the weighman.

It is a good idea to keep a record of complaints of deficiencies so that if the same driver's name appears at regular intervals a more intensive investigation can be carried out. With a suspect driver the object should be to catch him, rather than induce him to mend his ways temporarily, or he will remain a potential menace. If it is thought that he is carrying an object into premises on his vehicle to inflate the tare weight, observation should be kept upon him and if he is seen to throw material away prior to loading, after he is reweighed he can be stopped and he then has no defence at all to the charge of theft of the weight equivalent to what he has thrown off. Let him reload or at the best his offence will only be attempting to steal.

In conclusion, it must be remembered that wherever transport is concerned, either belonging to the firm, contractors, or carriers, this is the avenue whereby a maximum loss can be inflicted in a minimum of time and with a minimum of risk to those concerned. Precautions should be taken accordingly.

ADVICE TO DRIVERS

If valuable loads are carried and lost by hare-brained and irresponsible drivers, those who selected and instructed them are at fault and not the individual who cannot help his nature.

If the correct care has been taken in selecting and interviewing drivers, it must be assumed that they are appreciative of advice and guidance on circumstances which they may encounter. However, it must at all times be remembered that the prime duty of the driver is that of driving and it would be both unfair and unwise to lay down precisely what he should do in given circumstances. Their reactions will be as varied as their physiques and temperaments and if there is an obvious threat of serious injury by firearms no driver should be criticised if he puts personal considerations above all others.

Most drivers will appreciate that if the obvious intention is to obtain possession of a load come what may, he will be well advised to try to get away because he is likely to be subject to injury in any case. This would be particularly applicable in the rare instances where an endeavour is made to stop the vehicle by force. Ramming or threatened ramming are the obvious ways, with a further possibility of obstruction by a barricade. Most drivers will automatically think of escaping and if the opposition get hurt in the process that would be their occupational hazard. A further point would be that a badly dented and undrivable lorry is better than no lorry and load at all, interest in both might have lapsed if the vehicle had finished in a ditch or jammed into a blocking car. These are rare incidents and it is much more likely that an attempt will be made to stop by trick.

Stopping by trick

A driver cannot live in expectation of an impending attack at every routine happening. Nevertheless, there are several stereotyped approaches, of which impersonating policemen is the current favourite, particularly in the Metropolitan Police area; here lies the advantage in internal bolts. A driver cannot with impunity disregard signals from a person in police uniform unless the circumstances make it obvious that no reasonable person would have complied. The police do accept that a commercial vehicle driver can insist on staying in his cab and driving to the nearest police station for examination of his documents. Commonsense and discretion are the guiding factors on the driver's actions. Police uniforms are easily obtainable and there is a wide variety in the types of car in use; black is no longer the predominant car colour and the presence of a 'police' sign can be meaningless. However, police vehicles can be expected to be reasonably clean with absence of damage and rust, fully roadworthy, and carrying intercommunicating radio, and few forces operate foreign cars or anything above the medium price range. As for the officers themselves: black shoes are always worn, with plain white or blue shirts and black ties; untidy appearance and very long hair are rare. In case of doubt the driver should speak through the cab windows with the door kept locked.

Fake accidents are difficult to cater for—especially where they purport to involve the carrying vehicle. The locality and circumstances could decide a driver whether he has or has not been in an accident or whether a trivial accident has been deliberately contrived. A collision with a car containing several men in a quiet area is very much different to a bump in a traffic stream with a family car. In any case of doubt the driver should consider all possibilities before he leaves the safety of his cab. Where other vehicles are concerned in obviously genuine accidents, at the risk of seeming callous, a driver should again weigh the circumstances carefully before he obeys his natural instincts to stop and help; under no circumstances should he stop out of sheer curiosity. Another method of stopping a driver leading to the theft of a valuable load was recently experienced. The rear registration number plate of a trailer was unclipped while it was stationary, either in a traffic queue or on a ferry. Later, at a point in the lorry's journey suitable to their purpose, the thieves overtook the vehicle in an old van and the passenger beside the driver waved the plate out of the window at the same time pointing to the rear of the lorry. The driver recognised the registration number as his and pulled up behind the van to recover it. The man in the van's passenger seat got out, ostensibly to adjust the nearside mirror, and when the driver leaned into the van to get the plate from the other person, he was pushed from behind into the interior and overpowered without attracting any attention from passers-by.

Hitch-hikers

There is a simple remedy for the dangers that can accrue from these: simply do not carry them. This should be made a specific instruction from a firm and should be the subject of severe disciplinary measure if it is not obeyed. The danger is not always to the load: to a petty thief the contents of a driver's pockets are adequate—indeed it is a matter for speculation as to how often drivers have been threatened, blackmailed, or cajoled into parting with money and have never reported the matter for fear of ridicule. Female passengers are the greatest danger of all: the girl-tramp habituées of transport cafés can find it easy to persuade a weak driver to pull off into a lay-by where their accomplices could be lying in wait; alternatively, by virtue of their mode of life, they can offer a danger to health. Few respectable girls hitch singly and the female in distress is the most hackneyed way of inducing a stop.

Other methods

There is obviously considerable and continuous variety in the means that can be employed. Waving and flashing lights to indicate faults with the vehicle itself are common and diversionary messages have been left at cafés which are known to be regularly frequented by particular drivers, instructing them to divert to new premises. The latter should be queried if there is any doubt at all.

Checklist of instructions to drivers:

1 Do not park in quiet country lay-bys unless the presence of other similar vehicles gives a measure of protection by sheer weight of numbers.

2. Do not visit transport cafés where there are recurrent incidents or thefts.
3. Use guarded car parks, or well-lit ones when the former are not available. Under no circumstances leave vehicles in back streets near to lodgings for personal convenience.
4. When in doubt notify the police of the presence of your load.
5. Before leaving the vehicle for the night, check immobilisers, alarms, and all doors and locks, and sheet down loads tightly to conceal the contents and hinder removal; have a special arrangement in the knotting to ensure that if there is interference it will be detected on sight.
6. If obliged to stop at an accident or by signals from police officers, stay in the cab with the door bolted if you are in any doubt.
7. If your vehicle is involved in an accident you will have to use your discretion as to whether it is a deliberate or an accidental occurrence when you can safely get out of your vehicle.
8. Casual passengers and hitch-hikers must not be carried.
9. Waving and flashing lights to indicate faults with a vehicle and even diversionary messages left with cafés should be viewed with suspicion.
10. Any approaches from strangers suggesting collaboration in stealing the load should be reported to the police or employers immediately.

Insurance

Persons responsible for transport of valuable loads making overnight stops are recommended to look carefully at the small print of their insurance policies. Due to heavy losses sustained from vehicles parked unattended without satisfactory security precautions, insurance companies are insisting that the vehicles must be left in vehicle parks especially made for their security or, where one is not available, in others with at least first-class lighting. Where a security vehicle park is not used there is likely to be another requirement that the local police must be informed by the driver of the location of his vehicle and the nature of its load. Claims for reimbursement of losses are very likely to be disputed if reasonable precautions against theft had not been taken.

The degree to which, and the spheres within which, 'hired' labour is to be used should be carefully considered. Several organisations maintain pools from which their customers can draw temporary staff to substitute for permanent employees who are sick or on holiday, or to cater for a seasonal trade increase. The standards set by a company for its own drivers, and the checks it makes upon them, may not be met by these 'temporaries'. Fidelity bonds held by such organisations may not be considered by one's own insurers to be acceptable or an adequate basis for recovery in case of loss. If high-value loads are to be carried by such drivers, the insurers should be apprised and their prior approval obtained. They may well ask that a positive and satisfactory loss liability should be defined.

Hire transport is often verbally obtained without misgivings when it is known and locally based. On the vast majority of occasions this would be perfectly safe, but the precaution should be taken of ascertaining what insurance cover the carrier has. A case in point—a regular carrier for one unit of a large group solicited return loads from a second in the delivery area. The second, being of a cautious nature, and having materials averaging £1200 per ton, asked for assurance of cover; after long delay, a reluctant admission arrived that the maximum was for £100 per ton. The group's insurers would have been decidedly annoyed had they been called upon to meet the difference for a loss and may have refused to do so.

25

Traffic control on factory roads and car parks

The ability to travel in a private car and to have facilities to park safely close to his place of work is an amenity which the modern employee has come to expect. If his comings and goings are to be inconvenienced because he cannot do this it will be an adverse factor in deciding whether to take up employment with a firm or, for that matter, whether he will settle down once in employment.

In many industries a high percentage of all personnel do travel to work in their own cars; this is particularly true of the more skilled, and accordingly more valued, type of worker. It is therefore in a firm's own interests to make car parking space available and ensure that users can leave their vehicles in safety and get away later without inconvenience. This has been recognised by most large firms who provide specific marked-out car parking areas. It is important that representatives of the employees should participate in discussions concerning parking, traffic flow and general safety of personnel on factory roads or parks. This leads to goodwill and better acceptance of restrictions.

CAR PARK SECURITY

For all security purposes it is far better that employees' cars should be parked outside the factory perimeter fence so that they cannot go direct to their vehicles without passing a point of supervision. This segregation is virtually essential to ensure satisfactory control and direction of visitors. Moreover, if private cars are allowed to park inside the works and have to use the same entrances and exits as commercial vehicles there will be times of the day when searching or checking the documentation of the latter would be virtually impossible.

Internal parking

If the available space and the location of the premises are such that internal parking is inevitable, specific areas should be designated for the purpose and haphazard parking rigidly suppressed. This is not only in the interests of the motorists as production requirements demand the free movement of internal transport and this must have priority over personal convenience.

In selecting the areas their proximity to easily stolen materials should be taken into account. Carrying a large carton a few yards to a car boot is a vastly different thing from carrying it the length of a well-lit roadway in full view. There are advantages in splitting up available space into blocks and allocating them to particular departments or to different managerial grades. Where this is done, without making the point obvious, it may be possible to place those least likely to steal in the positions where there is most temptation to do so.

Where partial internal parking is a necessity the only way to do it without causing continual complaint is to make it a privilege of the higher managerial grades. The important thing is that every employee must clearly know where he can or cannot park. No motorist will ever willingly leave his vehicle any further from where he wants to be than is absolutely necessary, and he will find every imaginable reason or excuse why he should not. To be excluded from one's legitimate parking place by another, who has no right to be there, is most annoying. To prevent needless acrimony there must be some recognised disciplinary procedure against any individual who contravenes laid down instructions designed for the communal good. These normally take the form of a written warning in the first instance, followed by possibly a final warning, then exclusion of his vehicle from the car parks for a given period—if he refuses to comply with this, termination of his employment with the firm would inevitably have to be considered.

IDENTIFICATION OF VEHICLES

In a small firm this offers little problem; the owners of vehicles are known to the security staff and can be quickly traced if necessary. This is not true of the larger concerns where the sheer weight of numbers makes it impossible; moreover there is a regular turnover and interchange of vehicles.

Space is a valuable commodity and the size of car parks will rarely be so greatly in excess of needs that obstructive parking will not occur. Where all employees leave at the same time this has a limited nuisance value, but where shift work is carried on it is not hard to visualise circumstances in which a complete line of cars may be blocked in by one left hurriedly by a latecomer who has just commmenced work. If this individual cannot be immediately identified and his vehicle moved, not only will the irritation of those impeded be directed at the erring driver, the security staff and management will not escape criticism. Tannoy announcements may locate the driver, but in factory condiions this is by no means certain. It is far better to have some form of registration or identification whereby the driver can be traced to his department in the shortest possible time.

Such registration is also in the interests of the owners as a quick notification can be made in cases of damage, lights being left on, water or petrol leaking, etc. Nevertheless, reluctance to establish a system of this nature is occasionally met with from employees. It is hard to ascertain precise reasons; it would appear that the advantages to all concerned greatly outweigh the meagre inconvenience of notifying change of ownership and details so that a current card index can be kept. It may be that there is a basic thought that this is an extra piece of regimentation which should be resisted on principle.

If car parks are designated to particular departments or grades, quick identification is made easier by providing a colour sticker for the windscreen; the same colour can be used for the appropriate car park sign—this will at once show up use by unauthorised persons.

A card index system for vehicles can be very simply devised. Its sequence can be based either upon the numerals or the letters of the car's registration number; all that is required is that registration number, the name and department of the owner, and the telephone extension at which he can be reached—and the make and colour of the car could be advantageous to prevent mistakes. A simple card of this nature could be made out every time a change of vehicle occurs, or a new employee enters the firm, to ensure the index is kept up to date.

CAR PARKS

Design

It would be the exception rather than the rule for a firm to have an ideal car park design and ample available space in which to put it. It is far more likely that an irregular area will be allocated for the purpose. When this is so careful consideration must be given to making the best use of what is available.

Acceptable spacing to be allowed for each car has been found to be: 2½ yards (2.3 m) in width and 5½ yards (5 m) in length, with access lanes approximately 7 yards (6.4 m) wide.

No matter how good the plan looks on paper it is advisable to make the first markings so that they can be removed if necessary. Experience over a trial period can well lead to improvements in layout which will accommodate more cars and make entering and leaving the parking space easier.

Whether openings into the car parks should be marked specifically as exits or entrances will depend very much upon the manner of working of the firm in question. With most forms of day or shift working it is improbable that there will be incoming and outgoing traffic at the same car park at any one time. In these circumstances there is little point in designating entrances and exits as such, though it may be necessary to do so if there is any danger of collisions occurring inside the car park because of its shape. Directional arrows should then be used to indicate how the traffic should flow. When the final markings have been decided upon these should be made as permanent as possible and renovated when necessary.

If the parking area is well marked it is surprising how rarely bad parking and obstruction will occur. The offenders are apt to be persistent latecomers and it is reiterated that it is advisable that some agreed disciplinary procedure should be invoked against the regular offender. Where the parking is bad but the resulting obstruction limited a warning notice can be stuck on windscreen. This effectively delays the departure of the offender and the result of his thoughtlessness is quite clearly visible to the other users of the car park.

Illumination

This is a matter for the management since obviously considerable expense can be incurred in its provision. It is possible that the proximity of the park to the lighting of internal and external roads may give an acceptable minimum to locate vehicles and prevent accidents. From security viewpoints, quite obviously, the more illumination the better, but the responsibility for protecting the car's contents must rest with the owner and not with the employer. Notices to this effect, disclaiming responsibility, should be clearly displayed at the car parks indicating that users use them at their own risk.

Points where there should be additional lighting in the interest of safety are the car park entrances and exits. Also for safety considerations, the car parks should be split off from footpaths running alongside by a barrier, either in the form of a fence or a low obstruction, to prevent motorists from driving straight out over the footpath and onto the main road.

Supervision

Though the responsibility for good parking and the locking of cars must rest with the car owners, who in the main conform, there is nothing more annoying after a long day's work than being held up by tedious and unnecessary delays caused by obstructing vehicles. For this reason the car parks should be patrolled, soon after each shift has arrived in the works, to ensure that all obstructing vehicles are removed before any general exodus begins—this only takes a matter of minutes. Similarly, when making normal patrols an eye should be kept on car parks to observe the presence of unauthorised persons who may be intending to steal cars or property in them.

INTERNAL TRAFFIC CONTROL

Road markings and signs

All road markings and signs used inside any works area should be identical with those used outside for everyday purposes. It must be remembered that production has paramount importance. Any scheme involving control of the direction of traffic flow or parking and loading areas must be discussed fully with production personnel to ensure that there is no conflict with their needs. Before any scheme of road marking is embarked upon security, safety, and production representatives should meet to decide upon no-waiting areas, parking spaces reserved for works vehicles and equipment, where stacking may take place, the types of road markings which should be used, etc. Once decided upon these regulations should be enforced by the security staff in respect of visiting commercial vehicles as well as the firm's own vehicles.

Speed limits—In some large works, the stretches of carriageway may be such that it is necessary to set speed limits. If this is so, these should be clearly shown in the conventional way and occasional checks made, either by stopwatch or other means, to make it obvious that these limits are going to be enforced. Offending private car drivers should be banned from the works, after a preliminary warning; the firm's own internal drivers can be dealt with departmentally and visiting commercial ones reported to their employers with a request that the works rules should be obeyed during future visits.

Do not put up mandatory signs and then not enforce them. The vast majority of employees are appreciative of the measures that have been taken for their safety and convenience and lack of implementation would cause resentment and criticism. Offenders will constitute a negligible minority who receive little sympathy from their fellows.

One-way systems—The introduction of such systems should always be considered. Even though they may increase the distances vehicles have to travel their use is justified if a general improvement in the flow of traffic results; this, in turn, should lead to a higher efficiency in performance and a removal or, at least, reduction in road safety hazards.

Traffic control

It is again trite to say that the normal police signals for the control of traffic should be adhered to by security staff. These are those which are known and accepted and there is little doubt that, if requested to do so, the local police will be pleased to give guidance and instruction in their use.

White or yellow fluorescent gauntlets should be used when directing traffic under normal circumstances; in bad weather these should be reinforced by the wearing of mackintoshes of similar material. Where there is any danger to the man performing the traffic duty due to poor lighting, this must be rectified so that he stands out against his background.

Especially where the securiy officer is controlling traffic going out onto a main public road the question of insurance must be carefully looked into. Not only in the man's own interests, to ensure that he is adequately covered, but also ensure that the firm does not incur a gross liability should he give a signal which results in a major accident. One firm has been successfully sued for a very high amount indeed in respect of an incident of this nature.

Where an outgoing traffic flow crosses masses of pedestrians going to buses or other forms of transport, or walking home, serious consideration should be given to imposing a period of traffic standstill to enable them to leave in safety. This can be done by the use of stoplights or even by electrically or manually operated traffic barriers controlled by security officers. If barriers are used they must be painted so as to be clearly visible, preferably with black and white bands along their length, and carry a red disc with the word stop on both sides clearly painted on it or outlined with glass reflectors.

All employees are most appreciative of efforts to expedite their comings and goings and anything that security can do in this connection will be amply repaid by appreciation and cooperation.

Vehicular accidents

The reporting of these has been dealt with in detail in Chapter 6. It should be remembered that whatever the attitude of drivers might be at the time, subsequently insurance complications could give rise to ideas of suing the firm on whose property the accident has happened for some form of negligence. With the lapse of time, if there is no record for rebuttal, allegations which have little basis on fact find credence in legal proceedings.

A security officer on the scene of the accident should help employees in the exchange of the necessary particulars and get as many details as possible for future reference—especially if there is the slightest suspicion that negligence against the firm may be alleged or there is severe personal injury. A sketch plan and measurements are always useful; insurance particulars must always be obtained where the firm's own transport is involved or damage is caused to the firm's property.

CONTROL OF COMMERCIAL CAR PARKS

The demand for off-street parking has led to an unprecedented growth in the construction of multi-storey car parks and the utilisation of waste ground for supervised parking on prepayment.

Most have barriers and automatic ticket controls to prevent the owners being defrauded. This is not the case at many open car parks, be they operated by local authorities or by commercial interests. A vast amount of money changes hands at these but the attendant's job is one which is lowly paid and often allocated to disabled people. It seems to be accepted that his remuneration is likely to be augmented by 'perks'. Where he simply uses a ticket machine, the 'perk' takes the form of either the non-issue of a ticket, or picking up discarded tickets and re-issuing them rather than using the machine. These practices are known but tolerated by users on a policy of non-involvement and the lack of action by employers can only be explained as indifference stemming from either inflated profits or the knowledge of the paucity of the wages paid.

The only certain way to counteract fraud by the attendant is to establish some means of monitoring the vehicles that enter a park. This can be done by a drop arm barrier, electrically operated by the attendant and with pads coupled to an automatic recorder. An alternative is to dispense entirely with the services of the attendant substituting instead a self-contained unit at the point of entry with the drop arm functioning on payment of a fixed charge with an egress barrier pad operated. A yet more sophisticated unit issues a timed and dated ticket before the barrier is raised, this must be placed in the second control coupled to the egress barrier with the appropriate sum of money before that barrier is raised. Such installations which cater for a charge graded upon the length of time in the park are complicated and expensive, nevertheless they are worthy of consideration since they achieve both the saving of wages and the less determinate sum of 'perks' thereby repaying their cost within a forecastable and limited period.

Card-operated barriers are increasing in number where parking facilities for senior staff are limited. Hospitals are a particular example where each authorised user can be given an identification card to place in the mechanism in front of the barrier causing it to lift and the card is returned to him. Devices of this type save considerable controversy where precedence and dignity are symbolised by preferential parking.

PART SIX

FIRE AND ACCIDENT PREVENTION

26

Fire precautions

Fire prevention is a basic responsibility of all security staff and management, irrespective of whether permanent full-time firemen are also employed. To get this risk into perspective: losses to the country resulting from fires average at least twice the rate of crime losses; there is no doubt the former can more easily be controlled than the latter.

The seven days a week, twenty-four hours a day coverage given by patrolling security officers makes them the most likely agents for the detection and consequently the prevention and extinction of fires. This fact is recognised in medium-sized and small firms by making fire prevention a dual responsibility for the chief security officer and the tendency to do so seems likely to become universal.

LEGAL OBLIGATIONS

With increased immediate availability of units of the fire service, there is less need for firms to provide their own brigades and the trend, other than in large factories with high potential danger, is to rely on outside help rather than internal services. Despite this it is essential that all firms, be they large or small, whether they have a security force or not, should designate to a specific individual the responsibility for ensuring that legal obligations are observed, in respect of fire hazards and the safety of life from fire. These, in the main, are outlined in (*a*) the Factories Act 1961 and (*b*) the Offices, Shops and Railway Premises Act 1963. Provisions in both are similar but they apply to different types of buildings. Though these are the ones of main interest to security management, a recent Act—the Fire Precautions Act 1971—will progressively come into force and affect those with responsibilities for clubs, hotels, research and educational establishments, etc. The full range of premises covered is not at the time of writing at all clear, nor apparently are the means available in the form of trained manpower for inspection, supervision and enforcement. The main provisions of the Act will be outlined later but a check will be needed that the Secretary of State has made the required statutory instruments to bring the various sections into operation. Many of these premises will already be subject to control under the conditions of licences held in respect of them.

If there is any doubt about the effect of legislation on particular premises, immediate guidance can be obtained through the local fire authority which, under the Fire Services Act 1947, must maintain efficient arrangements for giving, on request, advice to firms in the area on fire prevention, means of escape, or restricting the spread of fires in respect of buildings and other property.

The 1961 Act applies when one or more persons are employed in manual labour in the premises. Similarly, the 1963 Act is restricted to premises 'in the case of which persons are employed to work'; where only partially used by employees, the test is whether more than a total of twenty-one hours is worked therein each week. Areas in which self-employed persons work are not covered by it.

The two Acts between them cater for most buildings in which people are employed, and the definitions of places to which they are applicable are given at considerable length. To abbreviate these—

Factory

This is any structure in which anything is manufactured, or

changed in shape or substance, and to which the factory inspector has the right of access—it can, under certain circumstances, include open-air premises.

Office premises

A building or part of a building, the sole or principal use of which is as an office or for office purposes. The Act is confined to the part used as office premises but will include portions which can be regarded as forming part of them—staircases, cloak-rooms and store-rooms.

Shop premises

This includes all shops in the everyday sense of the word. It also includes a building or part of a building which is not a shop but of which the whole or principal use is the carrying on there of retail trade or business.

Railway premises

This has certain exemptions from its definition, but these are covered largely by other definitions in the Act or by the Factories Act. The broad meaning is a building (or part of one) occupied for the purpose of the railway undertaking and situated in the immediate vicinity of the permanent way.

The most important provisions under these Acts which must be complied with follow on the next few pages.

FACTORIES ACT 1961

Section 40: Means of escape in case of fire

This section requires that no premises to which the section applies should be used as a factory unless there is in force, in respect of the premises, a certificate of the fire authority that the means of escape available in case of fire are adequate for the number of persons employed.

Section 41: Means of escape

As specified in this certificate the means of escape shall be properly maintained and kept free from obstruction. The fire authority must be given notice if it is proposed to make any material alterations to the premises or in the number of persons employed in any part of the premises which might affect the means of escape.

Section 45

This, combined with regulations made under section 50, applies section 40 to every factory:

1 Where twenty persons or more are employed.
2 Which was constructed or converted for use as a factory on or after 30 July 1937, where more than ten persons are employed above the ground floor.
3 Which was constructed or converted for use as a factory before 30 July 1937, where more than ten persons are employed in the same building above the first floor or more than 20 feet (6 m) above ground level.
4 Where explosives and highly flammable materials are stored or used.

Section 48

This requires:

1 Where any person is in the factory for purpose of employment or meals, doors shall not be locked or fastened in such a manner that they cannot easily and immediately be opened from the inside.
2 Doors opening into any staircase or corridor or any room in which more than ten persons are employed shall, with the exception of sliding doors, be constructed to open outwards.
3 All doors affording a means of exit from a factory should be constructed to open outwards.

4 Contents of any room shall be arranged and disposed so that there is a free passageway for all persons employed to a means of escape in the case of fire.
5 All exits affording a means of escape or giving access to a means of escape, *other than ordinary exits,* should be distinctly labelled with notices printed in letters of adequate size.
6 A suitable and distinctive form of fire-alarm should be installed which is audible in all parts of the building. (Section 52 requires that these shall be tested at least once every three months as well as on request of a factory inspector.)
7 Hoists, lifts, and other openings in floors must be enclosed or otherwise protected to hinder spread of fire and smoke.

Block letters of a distinctive colour at least 5 inches (127 mm) tall are recommended for fire notices.

The conventional red for warning and green for safety are the obvious colours to use—appropriate signs in plastic or aluminium can be purchased more cheaply from specialist manufacturers than by producing them internally.

Section 49

Applies to factories where more than twenty persons are employed in the same building above the first floor or more than 20 feet (6 m) above ground level or where explosives are stored or used; all such employees must be made familiar with the means of escape and the routine to be followed in case of fire.

Section 51

Requires that in every factory there is to be provided and maintained appropriate means for fire fighting, which should be so placed as to be readily available for use.

OFFICES, SHOPS AND RAILWAY PREMISES ACT, 1963

This Act regularised the legislation in connection with these types of premises which had been haphazard in contrast to that applicable to factories. In addition under the Factories Act 1961 (Extension of Section 40) Regulations 1964, section 40 is extended to any factory in or partly in a building housing any other factory (or part of a factory) or any of the office, shop and railway premises to which the Offices, Shops and Railway Premises Act of 1963 applies. In other words a certificate of the fire authority relating to means of escape is required if the total number of persons employed at any one time in the building or in the upper floors is in excess of the numbers mentioned in section 45.

Section 28

All premises must be provided with adequate means of escape, not only for employed persons but also in respect of others who may be reasonably expected to be on the premises.

Section 29: a fire certificate

This is required:

1 When more than twenty persons are employed to work at any one time.
2 When more than ten persons are employed, elsewhere than on the ground floor.
3 For any premises involved in the use of storage of inflammable materials or explosives.

In a similar manner to the provisions of the Factories Act of 1961, other sections deal with:

1 The labelling of means of escape.
2 Ensuring that all persons employed are familiar with the means of escape.
3 The opening of all escape doors in an outwards direction.
4 The arrangement of contents of rooms to allow free passage.
5 The provision of an *effective* form of alarm (not necessarily an 'audible' alarm) with a three monthly period at which it should be tested.

6 The maintenance and provision of adequate means of fighting fires.
7 Misuse or interference with fire equipment, which is an offence.
8 Notification to the local authority of any proposed material extension, alteration, or deviation in numbers employed, or materials of an explosive nature to be stored, for possible variation in the means of escape certificate.

It should be noted in both Acts that the refusal to grant a fire certificate may be the subject of an appeal to a magistrates' court. In the event of the magistrates' court finding against the fire authority there is no further right of appeal by them.

A fire certificate, once granted, must be kept on the premises to which it applies and be readily available for inspection at any time.

FIRE PRECAUTIONS ACT 1971

This is an 'enabling' Act designed to cover certain premises liable to risk which are not otherwise catered for. Its provisions may be brought into force piecemeal by the Secretary of State as and when he thinks fit to make the necessary statutory instruments. It is an understatement to say it is a complicated piece of legislation and it lacks the clarity of the other main Acts.

An explanatory guide has been published by HM Stationery Office, obviously the first of a series—*Guides to the Fire Precautions Act 1971. No. 1. Hotels and Boarding Houses.* This was necessitated by the first statutory order made—SI 1972/238 dated 21 February 1972—which made these premises subject to the provisions of the Act. The advice repeatedly given is that of consulting the fire authority for detailed guidance on all matters and especially before incurring expense in the purchase of equipment or altering premises.

A fire certificate is made necessary, subject to certain exemptions, for any premises which are being put to a 'designated use'. Broadly speaking, the designated uses involve numbers of the public being present or are circumstances of special fire risk. The classes of designated use and examples of buildings in each class are:

1 Use as, or for any purpose involving the provision of, sleeping accommodation (hotels, residential clubs, hostels, holiday camps, living accommodation at colleges and universities).
2 Use as, or as part of, an institution providing treatment or care (hospitals, nursing homes, residential clinics).
3 Use for purposes of entertainment, recreation or instruction or for purposes of a club, society or association (theatres, cinemas, billiard halls, dance halls, clubs).
4 Use for purposes of teaching, training or research (schools, universities, dancing tuition centres).
5 Use for any purpose involving access to the premises by members of the public whether on payment or otherwise (places other than buildings, e.g., circuses, beached or moored craft).

Where such premises are part of a building, any other part jointly occupied with them will be included for certificate purposes (section 1).

No certificate will be required where the provisions of other Acts apply nor for any premises used solely or mainly for public religious worship nor where they are occupied as a single private dwelling (section 2).

In certain circumstances the fire authority may serve a notice requiring a fire certificate to be obtained where part of the premises is used as a dwelling—there is a right of appeal within twenty-one days of service of the notice (sections 3 and 4).

On application containing the desired details the authority will cause an inspection to be carried out and, if satisfactory, will issue a certificate (section 5).

It shall specify the use or uses of the premises which are covered, the means of escape and the way it is ensured they can be safely and effectively used, the location, type and number of fire-fighting appliances and alarms. The certificate can impose requirements to ensure persons employed thereon receive adequate training for which records will be kept, and limit the number of persons who may be on the premises at any one time;

generally speaking fire certificates will be lodged in the premises to which they apply (section 6).

If it is proposed to make any material extensions or structural alterations to the premises, or any material alterations in their internal arrangement or their contents, or to alter the usage in so far as it is related to keeping explosive or highly flammable materials, the occupier before carrying out any of these actions shall notify the proposals to the fire authority (section 8).

Premises involving excessive risk to persons in case of fire

If the fire authority is satisfied that risk to persons in case of fire is so serious that, until steps have been taken to reduce it to a reasonable level, the use of the premises ought to be prohibited or restricted, the authority may apply to a magistrates' court for an order for that purpose (section 10).

The Secretary of State may impose requirements on the provision and maintenance of efficient means of escape, fire-fighting equipment and warning devices; internal construction and materials used; types of furniture used; training in the use of fire-fighting equipment; presence of attendants in adequate numbers in certain circumstances; the keeping of records of instruction on training given or other actions taken (section 12).

Enforcement (section 19)

Each fire authority must appoint inspectors who may do anything necessary for the purpose of carrying this Act and Regulations into effect and for that purpose have the power, at any reasonable time:

1 To enter and inspect.
2 To inquire as to ownership and responsibility.
3 To demand production of and inspect any fire certificate in force.
4 To demand such particulars and facilities necessary for him to carry out his duties from any person having responsibilities in relation to such premises.

The fire inspector shall have some duly authenticated documents to show his authority; it will be an offence to wilfully obstruct him in the exercise or performance of his duties or without reasonable excuse to comply with any requirement imposed. It is of interest that the inspector, if he discloses, other than in the performance of his duty, etc., information obtained by him in any premises, can be guilty of an offence.

Offences

Among the offences created are the forging and possessing of forged certificates, making false statements to obtain, giving false information to a fire inspector, and impersonating one.

It is a statutory defence to prove that all reasonable precautions and due diligence were exercised to avoid committing offences.

GENERAL OBSERVATIONS

These legal requirements are designed to limit loss of life, and to ensure that means are available for the extinction of fire. However, a large majority of fires are caused by the unpredictable careless acts of individuals for which no legislation can provide. It is therefore essential that precautions must be taken on an assumption that inevitably some fires will be caused by carelessness which may eventually produce a major conflagration.

The ultimate responsibility for extinguishing any blaze of consequence lies with the local fire authority and therefore the higher the potential risk the closer liaison that should be maintained between the Chief Fire Officer, the Fire Insurance Surveyor, and the Fire Prevention Officer. The latter should be encouraged to visit as regularly as possible so that the maximum advice can be obtained and the local fire brigade can be familiarised with the buildings and with the materials in use. Recommendations by the fire prevention officer as to the siting of appliances and the marking of hydrants should be followed meticulously. A plan of the building or factory area showing the location of all these and alternative sources of water supply should be kept preferably by the main entrance to assist the brigade.

There can be no single procedure recommended for action which is suitable for adoption at all premises in the event of a fire. This is inevitable owing to the divergent number of storeys, structural standards, processes, and materials, and the number of people employed, as well as possible means of escape. One thing is common to all: comprehensive instructions must be available to all employees, clearly displayed, and easily understandable; no room for doubt must be left by them as to what action should be taken in the event of fire.

IMMEDIATE ACTION ON FINDING FIRE

1 Raise an alarm and make sure that the public fire brigade is informed.
2 If there is reasonable hope of extinguishing the blaze, attack the site immediately.
3 Put into operation a pre-arranged plan for evacuation of personnel, notifying the management and the trained staff required for action in case of fire (see Appendix 15).

The magnitude of an outbreak must dictate whether attacking should take a priority over reporting—it is obviously foolish to allow a small fire to spread by spending time on reporting; it is equally foolish, in any case where there is doubt, to delay reporting whilst making an abortive attempt to put the fire out. Normally there will be other persons at hand to assist and to notify the switchboard, but a patrolling security officer at night would be well advised not to over-estimate his capabilities.

Providing there is no danger to the persons concerned, every effort should be made to contain the blaze pending the arrival of the fire brigade. All staff other than those actually engaged in fighting the fire should vacate the surrounding area. Onlookers should keep out of the way and doors and windows as far as possible should be kept closed to prevent a quick spread of the fire.

It would be unrealistic to stop all work because of a fire in one isolated department and only the alarms in those most likely to be effected should be sounded. These can take the form of:

1 Manually operated bells—these are only suitable for small single storeyed buildings with a low fire risk.
2 Electrically operated sirens, actuated either from points throughout the factory or from a switchboard.
3 Automatic electrically operated alarm systems activated by heat or smoke. Such systems would normally incorporate manually operated points as well.

Where a mass evacuation is necessary it should be done with a minimum of fuss and panic and the employees assembled at a pre-arranged point to check that all have left. A search for stragglers should be made before the building is vacated, if this can be done with limited risk to the searchers. The head of each department should be responsible for checking that all his personnel are out.

It is essential that the fire brigade should have been directed in the first instance to the entrance nearest to the fire. They should be met and the person meeting them should be fully aware of the location, the best route of approach, whereabouts of water supplies, special hazards, and any factor which might need priority action, for example, trapped persons.

REPORTING AND INVESTIGATING

All too often these are matters which are neglected, with the inevitable results that lessons are not learnt and used equipment is not replaced or refilled before it is needed again. There are really two separate groups into which industrial fires will fall for reporting purposes: the first, minor fires where damage is limited and there is no personal injury (these will include the recurrent trivial ones which may be almost inevitable, perhaps because of the nature of a manufacturing process being used);

the second, larger fires where insurance claims are likely to follow.

The first group must be reported and recorded if only to show up any pattern whereby preventive measures can be strengthened, also to ensure that extinguishers which have been used are replenished. Experience shows that at least as many extinguishers are used without notification to the fire officer as are reported to him. A simple card system could be used for this type of incident giving date, time, department, installation involved, apparatus used for extinction, nature of fire, and signature of supervisor.

With a more extensive fire, every effort should be made to establish the cause, irrespective of whether this might cause embarrassment to particular individuals or departments. This will necessitate a comprehensive report compiled as soon as possible, whilst memories are fresh, preferably on a printed form to ensure that details are not overlooked. A suggested list of items for inclusion is:

1 Time, date, general area of outbreak and person finding.
2 Time, date last seen in order, and by whom.
3 Precise location of fire.
4 Times of notification, arrival, and departure of the fire brigade.
5 Identity of officer in charge and means of contacting him.
6 Cause of fire, if given by him.
7 Senior personnel notified of outbreak and times of doing so.
8 Appliances used that need replenishing.
9 General description of fire, including structures and equipment damaged, factory services affected—electricity, gas, etc.

The foregoing details can be filled in by the senior security or fire officer attending the fire but further space should be then available for the chief fire officer to complete with:

1 Time and date of notification to fire assessor.
2 Appreciation of effect on production.
3 Comments, and observations to prevent a recurrence.
4 Possible causation.

The completed form should be circulated to the manager and heads of departments concerned. Fire assessors do not take kindly to being informed of a fire after everything has been tidied up before they have an opportunity of inspecting the site—hence inclusion of that item. (Specimen report form, Appendix 16.)

Even in the case of a minor fire, a note should be made in the security department's occurrence book (see page 56) so that security personnel subsequently coming on duty can pay especial attention in case smouldering material has evaded detection.

Where the fire is of such magnitude that the management decides that an official inquiry should be held upon it to establish the cause and to consider future prevention, and desirable variations in procedures, it is advisable that the composition of the panel should be independent of the departments concerned and the services affected—if there is blame to be ascribed it is unlikely they would be unbiased.

Arson

Arson is not a frequent offence but the possibility should be borne in mind where there is:

1 Fire at several original sites.
2 Simultaneous outbreak in separate places.
3 The presence of flammable material foreign to the area concerned.
4 The presence of burnt-out petrol tins or similar containers.
5 The smell or indication of flammable fluids which should not be there.
6 Indication that the premises have been broken into by intruders.

In these instances the police should be notified for inquiries to be made and there is no harm done if the suspicions are unfounded.

MEANS OF ESCAPE

There are precautions imposed by the legal requirements which should help in ensuring the safety of employees. First of these is the means of escape. Principles to be applied when assessing the adequacy of these are:

1. No one should have to go towards any fire in order to escape.
2. Each route should be as short as possible and of adequate capacity to allow the speedy passage of the number of persons who may have to use it.
3. Each route should lead to the open air at ground level either directly or through a fire-resisting barrier.
4. Enclosed parts of escape routes should be protected against penetration by smoke or fire.

These principles can be met by:

1. Each occupant having a choice of escape routes.
2. No occupant should have to go more than about 100 feet (30 m) to reach open air or a smoke-free fire-resistant stairway, corridor, or lobby (lifts should not be considered as escape routes).
3. All corridors, stairways, and exits should be adequately wide to allow occupants to leave quickly without panic.
4. The walls, floors, and ceilings of any passageway forming part of an escape route should have a fire resistance of at least half an hour with self-closing fire-resisting doors at every entrance to the enclosed area.
5. Exits from escape routes should be sited so that people can disperse from them in safety without difficulty and without being confined in yards near the building.
6. External escape staircases should be provided where there is no possibility of using enclosed fire-proof stairways.
7. All apertures, stairways, lift shafts, and hoists, whereby smoke and fire could spread rapidly through a building, should have been surveyed and enclosed as far as possible.
8. All escape routes should have been clearly marked and all the doors on them should open outwards to avoid any confusion.

FIRE-FIGHTING EQUIPMENT

This should be present in a variety of forms dependent upon the type of risk incurred in the particular premises. Portable equipment should be placed in a conspicuous position which is readily accessible at a properly maintained fire point. At the same point there should be a fire-alarm contact for manual operation with fire instructions clearly displayed. Whilst it is essential that it should be near to possible sources of ignition a user must have a safe route of escape. Where different types of extinguishers are sited together there should be some clear form of identification to prevent those being used to attempt to extinguish sources of fire for which they were not intended, and against which they may be dangerous to the user.

Installations using water, foam, carbon dioxide, or dry powder are those most commonly in use. All portable equipment will normally be clearly and simply labelled with instructions and details of the contents; a warning may be shown of the type of fire on which they should not be used.

Briefly, fire has three needs: fuel, air, and heat—removing any one these will put it out. Quenching can be effected by: starving—removing the fuels; smothering—removing the air; or cooling—removing the heat. Starving is rarely feasible and the other two methods are commonly used. For the purpose of establishing that the most efficient fire-fighting agents are used, risks are normally classified into three main groups:

1 Class *A* risks: carbonaceous materials like wood, textiles, paper, and things manufactured from those materials in any form.
2 Class *B* risks: flammable liquids.
3 Class *C* risks: all risks where the problem of fire fighting is complicated by the presence of live electrical equipment and the consequent danger of shock.

Class *A* risks are usually extinguished by cooling—water is the obvious choice for this and can be used in the form of a jet or spray according to the type of fire encountered. Sprays are most effective on surface fires and a jet for more deep-seated ones. A wetting agent is sometimes added to reduce the surface tension of the water in the container.

Class *B* risks are extinguished by smothering—excluding oxygen. This may be done by a variety of methods:

1 Powder—sodium bicarbonate, for example, with a free-flowing agent added.
2 An inert heavier-than-air gas to settle on the site—carbon dioxide.
3 A foam to float on the surface of the burning liquid.
4 A non-toxic vaporising liquid extinguishing agent.
5 An incombustible sheet—asbestos.

Class *C* risks are countered by smothering with a non-conductive agent; 1, 2 and 4 listed under Class *B* risks are suitable.

Equipment using water

Static equipment

1 Automatic sprinkler systems. These are brought into operation by an automatic fire detector which gives an alarm and delivers water in spray form to the seat of the fire.
2 Hose reels. These are adequate lengths of non-kinking rubber tubing wound on a reel and permanently connected to a mains water supply. (There are recommendations for the length and diameter of these tubes and for the water pressure, nozzle bore, and siting. Advice should be sought from the fire prevention officer on these.)
3 Wet or dry hydrant systems or rising mains (in buildings). These are to provide a substantial supply of water close to where it may be needed by the fire brigade. In the 'wet' systems the main is charged with water permanently up to each outlet valve; in the 'dry', connection is made by the fire brigade on arrival.
4 Special water spray systems. These are fitted for the protection of specialised plant and each system is designed to meet individual requirements.

Portable equipment

1 Extinguishers. To simply supply water in either jet or spray form by expelling water from a container by virtue of pres-

sure, either generated by gas pressure, soda acid reaction, or by pumped pressure.

2 Buckets. These should be kept filled and lids fitted to avoid wastage and contamination.

Equipment using foam

Fixed systems are usually individually designed to protect large flammable liquids such as oil storage tanks, they may be automatically or manually operated. They can be from self-contained systems in which the foam supply is part of the installation or from systems of pipework to which a fire brigade may connect foam-making equipment.

Portable equipment is of two types, both extinguishers: in one the foam is generated by chemical means, in the other by mechanical. These are specially suitable for use on small fires—deep fat frying ranges in kitchens or oil fired boilers.

Carbon dioxide equipment

Fixed equipment is individually designed and may be automatic or manually operated. There is danger to employees of suffocation from carbon dioxide if they are trapped in a confined area.

Portable equipment is for use on small fires of flammable liquids, electrical equipment which may be live and where it is necessary to avoid damage or contamination by powder.

Dry powder equipment

Fixed systems can be automatic or manually operated. The powder is blown out in a cloud from outlets in a system of piping by pressurised gas, usually carbon dioxide. Installations are individually tailored for dealing with fires in flammable liquids and electrical equipment.

Portable equipment is in the form of extinguishers. These are most useful for all-round purposes in extinguishing electrical fires and fires of flammable liquid which have spread over a large area.

Equipment using vaporising liquids

These are portable and mainly used to apply a non-conductive extinguishing agent to fires in live electrical equipment. They can be used for other types of small fires but have no advantages then over the conventional means. The toxicity of their vapours, so far as early types were concerned, restricted their application in confined spaces and indeed all CTC cylinders should now have been phased out. BCF (bromochlorodifluoromethane) is generally regarded as the form giving satisfactory efficiency coupled with acceptable toxicity but care should be exercised if used indoors. Practically all cylinders have a controllable discharge. Apart from their suitability for fires coupled with electrical danger they are valuable for dealing with small flammable liquid fires and vehicle engine fires in particular.

ESSENTIAL PRECAUTIONS

1 Water jets must not be used on electrical fires unless it is certain that the current has been cut off—otherwise the jet could conduct electricity to the person holding the hose.
2 Water jets should not be used on oil fires unless they are fitted with a foam attachment—otherwise the oil would float on the water and continue to burn.
3 CTC (carbon tetrachloride) cylinders should be treated as obsolete, they may be dangerous in confined space and care must be exercised when using any vaporising liquid extinguisher in such circumstances.
4 All extinguishers must have the last date of recharging and the last date of pressure testing recorded by fitting tags or by painting on them—an office record should also be kept of these.

With regard to the testing of extinguishers, full information about these and inspection can usually be obtained from the manufacturers and from the Fire Service Drill Book; the latter also gives instructions concerning maintenance of fire hose and hydrants.

PREVENTION OF FIRES

There is no reason why this should not commence in new buildings before a stone is laid. If the elimination of fire risk is laid down by prospective owners as a basic consideration to be taken into account by their architects and planners at the design stage, the latter will have to consult with the local fire authorities from the outset to exclude unsuitable materials and potential fire hazards. Access for mobile appliances can be assured, adequate hydrant points dispersed to give full coverage, and optimum means of escape incorporated. The siting of storage areas for high-risk materials and fluids can be discussed, as well as their isolation in the event of outbreaks. Economies in installation cost and improved appearance can be achieved by incorporating permanent installations of the sprinkler type during the actual construction work—it is considerably more expensive to add these subsequently.

Sprinkler systems

Sprinkler systems are very effective in operation and rarely subject to fault or malfunction; they immediately cover areas which may be difficult to reach with hose jets and to a degree that is not possible with portable equipment. They are efficient in extinguishing fire in all normal combustibles including those used in the construction of the building. The fear of accidental water damage, or unnecessary damage in the case of fire is grossly exaggerated—only those heads in the region of the fire operate. It is advisable to protect all the building when fitting an installation otherwise an outbreak in an unprotected part might interrupt the water supply to the other or become intense beyond extinction.

Insurers will insist on sprinklers where their potential loss is high but it should not be forgotten that reductions in premiums ranging from 50 to 70 per cent may be made where an installation is put in, also, that automatic sprinklers are classified as plant and may qualify for allowances against taxable profit. To obtain premium discounts, the systems must conform to rules made by insurance companies which govern density of 'heads' type of water supply, rate of discharge, etc., it follows that no action should be taken without prior consultation with the insurers.

Automatic detection devices

Damage will have been done by the time sprinklers operate and there is good reason to incorporate electronic warning equipment which is much more sensitive and of course can be fitted in premises by itself. Where the main purpose of having security staff on premises is the fire risk, the job might more efficiently be done by a warning system with or without sprinklers; the cost will soon be written off against that of manpower.

Operation is similar to that of burglar alarms; heat or smoke sensitive detectors are incorporated in an electric circuit with audible and visual warning devices. It is important that the circuit is 'closed' so that it is broken when a detector operates—this enables it to fail safe and operate the alarm if a wire is cut or a detector broken.

Planning

Fortunately, legislation now ensures that the fire authority has ample opportunity to inspect new planning and to make recommendations—if these are disregarded the granting of a certificate would be justifiably withheld.

In connection with materials and construction, this is a specialist subject and the Fire Protection Association has produced a large number of technical information sheets which are invaluable; a full list of all publications can be obtained on application to the Association's office at Aldermary House, Queen Street, London EC4, and will be found to give expert detailed guidance on every aspect of fire prevention.

It is at this planning stage that the fire and security officers should get together to ensure there is no clash of interest—it is the former's responsibility to enable people to get out of buildings in emergency, the latter's to prevent other unauthorised persons getting in. A satisfactory compromise is not impossible provided discussions start early enough.

With regard to the equipment required to be installed and its siting, again the fire authority will make detailed recommendations which should be followed—the only one which may be in dispute is that of sprinkler systems which are an expensive item. In the event of opposing opinions on the necessity of the installation being irreconcilable, the matter may have to be resolved before a court if the certificate is withheld.

For the maintenance of equipment, contracts with manufacturers offer the best guarantee of skilled checking but security staff, in addition to receiving instruction in the use, should have at least an idea of the rudiments of servicing and refilling. One precaution which must be taken in respect of water-filled extinguishers is that of protection from frost for obvious reasons. The necessity of replacing expended extinguishers must be re-emphasised; any user is likely to put the empty one back from where it was taken and if the reporting system has not functioned properly this may escape notice; if this is a recurrent feature the solution will be to devise a system of storage or suspension to which the extinguisher cannot be returned after withdrawal. This again can be expensive and it is far better to educate staff to a consciousness of fire prevention's importance, so that they will meticulously follow a laid-down system to ensure that all equipment at their disposal is always serviceable.

When all the precautions of a permanent nature by way of construction, design, and provision of equipment have been met, there still remain the dangers which accrue from human fallibility and unpredictable breakdowns in services. These can only be countered by: enforcing good housekeeping standards to prevent the accumulation of flammable material to form fire hazards; instituting a routine of checking when premises are vacated and systematic patrolling thereafter; and, perhaps most important, the instruction of all new employees or transferees into different departments in:

1 The fire alarm signal.
2 The whereabouts of the means of sounding the alarm and how to do so.
3 The whereabouts of the equipment with which to put out a fire and how to use it.

4 The means of escape from where they work and where to go if they have to get out.

This form of instruction is most important where industrial processes are in use where, by their very nature, there is inevitable risk.

FIRE DRILLS

The question of fire drills is a difficult one. The object is praiseworthy and advocated by all responsible bodies; it ensures that everyone knows: how to leave in an orderly fashion without fuss and commotion which could engender panic; the alternative means of escape in the event of the obvious being blocked; and where to go to assemble for a roll call to check that no one has been trapped in. But, and this is a major 'but', in many large industrial concerns such a drill would entail the closing down of machinery and processes which would have an effect on production perhaps equivalent in financial loss to that caused by a sizeable fire . . . the views of management in such instances are obvious! Where this applies efforts should be made to carry out drills piecemeal by departments; the instructions on what should be done in emergency should be reiterated at every opportunity and the printed directions prominently displayed (see Appendix 15).

A form of 'dry' or 'static' fire drill has been evolved and practised with success by a number of firms to whom the holding of a full-scale exercise would prove prohibitively expensive. This goes through the complete sequence which would be followed in the event of a genuine outbreak, but without the movement of personnel from their working positions—with the exception of designated fire wardens (or supervisors or marshals, whichever terminology is used). A self-explanatory documentation for such an exercise is shown in Appendix 17. The cooperation of the local fire brigade can be solicited, and will no doubt be readily given, to furnish an air of reality and also provide opportunities to form an appreciation of delays and difficulties which may arise.

Fire wardens

An adequate number of willing and responsible employees should be designated and trained as wardens so that there is representation in every department. The names of these should be shown on fire notices so that everyone is aware of their identity, can approach them with queries, and will respect their orders in the event of an emergency. Notices marked 'fire warden' are recommended outside applicable offices.

Written instructions should be held by the wardens showing precisely their duties and areas of responsibility. Dry drill is ideal for ensuring that these have been given and comprehended. The wardens should endeavour to keep an up-to-date list of the employees for whom they are responsible.

Checks should be made at frequent intervals on the movements of personnel so that any necessary replacement wardens can be appointed and trained.

CAUSES OF OUTBREAKS

Knowing the main causes of fires is the first step in knowing how to prevent them and what to look for if there is suspicion that a fire is smouldering in a given area. No effort has been made in the list hereunder to place these in priority, obviously they will vary from industry to industry.

Electrical faults

These can be of a variety of causes:

1 Overloading of circuits beyond capacity.
2 Short circuits due to wear or damage to insulation.
3 Careless maintenance.
4 Electrical equipment overheated due to ventilation failure or overloaded mechanically.

Heating appliances

Portable heaters, gas and electric fires, stoves, open fires, oil burners, steam and hot water pipes, etc. All harmless if correctly installed and maintained but dangerous by proximity to flammable material, or by generation of sparks or breakdown. Fixed fires and stoves should be on non-flammable bases.

Process dangers

Usually accidental or due to carelessness in operation as safety from fire risk is taken into account in designing any process. Typical dangers:

1 Accidental ignition of liquids or flammable gases (including spillings of liquids).
2 Accidental overheating of substances under processing.
3 Flame failure in heating equipment causing explosion and fire.
4 Frictional heat and emission of sparks.
5 Breakdown in ventilation or cooling devices.
6 Chemical reaction getting out of hand causing explosion.

Static electricity

Particularly dangerous where solvent vapours are in use, all fixed machinery has to be bonded to earth and increased humidity will reduce the chance of an unexpected spark. Instruments are made to detect and measure the presence of static charges.

Flammable dusts

These are both an immediate explosive risk and a long-term one. In their finely divided form, and adequate concentration, certain metallic and carbonaceous dusts can form what is in effect an explosive cloud. The remedy is regular overhaul of dust extraction installations. Deposits of dust in roof voids can smoulder unobserved for lengthy periods before breaking into flame. In an enclosed area this can lead to a progressive build-up of heat which will cause a sudden flare-up over the entire area.

Spontaneous combustion

Certain substances, by decomposition or chemical reaction, gradually heat up to ignition point without any outside agent other than the presence of oxygen. Oily rags, sacks, and oil seeds are examples. These should be cleared regularly and, if they have to be retained, kept in fire-proof containers.

Rank carelessness

Smoking heads the list here. Works regulations should be strictly enforced in prohibited areas. Wastepaper baskets are the most prolific source of fire in offices. Repair work with blow lamps and welding equipment can also leave smouldering material if care is not taken. Means of extinction should be to hand while the work is in progress and security staff should always be notified for special attention to be paid subsequently.

FIRE PATROLS

A fire during working hours is unlikely to escape detection for more than a limited time and consequently has restricted opportunity to do real damage, as opposed to those which break out in the absence of the main body of staff. Even those in daytime are concentrated in the areas least frequently visited. The moral is obvious: fire patrols must be worked to a specific system which ensures that all parts are visited at regular intervals. This does not conflict with the principles of security patrolling and there is no reason why the two functions should not be combined with correctly trained personnel.

Again, the practice of employing aged watchmen is to be deplored; to be of any value at all for fire prevention, the individual used must be in full possession of all his faculties and of sufficient intelligence to use his own initiative in deciding whether immediate action by him can prevent an outbreak developing or whether he should seek outside help. An incompetent man is more dangerous than no man at all and by his own carelessness he might even start fires.

Familiarity with the premises and with the processes carried on therein are essential assets of the fire patrol. For this reason, staff specially employed by the firm are preferable to engaging the services of an outside professional agency; these do compile a very detailed list of instructions for their employees to follow but there are unavoidable changes in those detailed to supervise various premises which must reduce the efficiency of their patrols. Nevertheless, where a firm is small, it may well be economic to engage the services of an agency to give periodic visits; the cost of these will vary appreciably with the area and the commitments of the particular agency in the near vicinity—competitive quotes should be obtained. It should be possible for several firms to reach agreement to jointly employ staff to patrol their combined premises—this would certainly reduce costs and improve the quality of the patrols—but agreement seems rarely to be reached in this manner.

During daytime the necessity for patrols should be a strictly limited one, other than to those areas where either there is a regular incidence of fires or where few people are employed. After working hours, however, a full inspection of the whole premises should be made as soon as possible. The object of this is not just to detect any incipient fires but also to eliminate factors which may lead to an outbreak. A full inspection should imply actually entering every room—a good sense of smell is as valuable as any in the detection of smouldering material. Clocking points can be incorporated and in cases of high risk, an insurance firm may insist on this, though a degree of flexibility should be insisted on in the interests of security. Where they are used, they should be sited at the far extremity of any area to be visited to make certain the whole is seen. Too often the point is placed beside the entrance door and the lazy or hurried patrol may go no further in.

A form of checklist to be borne in mind by a fire patrol should include :

1 All electric fires or heaters left on, other than those essentially needed, should be switched off.

2 Gas and electric cooking facilities should be checked off.

3 Plant running but not in use should be switched off, and checked if cooling down.

4 Heaters, obstructed by overalls or other flammable materials left on them should be cleared, missing fire-guards should be replaced.

5 Doors and windows, internal and external, should be closed —the external locked against intruders and the internal closed to prevent possible fire spread, and also to give indication of the presence of intruders if subsequently found open.

6 Flammable materials left near any source of heat should be moved to a safe position.

7 Any leakage of oil or other possibly flammable liquid should be investigated immediately.

8 Check that all fire-fighting equipment is present, serviceable, and unobstructed; that access is available to all hydrants; and that fire alarm points are intact.

9 Smouldering fires due to electrical shorts or cigarette ends are not existent.

10 Any naked flames should be extinguished and soldering irons checked that they are disconnected.

11 Recognised avenues of access for the fire brigade must be unimpeded.

12 Sprinkler heads are not obstructed by piled goods.

Inspection during a patrol should not be confined to floor level —ceilings and roofs are equally important as much electrical cabling and ventilation conduits are sited in the roof and the first indication of trouble can be when fire breaks through the exterior of the roof.

Where automatic processes continue to function during non-working hours, or heating devices have to be switched on at given times, details of dial readings that have to be checked and other actions that have to be carried out should be in printed form to prevent mistakes. Though not precisely a fire problem, where a computer installation has to be visited the patrol should be conversant with the room temperature that is specified—overheating, without any suggestion of fire, can cause massive damage to tapes if the ventilation and air temperature control fails.

Finally the exterior of premises should not be neglected, particularly where timber for case-making is stacked or tarpaulins or sacks are stored or waste material has accumulated. In cold weather there is a temptation for workers to use open braziers for warmth; these are doubly dangerous if left burning when work has finished—a sudden gust of wind can generate a swirl of sparks. There is the ever-present danger from irresponsible children or vandals and perimeter fencing inspection has a fire as well as security purpose. Flammable goods should be stacked a safe distance from fences to prevent accidental or deliberate fires being started by outsiders.

It is a waste of time simply correcting risks which result from carelessness; a report should be made of each of these so that they can be drawn to the notice of those responsible to prevent recurrence.

GENERAL FIRE PRECAUTIONS

In conclusion a checklist of general precautions, other than concerning equipment and patrolling, and perhaps under a form of good housekeeping heading:

1. Clearly demarcate areas where smoking is not permitted.
2. Provide adequate non-combustible ash trays where smoking is permitted.
3. Enforce a period, say half an hour, of non-smoking before the end of each working day everywhere in buildings.
4. Where cleaners are employed in offices each evening, make the emptying of waste-paper baskets a priority. Use metal bins instead of the conventional baskets.
5. Provide metal lockers for clothing and overalls to counteract the cigarette or pipe left in a pocket.
6. Have a regular and frequent system of floor sweeping and waste collection and removal.
7. Store collected waste at a safe distance from buildings and either dispose of it in an incinerator or have it removed by outside contractors.
8. Oily rags and items that might be the subject of spontaneous combustion should be collected separately.

9 Try to convince the head of each department that he should be the last to leave and to give a quick look round to see that all is in order before he does.

It might well be thought that the most difficult matter has been left to the last item of these precautions and indeed, this is so. The whole essence of fire prevention is that of educating staff at all levels to the understanding that fire can kill, can deprive of livelihood, and, whilst it is a good servant, it is an unreliable and treacherous one which needs constant supervision. Once this is instilled in everyone its danger is reduced to a minimum.

27

Accident prevention

It is not unusual for the chief security officer of a small or even a medium-sized factory to combine a safety function with that of security, but for rank and file members of his staff it is a purely ancillary matter in which the twenty-four hour daily cover that they give provides a potential field of value to their employers.

Normal risks and incidents will be dealt with by the safety officer, it will only be those rising outside normal working hours which should receive attention as soon as possible to prevent injury or damage which really affect the security officer. There must be no suggestion of interference in this field; it is the responsibility of specially trained staff—security's role is a purely helpful one, consistent with the overall mandate to protect the interest of employers in every possible sphere. But to keep this in true perspective, it should be remembered that an estimate of man-days lost annually in factory accidents alone exceeds 20 000 000 with a corresponding overall cost from all industrial accidents approximating £300 000 000.

HOW THE SECURITY OFFICER CAN HELP

It would be unrealistic to suggest that security staff should be given special training in this subject except in very exceptional circumstances where recurrent risks may arise. Nevertheless, there are a number of things which may be temporarily overlooked by personnel intent on production where commonsense observation by a patrolling officer may be utilised to advantage. These, when observed, should be drawn to the attention of the supervisor within whose jurisdiction they occur; where repairs are necessary, it is his responsibility to requisition the work and then ensure that it is done as expeditiously as possible.

If a security officer sees what he considers to be unsafe practices, he should draw them to the attention of the supervisor for his action—he should not approach the worker. It might be that a procedure that appears unsafe to the uninitiated is one which is acceptable. In all cases, a note should be left for the safety officer and a record made in the occurrence book for the information of other staff, to prevent duplication of reports and to ensure that attention is in fact given. No doubt items will be reported that are already the subject of action, or are known and catered for, but the residual balance will justify the practice.

Items which should attract the attention of a patrolling security officer are perhaps best given in the form of a checklist.

1. Patches of oil or grease which constitute a danger on roads or footways.
2. Protruding slabs or cavities in footways, broken tiles or holes in tiled floors, loose floorboards.
3. Defective lighting over staircases or any place where people may have access after dark.
4. Defective stair-treads or where the stair edges are bady worn or broken.
5. Broken or defective handrails.
6. Damaged ladders.
7. Obstruction of gangways, fire points, and exits.
8. Dangerous stacking of materials.
9. Leaking valves, joints, etc.
10. Repeated dangerous parking of vehicles.
11. Reckless or dangerous driving of vehicles.

12 Unauthorised riding on fork-lift trucks, etc.
13 Failure to use protective equipment in dangerous areas, where specific requirements are laid down in factory rules.
14 Leaving unattended loads suspended on overhead cranes.
15 Horse-play by employees, anywhere in the premises but especially amongst machinery.
16 Deliberate interference with anything provided for first aid, welfare, fire or safety purposes.
17 Unauthorised personnel interfering with electric services and switch-gear.

The foregoing is really only a sample list, it should be reiterated that anything which might adversely affect the well-being of employees should be commented on, if observed by the security officer in the course of his patrol. Again it is not a matter where he himself should take immediate action, but rather that he should draw the attention to those who have a direct responsibility.

LABORATORY TESTS AND EXPERIMENTS

In many industries, and frequently under laboratory testing conditions, machinery or equipment is left running and unattended. This places the patrolling security officer in something of a quandary if he has received no prior notification. If the running is unintentional he has a potential safety/fire hazard coupled with power and other wastage. If intentional and there is a breakdown he would be bereft of instructions; if he uses his discretion and switches off that which should continue to operate he could cause delay, spoil a long-standing experiment or interfere with the production programme.

All such usages, of course, should be notified and logged but if this is asking too much the use of a standard informatory notice attached to the equipment or machinery should be introduced. A recommended printed form which covers all the contingencies is shown in Figure 26:1.

FRONT

Operating overnight
and
outside normal working hours

Apparatus

Running conditions

Instructions and/or special hazards

Action to be taken in event of supply failure

Water

Gas cylinder

Electricity

Air

Contact (in emergency only)

Name | Telephone | Date

Address | Signed

BACK

Instructions

1 Essential instructions must be clearly written, in ink, on the reverse side of this card

2 This card must be displayed in a prominent position on or near the equipment left running

3 The location of the equipment left running must be entered in the Safe Running Permit book held by the Laboratory Safety Officer. One entry in this book will cover a maximum period of one month

4 Equipment found running which is not recorded will be promptly shut down

5 Please remove this card when apparatus is not running

Fig. 26: 1.

SAFETY REGULATIONS

Regulations to enforce safety precautions are made by law, under the Factories Act 1961 and the Offices, Shops and Railway Premises Act 1963; if these are not complied with a prosecution by the Factory Inspectorate may follow. However, an employer may be subjected to a common law claim by an employee if he is neglectful of the employee's interest in such a way that injury results. Precedents under common law establish that an employer is deemed to have a legal obligation to provide a safe workplace, safe plant and appliances, a safe system of working, proper job instructions, and competent workers, in order that all his employees may carry out their work in safety.

A legal duty is also laid on employees by section 143 of the Factories Act 1961 which makes it an offence for any person employed in the factory to wilfully interfere with or misuse any means provided for securing the health, safety, and welfare of employees. The section also goes on to say that no person employed in a factory shall wilfully or without reasonable cause do anything which may likely endanger himself or others. The security officer, acting as always in his employer's interests, should do what he can to prevent his employer incurring liability of any kind for injuries or laying himself open to prosecution for some breach of regulations.

POWERS OF FACTORY INSPECTORS

HM inspectors of factories have exceedingly wide powers of entry and action in connection with premises for which they have legal responsibilties. There is no obligation upon them to notify the management of an intention to visit, and in these circumstances it is incumbent upon security officers to know that they cannot obstruct the movements of a factory inspector. This was recently emphasised when two factory inspectors visited garage premises; the proprietor's son called employees and closed and locked the doors, for which, in the course of a subsequent prosecution, he was fined. No routine warning of visit to the garage had been given so the defendant claimed that the inspector was trespassing and that the warrant, signed by a minister no longer in office, was out of date. Neither point was valid.

The main powers of inspectors are given by the Factories Act 1961, section 146. For the purposes of enforcement of the Act, a factory inspector shall have power to do all or any of the following things:

1 To enter, inspect, and examine, at all reasonable times by day or night, a factory and every part thereof if he has reasonable cause to believe that any person is employed therein. Also to enter by day any place he has reasonable cause to believe to be a factory in which he has reasonable cause to believe that explosives or highly inflammable materials are stored or used.
2 He can take with him a police constable if he has reasonable cause to believe he may be seriously obstructed in the execution of his task.
3 He can call for the regular production of registers, certificates, notices, and documents kept in pursuance of the Act, inspect, examine, and copy them.
4 He can make such inquiries as are necessary to ensure that the provisions of this Act and others concerning public health are complied with, in connection with the employment of any persons in the factory.
5 He can require any person he finds in a factory to give such information as it is in his power to give as to who is the occupier.
6 He can question anyone he finds in the factory or whom he reasonably believes to have been employed in the factory within the preceding two months and require the person examined to sign a declaration of the truth of the matters about which he has been questioned (no one is required to answer questions which are incriminating against himself).
7 If the inspector is a registered medical practitioner, he can carry out such medical examinations as may be necessary under the Act.
8 In addition to the foregoing, he is given adequate power to exercise such other steps as may be necessary for carrying the Act into effect.

An onus is also laid by the same section of the Act upon the occupier, his agents, and servants; they must furnish the means requested by the inspector for entry, inspection, examination, inquiry and the taking of samples or any other exercise of his powers under the Act.

Obstructing an inspector

The definition of this covers: wilfully delaying, failing to produce documents or registers, withholding any information as to who is the occupier of the factory, and concealing and preventing (or attempting to conceal and prevent) a person from appearing before or being examined by the inspector. The offence to obstruct an inspector in the execution of his powers of duty is created and also the occupier of the factory is made guilty of an offence where an inspector has been so obstructed.

Certificate of appointment

A point that should be known to all security officers: an inspector will be furnished with a certificate of his appointment; when visiting a factory or any place to which the provisions of the Act apply, he must produce this certificate on request from the occupier or other person holding a responsible position in management of the factory. A security officer is in such a position so he can request the production of this certificate before he permits entry. He must not impede entry once he is satisfied that the certificate is genuine.

Generally speaking, factory inspectors will operate continually in a particular area, but there is no reason why they should not operate elsewhere at the request of the Superintending Inspectors of adjacent areas. For this reason, security officers may encounter unfamiliar faces acting in this role. It is somewhat unusual for inspectors to visit without prior notification, and when this is done outside normal working hours the chief safety officer could well wish to be notified by the security staff.

DEFINITIONS OF ACCIDENTS

It is always useful to have knowledge of the terms used in

common parlance concerning any subject. In safety for the purpose of recording, there are degrees of severity of accidents, these are as follows:

No lost time or slight accidents

These involve no loss of time beyond the shift on which the accident occurs.

Lost time accidents

These involve losses of working time *beyond* the shift on which they occur. They are further classified into 'moderate' and 'severe' according to the degree of severity.

Non-reportable or moderate accidents—These involve not more than three days loss of time beyond the shift on which they happen.

Reportable or severe accidents—These involve more than three days loss of time beyond the shift on which they happen. All 'severe accidents' are also known as 'reportable' or 'notifiable' accidents. This means that they have by law to be reported to the factory inspector.

Frequency rate

This is a figure which is calculated from a formula to show the number of accidents occurring during each 100 000 man-hours work. The formula is simply the number of accidents a month (or a year) divided by the hours worked in the equivalent month (or year) and multiplied by 100 000. Having records of frequency rates enables the comparison of accident records between similar factories, departments, or indeed industries. If prepared in a graph form of frequency rate against monthly periods they will show tendencies to improve or worsen which in turn may reveal the adverse or beneficial safety values of new techniques.

Accurate and comprehensive recording of accidents is essential if they are to be reduced. A statistical examination of causation factors in a particular department, or during a particular operation, can lead to the elimination of a hazard which it might not be possible to visualise at first glance.

ACTION BY SECURITY STAFF AT ACCIDENTS

Security attendance at the scene of an accident will almost invariably be to render first aid. Whilst this is the first essential, the cause of the accident should be ascertained from the injured person, if possible, whilst treatment is going on. The knowledge of this is desirable from a medical point of view, and it has two further advantages; first, it may prevent an immediate recurrence to someone else, second, it might obviate a subsequent unfounded claim for compensation on the grounds of negligence by the employer.

The practice of many firms in having a form which must be completed when treatment is given, or when an accident, not apparently requiring treatment, is reported, is worthy of being copied. This enables information to be passed to the safety section immediately, facilitates their records and, if the form is correctly made out, gives a basis for action to remedy possible defects immediately. It also compels the supervisory staff to take cognisance of the accident and consider preventive action. A sample form is shown in Appendix 18.

Security officers should familiarise themselves with the manner of reporting and recording in use in the relevant department of their particular firm as part of their general knowledge of works techniques which they themselves might have to use. They should also familiarise themselves with regulations which have a safety application. One of their duties is reporting contraventions of the regulations; in doing so they are not only upholding the authority of the management but are doing a service to the individual concerned.

FIRST AID

Even if management have not incorporated first aid as a part of their security officers' job description, everyone who undertakes that type of employment should consider he has an obligation to acquire proficiency in it. The wearer of a uniform which is associated with both authority and the giving of assistance will be expected to act with those characteristics in an emergency and injury is an emergency. The efficient handling of casualties can lead to improved goodwill and cooperation from the general workforce. Superficial knowledge can be dangerous; so far as legislation is concerned, and for all practical purposes, anyone who is termed a person with first aid responsibilities must be a holder of a current certificate from a recognised organisation.

LEGAL REQUIREMENTS FOR FIRST AID

Both the Factories Act 1961, section 61, and the Offices, Shops and Railway Premises Act 1963, section 24, lay down similar requirements which must be satisfied by employers.

First aid boxes

Adequate first aid facilities in the form of first aid boxes must be provided on the scale of one for every 150 persons who are employed at any one time and an additional box or cupboard for every additional 150 persons working or a fraction of 150. This means that if 160 persons are employed on a shift, two first aid boxes must be available. Where there is a surgery or first aid room where people can be treated immediately the requirements with regard to first aid boxes may be waived and the factory be given a certificate of exemption by the factory inspector.

Minimum contents for a first aid box, where more than fifty persons are employed at any one time are laid down by the First Aid Boxes in Factories Order 1959; these include small, medium and large sterilized dressings, adhesive wool dressings, triangular bandages, eye pads, rubber bandages or pressure bandages, adhesive plaster, cotton wool, etc. These are minimum contents, the mandatory contents should be studied and added to as advised by a firm's medical officer, having regard to the special requirements of the industry or offices.

First aiders

On all occasions where more than fifty persons are employed in a factory, the first aid box or cupboard must be under the charge of a responsible person trained in giving first aid treatment. This person must always be readily available during working hours and a notice must be displayed for the information of employees, quoting the name of the trained person or persons available. In the case of fuel storage premises in the open, with a greater element of danger, a notice bearing the same information must be given to each employee. Where office buildings form part of a factory or an engineering site of any kind they will be considered as part of a factory for numerical purposes in respect of requirements.

It is the practice in many factories to nominate shop floor or clerical employees as first aiders. Indeed many firms have built up highly competent and competitive first aid teams which have entered national competitions. Where shift work is a regular feature ensuring that an adequate number of trained employees are always available can cause administrative headaches. If it is a requirement that all security staff are trained to acceptable standards they can take over the responsibility and their duties make them always available. A security department which is always manned and has telephonic communication with the outside will always be a place to which persons who are hurt or in need of assistance will gravitate. For this reason alone there should be a fully equipped first aid box in the security department and staff trained to use its contents.

Training

First aid training is done by three recognised societies: the St John Ambulance, the St Andrew's Ambulance Association, and the British Red Cross Society; all three societies now use the same *First Aid Manual*. Certificates granted by them are for a duration of three years, after which they lapse if the holder does not undergo a course of revisionary training and a further examination. Where security personnel are concerned they are likely to be more regular practitioners than the average first aider and a revisionary course every two years should be regarded as a minimum requirement. First aid is a practical subject and it requires practical training and instruction—not just the reading of the *First Aid Manual*. It must be borne in mind that the first aider, no matter how well he is qualified, must never take upon himself the responsibilities or duties of a doctor.

The First Aid Manual

This is issued by the St John Ambulance Association and is regularly revised and very comprehensive. Slight variations in techniques are progressively introduced with experience and an up-to-date copy should always be held available for reference by the security staff. In view of the availability of such an authority there is no object in giving other than essential points for the treatment of injured employees.

FUNDAMENTAL PRINCIPLES OF FIRST AID

1 *Prevent further injury*. This can be done by either removing the source of danger, or removing the casualty, for example:
(*a*) Switching off electrical current.
(*b*) Stopping machinery in motion.
(*c*) Moving casualty away from fire or falling debris.

2 *Restrict movement*. Do not move a badly injured person unless it is absolutely necessary to prevent further injury to him. If he has internal injuries or a broken spine unnecessary movement could cripple him permanently or kill him; do not move him before the arrival of fully trained and fully equipped personnel, nor allow anyone else to do so. Do not give him anything to drink if there is any suspicion of internal injuries, even if he is fully conscious.

3 *Artificial respiration*. If there are no signs that the casualty is breathing but if there is any possibility at all that he is alive artificial respiration must be attempted. Mouth to mouth respiration is now accepted as the easiest and most effective

method to apply. Unless given artificial respiration a person who has stopped breathing may die within three minutes of doing so.

4 *Stop severe bleeding.* If the rate of loss of blood is obviously causing danger to life apply a clean dressing pad or anything sterile and press firmly on the wound. Do not waste time cutting clothing unless this is essential and, in the absence of sterilised materials, with blood gushing, stop the bleeding with anything which is available.

5 *Send for assistance.* Get assistance by any means possible which does not involve your having to leave the injured person to his detriment; get a reliable person to telephone for help. The message he sends should mention the precise position, some idea of the extent of the injuries, and an indication as to what has happened.

6 *Alleviate pain.* With the minimum of movement manipulate the casualty into a comfortable resting position (see paragraph 2 above). Talk quietly to him in a confident and reassuring manner until skilled help arrives. Warn onlookers to keep well away, leaving a passage for a doctor or ambulance. Do not discuss the accident with the injured person other than to ascertain how it happened, and do not discuss it with other people in his hearing.

7 *Minimise effects of shock.* If there is to be appreciable delay in the arrival of skilled help, do not forget that shock can have a serious delayed effect. This can be recognised in a casualty by a pale, cold, clammy skin, and with a rapid pulse and breathing rate. The treatment: keep the casualty warm but without overheating. The object is to conserve his body heat. Loosen his clothing and reassure him.

8 *Assist skilled help.* Give those taking the casualty away as much information as you can about how he received his injuries. If he is unconscious tell them anything of value he may have said before he lost consciousness.

Movement of a severely injured person

Unless movement is necessary to prevent further injuries do not change the position of the casualty until the nature of his injuries are known. If he has to be moved do so by moving his body lengthwise, preferably by supporting and pulling from the shoulders—not sideways. Slip a blanket or other material under him so that he can ride upon it without direct force being applied to his body. If he has to be lifted do not 'jack-knife' him by lifting heels and head only. Obtain assistance and support each part of his body by persons kneeling alternately at either side and lifting his body in a straight line. If he then has to be transported, for any reason at all, endeavour to do so on as rigid a stretcher as can be improvised, fastening him to it. Several forms of improvisation are laid down in the St John Ambulance *First Aid Manual.*

CRIMINAL INJURIES

In most cases the facts will be obvious but if there is an injury for which there is no apparent logical explanation, the possibility of it being the result of violence from another should not be overlooked. Facial or head injuries may leave the sufferer unable to tell how he got them so look around for the presence of a possible weapon which has been used. A mental note of persons in the vicinity who may have seen something should be made. In all instances of serious injury, whether criminal or accidental, it is advisable to seal off the immediate area to prevent interference by unauthorised persons pending examination by those appropriately qualified.

28

Emergency plans

In many commercial and industrial concerns the risk of a major emergency is so remote it will be difficult to convince management of the advisability of formulating a plan in readiness for such eventuality. On the other hand there are processes which have recognised risks of disaster the effects of which could very likely extend beyond the perimeter of the premises concerned. Industries involving explosives, chemicals and high concentrations of petrol or fuel oil spring immediately to mind and where it is to be expected special precautions and plans to deal with emergencies will have been drawn up.

In this connection we have been given permission by the Chemical Industries Association Limited to quote extracts from their publication *Major Emergencies* and these will follow.

The possibility anywhere of a serious fire is never absent but there are other catastrophes which can occur such as serious flooding, electric storms, gas explosions and thought must be given to what must be done to make the task of the emergency services easier and more effective in those circumstances.

These can happen in commercial and industrial concerns of any size and criticism can be expected from official sources if the consequences of one were more disastrous than would have been if it had been prepared for.

It is essential that plans be prepared in readiness to deal with any major emergency which may arise so that immediate correct action can be taken on its occurrence to remove further danger or loss and cause for panic.

DEFINITION

A major emergency is one that may affect several departments within a works and/or endanger surrounding community. It may be precipitated by the malfunction of the normal operating procedures, by the intervention of some outside agency, such as a crashed aircraft, severe electrical storm, flooding, a deliberate act of arson or sabotage or by civil disturbance.

EMERGENCY CONTROLLER

A senior member of staff with a thorough knowledge of all the processes and associated hazards should be named as emergency controller. He should have an elected deputy to take account of his absence from the premises in any circumstances. Out of normal office hours the senior member of management on the site should take initial control until relieved by the emergency controller.

The list of duties to be performed by the person in charge must be laid down in writing and be available immediately in any contingency.

EMERGENCY CONTROL CENTRE

This must be designated within the plans. It must be adequately equipped with means of receiving information and giving directions. Alternative means of communication must be available in the event of the main telephone system being rendered inoperative.

If personnel radio equipment is available, this can be of great initial assistance in communications and as a second line should the telephone fail.

The centre should have a map of the site which can be used to indicate traffic hazards in consequence of the emergency, and locations of emergency equipment, e.g. fire hydrants and extinguishers, protective clothing, and breathing apparatus. Special risks such as radioactive materials should be shown. A list of names, addresses and telephone numbers of key personnel likely to be required in an emergency and the action taken should be kept and be used in a review of the procedures carried out in the light of the experience gained.

EVACUATION

In the immediately affected area it will almost certainly be necessary to evacuate non-essential personnel. The extent of evacuation of personnel in areas adjacent to the affected area will depend on circumstances which are too many to be described here in detail. A decision will be the responsibility of the emergency controller.

Assembly areas to which evacuated personnel will report should be chosen beforehand. Each should be clearly marked and known to all employees. An appointed member of the staff should be in charge and there should be means of communicating with the emergency control centre.

A procedure will be required to ensure that everyone on site can be accounted for. In practice the most workable arrangement is to provide a nominal roll at each assembly point which lists employees by their department or section. These have to be kept up to date by personnel department but inevitably this will mean that for most of the time they will not be completely accurate. Each company should devise a scheme which is practical and workable in the light of their own circumstances.

EXTERNAL AUTHORITIES AND OTHER COMPANIES

The closest contact must be maintained with the local police, fire and medical authorities. The development of a mutual aid scheme with neighbours should extend to the type and quality of emergency equipment each participant has available, and the number of trained personnel. Early involvement of the local authorities' services will allow interchange of experience and suggestions.

It should be borne in mind that a major emergency may involve a failure in the supply of electricity, water, gas and/or telephone communications. Discussions with the appropriate authority will help to determine priorities in re-establishing supply.

These notes are intended only as guidelines in the drawing up of emergency schemes which should be prepared in detail and issued to all those required to take action. Exercises are recommended from time to time to ensure all key personnel are up-to-date with the latest information.

BOMB THREATS

The increasing incidence of hoax messages to offices and factories spotlights the need for precise instructions to all who may be directly concerned with making decisions as a result of an alert.

Threats may come from:

1 Misguided practical jokers.
2 Malcontents presently in the company's employment or disgruntled former employees deliberately causing inconvenience and dislocation in production, etc., without sinister motivations for damage or injury.
3 Extremist organisations operating primarily in the fields of local, regional or national politics with malicious intentions.

Threats of this nature are usually made:

1 By telephone direct to a location, in which case they would usually be received by a telephone operator, or
2 By telephone through the local police who may have received a message direct, or who may be repeating a communication to the Press, radio or television authorities, or to the local

telephone exchange from which it has been passed to the relevant police station, or

3 By anonymous letter.

Letters received containing information on the alleged placing of a bomb should be handed to police for any action they consider desirable. They should be handled as little as possible and by a minimum number of persons.

General

The objectives of the guidelines which follow are:

1 To ensure maximum safety of personnel.
2 To protect company property.
3 To minimise interference with normal production, business, etc.
4 To enable an early appreciation of the situation to be made and to arrive at correct decisions without anxiety and confusion.

Instructions should be distributed on a need-to-know basis. It has been found useful in discussing the drawing up of these to confer with representatives of trade unions or staff associations so that their cooperation is obtained when they have to be implemented.

Telephone operators

Experience in this and other countries has been that when a bomb is placed in premises, it is usually with the intention of creating damage and not personal injury to occupiers/employees. Hence warning of the location of any explosive device and/or the planned time of explosion is very often given or can be obtained by questioning to allow time for evacuation of staff.

Since the warning is almost certain to be received by the telephone operator, her actions are of major importance and trouble should be taken to explain the important part such operators can play. The instructions to telephone operators, therefore, must be comprehensive so that by keeping the caller talking every scrap of information can be obtained to assist in any subsequent decision. Details of the part they can play can be found in Appendix 29. Immediate transfer of a call to an extension is not recommended as it could result in the loss of the connection through mechanical failure or because the extension is not attended or engaged.

Management

The decision to evacuate all or part of company premises must rest with management present when the message is received. The time factor may not permit reference to higher authority. Therefore, the instructions should provide for a selected number of management to be designated, by name, with their respective telephone extension numbers so that the recipient of the message or letter knows immediately who to contact. The decision on evacuation of personnel and return after evacuation will be delegated to these managers by seniority. They could be described as 'controllers' or some other title. On an alert they should automatically gather together at a predetermined control centre from which to direct operations or to receive communications. The controllers should have instructions recording their responsibilities including the communications they must make to those persons who should be informed of the emergency and the action taken.

Factors bearing on a logical decision to evacuate the building

1 Provenance of the call—e.g. direct, via police, etc.
2 Nature of the call—apparent age of the caller, speech, attitude, approach, etc. (See Appendix 29.)
3 Recent history of such threats in the locality, real or hoax.
4 Prevailing conditions in the locality as they affect the firm, industrial tensions, strikes, employment problems, political unrest, threats against industry, etc.
5 Company experience with any previous calls.
6 Guidance from the police who will advise only with relevant information but will not accept responsibility for making a

decision in respect of evacuation which is that of management.

7 The number of persons working and the general implications of a large-scale evacuation.

When evacuation is ordered, employees should take with them readily available handbags, personal cases and parcels to facilitate early identification of any suspicious objects.

If a bomb threat is received when the premises house only a few people such as cleaning or maintenance personnel, it is always safest to evacuate and search and wait at least half an hour after detonation time before re-entering the building.

Where evacuation is ordered during inclement weather, it may be desirable to go to a location within the building concerned. It is imperative that these shelter areas be searched before people arrive there. These locations must be known only to a select few.

If evacuation is to be outside, move everyone away at least 300 feet from the building where possible.

Reciprocal arrangements might be possible with neighbouring businesses to accommodate evacuated employees.

Lifts should not be used for evacuation but brought to ground level. However, special arrangements will be necessary for disabled persons who normally use lifts.

Genuine bomb users want the maximum publicity and it follows that calls through the local police are potentially more likely to be real and should be considered accordingly when deciding whether to evacuate a building.

Searches

Arrangements should be made for searches to be made. For details see Appendix 30.

PART SEVEN

AIDS TO SECURITY

29

Alarm systems

From small post-war beginnings the burglar-alarm installation industry has grown at a rate which more than equals that of the incidence of crime. About a dozen firms now operate on an almost nationwide basis with numerous smaller local equivalents, several of which have combined into a consortium to combat their larger rivals. There is comparatively little variation in the types of equipment used but expertise in installation, and for that matter workmanship, can differ between firms and indeed between branches of any one firm—efficiency in one area does not automatically mean efficiency elsewhere.

When the installation of an alarm system is contemplated quotations should be obtained, ideally from several firms; their specifications will correspond but prices may vary widely and the most expensive is not necessarily the best. Police crime prevention officers cannot be expected to recommend any firm in preference to the others but may be prepared to say which has the greatest incidence of false alarms in that particular area. This will reflect upon the workmanship and service which is given. Whatever else, a well established company should be used. This is a popular field for small local firms based upon electrical contractors, which set up hopefully and then disappear, leaving their installations to be serviced as best they may.

It is advisable to look closely at the small print on agreements before completing—some endeavour to exclude every conceivable form of responsibility and may make the user pay for equipment replacement and services that he believes would be included in rental/servicing. When in doubt about the equipment which is being suggested, ask to see a similar installation that is already functioning—then ask that user what snags, if any, have been encountered. A highly expensive linked system of rays to cover a very long factory perimeter was 'sold' and on the point of signature by the local manager when a gentle request 'to see one that's working' produced a reluctant 'well there isn't one yet' and a further voluble attempt to persuade 'a firm with so much at risk to experiment'—at the user's expense!

METHODS OF OPERATION

There are four main methods of operation for alarm systems:

1 By Post Office line on the 999 system to the police (the most numerous at the moment).
2 By direct line to the nearest police station.
3 By direct line to a central station or to an automatic satellite linked to a central station manned by the alarm firm.
4 Local audible bell type from which no telephonic message is sent.

A further variation has not really got off the ground. It is that of causing the transmission of a wireless signal in lieu of the telephonic link. A pilot scheme using the permanently manned office of an alarm company did not come up to expectations and met other opposition. Providing congestion of available channels and the traffic upon them were not insuperable obstacles and equipment were reasonably priced, there seems no reason why this should not be developed in special circumstances where no telephonic link exists. The police have at their disposal means which can be installed on a temporary basis to provide for a positive risk; triggering off of this alarm superimposes a coded bleep on a wavelength used by them. The number of such devices that can be used in a given area is limited and their loan will only be made where substantial grounds to anticipate attack have come to light.

Another scheme which is prospering is that of satellite automatic stations. These are of the direct line type and radically reduce Post Office charges on their customers. The satellite unit, which is unmanned and comparatively small, is strategically placed amongst a number of installations which are connected directly to it. It is then linked to a centrally manned station; Post Office charges are on a per mile basis and could be prohibitive when the distance to the station is great; by this system the cost is split between the users by virtue of the single line needed. This is ideal for a small industrial estate and enables the best form of protection to be provided at reasonable cost.

Electronic message-transmitting systems

These all work on the principle of interference with a warning circuit which activates a tape or disc mechanism, automatically dialling 999 and conveying a pre-recorded message which indicates the location of the premises and the presence of intruders; or, in the case of a direct line, causes the illumination of a light and the ringing of a bell on a panel at the terminal station. This alerts the police to put a pre-arranged plan into operation to assemble an adequate number of men at a fixed point near the premises, but out of sight and hearing; the area will have been previously surveyed and when sufficient men are assembled they move simultaneously to selected places to entirely surround the premises; keyholders from the firm are notified in accordance with the list supplied to the police—but, to save time, firms are often agreeable that keys shall be lodged with the police. Where this is done the keys are retained in a sealed envelope at the police station and are only withdrawn on the sounding of an alarm. Where a central station is in use the operator notifies the police, either through a direct connection or by dialling 999; apart from this brief delay such listening posts have advantages in picking out false alarms due to faulty opening and closing down procedures, thereby saving police time. On the adverse side, the calibre of the employees in experience and perhaps in resistance to external pressures is scarcely likely to be comparable, and the regular change round of police staffing restricts in comparison the possibility of conspiracy with thieves.

The permission of the local chief constable is necessary for a direct line to a police station. There is no general practice on the granting or refusal despite the fact that the Home Office have approved a standard form of indicator for use in police stations. It seems invidious that instances can arise when the police apply pressure for an installation to be put in only to be followed by refusal to allow the most efficient and thief-proof type to be used, where no commercially operated central station exists but a permanently manned police station is available.

Audible bells

Many electronic alarms are fitted with audible bells functioning on a delay mechanism at a fixed time after the initial telephonic message has been transmitted. This is a precautionary measure, often insisted on by insurance companies, but disliked by the police, who thereby have a limited time in which to carry out their plans before the criminals are alerted. The insurance view is that the alarm will make the intruders decamp before they can steal anything, if the arrival of the police should be delayed for any reason. However, if the buildings are at such a distance from normal police coverage that it is not feasible to assemble the required number of personnel in less than ten minutes but the bells work six minutes after the alarm is activated, the police cannot be blamed for having a jaundiced view since they will obviously arrive when the premises have been vacated.

Once premises are known to be protected by alarms, as they must be if the bell sounds, thieves can either leave them strictly alone or make efforts to see how this can be circumvented. There is a growing tendency to try to defeat alarm systems and the course that is subsequently followed will depend purely upon the determination of the thieves.

Audible bells have outstanding disadvantages—their only virtue is that of cheapness. Experience has shown that people other than police either take no notice or do nothing about it on hearing one. Frequently such bells are fitted where no one can hear them in any case. They do not catch criminals: they may chase them away—or they may induce them to take a look and see how they can stop the bell next time and work in peace. Such is the oldest type of alarm, cheap and widely used on small shops and premises where expense is a major consideration. Unfortunately, most thieves will have little trouble in putting the bell out of action—this can easily be done by either cutting the wiring to it, filling it with cotton wool, or even discharging a fire extinguisher into it. A bell is only effective when it causes thieves to leave the premises immediately.

SURVEYING PREMISES FOR ALARM INSTALLATION

All commercial firms supplying alarm systems will survey the premises of prospective customers without any charge or obligation. They will specify the most suitable type of protection which could be installed and submit estimates of overall cost. Insurance companies employ surveyors who work in a similar fashion at the request of applicants for insurance cover, and in this instance there is no charge for the survey. Police crime prevention departments also furnish free and detailed advice on a firm's problems; it must be appreciated that they will not be able to detach a man for a prolonged survey extending over a period of days—this is the task for a professional firm. Cost considerations apart, the practice of requesting quotations from several alarm firms has an additional advantage in that the coverage proposed by each firm can be compared and the best system selected to conform with the firm's usage of the premises.

Where experts are called in to advise and make surveys they must be informed, in confidence, of every risk that may be encountered so that they can include it in their considerations. Justifiable standards of protection will vary according to the risk attached to the particular premises: a jeweller's shop would require much more detailed protection and consequent cost than an ordinary grocer's shop. Similarly, a tobacco warehouse would justify provision far in excess of a furniture equivalent. The location of the buildings is a further factor; if these are sited within a works, the perimeter of which is well lit and patrolled by security staff, there is little object in having detailed protection, other than at an essential point. If an office block in the dimly lit back street is concerned, it will obviously need far greater coverage.

Overall protection of premises may not be required; at some points nothing of value will be stored; other areas may contain safes, valuable metal stores, or office equipment, all of which merit security. A single contact at one particular point may obviate the need for six at other locations; certain parts of the building could be strengthened by physical means to make an intruder's task almost impossible. These are all matters for the surveyor, who must be fully qualified in every way, not only in connection with equipment but in knowledge of the probable modes of operation of prospective thieves.

Two separate main schemes exist for the protection of premises: the first is that of protecting the perimeter so that no one can get through the vulnerable interior; the second is that of 'trap' protection, which, in effect, allows the intruder in but ensures that he subsequently trips an alarm in endeavouring to reach his objective. The perimeter variety can be avoided by attack over roofs and dropping through; this of course limits the quantities of material which can be taken out, but, where values are high, this can still be productive for the thief. The ideal is a combination of 'trap' and 'perimeter'; this allows the police a maximum warning before the thief can get to his main objective and also ensures that should he get past the first line of defence he is caught with a subsequent line which he may not expect.

It is not always easy to disguise perimeter protection since this involves windows, where tubing and wiring will be visible, as could be vibratory contacts. Not only that, but should a thief carefully examine the outer doors he will probably notice the various contacts which may be fitted there. Trap protection usually takes the form of invisible rays, pressure pads, pressurised systems, etc., which on the whole are less obvious than the perimeter protection.

Where a number of surveys of the same place are made, it should be borne in mind that it is likely that the insurance surveyors will specify a much higher degree of protection than either the police or the alarm company (because their own finances are directly involved); the police will advocate the next;

and the alarm company, not wishing to price themselves out of work, will hit a medium which will preserve their reputation and supply an acceptable degree of security.

MEANS OF DETECTION

There are a variety of devices; some of them have a multi-purpose nature which may be utilised by installation companies. The main ones are described below in non-technical terms.

Door contacts

These are the simplest yet most effective and trouble-free devices. There are two types: the mechanically plunger-operated type and the magnetic type.

Mechanical. The first, sometimes called the switch type, is normally fitted in the door jamb. The plunger is depressed when the door is closed and when it is opened about 4 inches (10 cm) the plunger, which is spring loaded, operates and the alarm system is completed. The mechanism is micro-switched and the plunger should be adjustable to take into account variations in the fittings of different doors.

Magnetic reed switch. The second type consists of a combination of a magnet and a glass capsule containing two contacts which are closed by bringing the magnet into close proximity to them. The magnet and the switch are housed separately, one in the door surround and the other in the door itself. New varieties of this magnetic contact have been devised which are resistant to any attempts at interference by other magnets.

This type has advantages over the plunger type as they are easily rendered invisible in wooden doors, there is no corrosion effect upon them, and they are perhaps more difficult to counteract.

Heavy-duty contacts

These are used for larger doors and again may be either of the mechanical or magnetic type. They are used on garage and factory doors of the large roller-shutter and concertina type. They are basically the same as the smaller contacts but more robust. The disadvantage of the non-magnetic variety is that they are clearly visible and, once seen, it may be possible to circumvent them.

Closed-circuit wiring

This again can be in two types: tubular wiring and simple wiring.

Tubular wiring—This is widely used for the protection of external windows and consists of light steel tubing mounted on a wooden frame, shaped to fit the whole vulnerable space of the window. This of course gives a bar effect to the windows which cannot be concealed; it is obvious what they are to anyone who knows anything about burglar-alarms.

The tubes themselves each contain a fine non-stretch wire throughout their full length, and this is connected into the circuit wiring. After breaking the window any further entry must be made by forcing the tubing; cutting the tubing breaks the wire and interrupts the circuit, activating the alarm. Stretching the tube, forcing the tubing apart, will similarly break the wire. This is very effective but also very obvious.

Simple wiring—Not contained in tubes, this is used for protecting doors, vulnerable walls, ceilings, and forming protective enclosures around safes. Wire of the same type is stretched over the area to be protected in the form of a network and similarly linked into the circuit wiring. The whole is then covered over by hardboard or some similar material to blend into the surroundings. Any strain upon the wire caused by physically breaking through will break the wire and activate the circuit.

This form of protection is ideal for lightweight doors, in combination with an ordinary plunger or magnetic contact. The object of this is to defeat the criminal who has some knowledge of alarm systems and will break a hole in the door, rather than open it, or force a hole in a plasterboard wall. This is a comparatively cheap and certain means of protection, unless there is expert interference with the circuitry.

Window foil

An alternative to the tube and wiring of external windows is the use of foil, which is by no means as obvious to the intruder. Very thin lead foil is stuck firmly to the inner surface of the window, completely around its perimeter, and indeed, if required, a form of pattern could be included. The foil is connected to the circuit wiring in the same way as the previous methods. It follows that if a crack is made in the window the lead foil likewise will be broken and the circuit interrupted.

The foil is normally given a coat of clear varnish to protect it and while this is a good method of window protection it requires expertise in application. It can also suffer from the efforts of enthusiastic window cleaners!

Floor pads (pressure mats)

These consist of two layers of conducting material separated by an insulator, all of which is enclosed in a waterproof and non-conducting cover. The idea is to form a very thin sheet which can be put underneath a carpet or other floor covering so that pressure upon it will cause the closing of the circuit and activate the alarm.

These can be used as a form of trap protection on stairs or landings, in front of safes, behind doors or beneath windows. Care must be taken on their installation so that they do not show their presence after a time by an outline on the surface above them—this may involve the removal of carpet underfelt, or the placing of some suitable material around the perimeter of the pad and under the floor covering so that the outline of the pad is lost. It is advisable not to use these on concrete floors or under circumstances where they may be subjected to chafing or the effects of condensation.

Ray-operated alarms

These operate on the principle that anyone or anything passing through the beam of the equipment will cause a break in the circuit and activate the alarm. Infra-red light is used, which is invisible to the human eye, though by careful inspection a dull red glow may be seen from the transmitting bulb. This discovery is of no value to an intruder since he will have to break the beam to see it.

Rays are considerably more expensive than the equipment which has been previously mentioned. The apparatus contains a system of lenses for focusing and may have a fan shaped rotating disc to modulate the ray. The cost of individual rays will depend upon the distance which has to be covered. These prices vary from company to company, and the user must install a power source adjacent to the rays. Mirrors can be used to take the rays round corners, extending the range of a single ray, but this has complications and entails regular cleaning of the mirror surfaces.

Rays can be a nuisance in so far as they are extremely sensitive and the flight of a bird through a single ray can activate the alarm. By way of example of such sensitivity, a ray was fitted across the back of a pair of Bolton doors which did not meet completely near the top. An illuminated sign was sited directly over the doors and for some time a series of inexplicable false alarms were caused; the illuminated sign was switched off and these ceased. It can only be assumed that the intense beam of light attracted large moths which passed through the ray. The remedy for this, in premises or places where birds or large moths can reasonably be expected, is to fit twin rays adequately spaced to ensure that a sizeable object is needed to break both at once; this will naturally step up the price of the installation. The alternative is to site the ray so closely to fixtures as to make it unlikely that birds will cross it. For very high risks, 'barrier rays' can be provided which in effect form an invisible fence round the area—needless to say they are costly.

Infra-red bulbs need regular replacement in older installations, but longer lasting alternatives have now been invented. One disadvantage in warehouses, or wherever materials or vehicles may need to be moved about, is that certain areas must be kept clear, thus reducing expensive storage space, or the rays will have to be cut out of the alarm circuit.

Vibro detectors

These are mainly used in connection with windows and fences, though efforts have been made to extend them to wall protection with a certain amount of success. Basically, they are devices which, when subjected to vibration or shock wave, cause a breakage in the circuit running through them. Most types are adjustable for local conditions to prevent false alarms, and variations of such devices have been used in the protection of vehicles. It is this need of adjustment which makes them unreliable: if too finely adjusted they might react to children banging against windows; if too coarsely adjusted, it might even be possible to cut a complete square out of the window with a diamond without upsetting the vibrator contact.

These can be used for the individual protection of filing cabinets and safes to which they can be clamped by means of powerful magnets. Any effort to remove them will of course set them off as will any attack upon the container to which they are attached. An obvious disadvantage is that they are immediately visible and any expert thief will endeavour to find other ways of counteracting them by interference with the circuitry. The varieties used on safes can contain a heat-sensitive device to ensure that the back is not carefully cut away with oxyacetylene or other cutting gear. These could have a fire-prevention value too.

A variation which has had a military application is that of utilising vibro/sound detectors for outdoor purposes. It is unlikely to be practicable for ordinary commercial or industrial purposes but the principle is to set a chain of weatherproofed units in gravel or other suitable media round the area to be protected so the sounds made by the passage of an intruder will be picked up—very expensive, and avoidable if the presence is known.

Sound detectors

These can only be used satisfactorily where there is either a complete absence or an almost constant level of sound to which the apparatus can be adjusted. Essentially, a sound detector is a microphone and amplifier system connected into the circuitry; where a sudden increase in sound is detected this circuit is broken.

These are primarily for special risks and may be combined with a heat detector which could have a dual purpose as a fire alarm and also a means of detecting the use of a thermal lance or other heat-using cutting device.

Sonic detectors are of a special value for retail premises with their plate glass windows and display showcases in that they react to the sound of breaking glass before the actual theft takes place. Indeed, sensitivity can be adjusted to pick up the first attempts to break in. There are good grounds in this application for coupling to instantaneous bells since the intruder is likely to be on the scene in any case for the least possible time.

Pressure differential systems

These are further specialised systems which can only be used in comparatively small self-contained areas. The principle is that a fan is installed in the wall of the compartment which can blow air either into or out of the area to be protected. A diaphragm constantly monitors the pressure difference between inside and outside; this is connected to a pendulum type contact which in turn is included in the alarm circuit. Any additional aperture made into the protected area causes a variation in pressure which is registered by the diaphragm and transmitted to the pendulum contact.

The use of this device is almost entirely restricted to strong rooms and the like. It is an expensive and sensitive installation which, by virtue of its sensitivity, can be subjected to false alarms. It has a drawback in that no warning is given until the protected area has been violated and the intruder may have time to escape profitably if he has planned carefully and has knowledge of the system in use.

Ultrasonic systems

The general principle is that high-frequency pulses are transmitted by the apparatus which are inaudible to the human ear.

These are reflected by objects in the surrounding area and are picked up by a receiving unit which adjusts to a pattern when switched on. A change in the pattern will cause an alarm condition—any movement in the area covered by the apparatus will do this but sensitivity can be controlled.

There have been considerable improvements in the efficiency of this type of equipment during recent years. It has benefited by 'spin-off' from the American space industry; the field each 'head' can protect is limited but it can be adjusted to exclude areas likely to cause faults, some types will accept regular movement such as rotating fans or constantly working machinery, without going into an alarm state.

These are still expensive forms of equipment, but can be very effective. They need careful and expert installation, the optimum positions being found to some degree by trial and error.

Radar detectors

These too can be extremely efficient but are also expensive. The principle is widely known nowadays; a radar pulse is sent out from a combined receiver/transmitter which is reflected by all stationary objects in the immediate area at the same frequency, movement will trigger off the alarm by causing a variation in the reflected frequency. The sensitivity of the equipment can be adjusted to exclude the effects resulting from the presence of birds or rats, etc., and portable forms are available.

The apparatus can be used directionally and miniaturisation has reduced the size of the equipment needed to the extent that it is possible to build it into the base of a telephone handset. Like the ultrasonic system, the area each unit will cover is limited and the optimum placements would have to be ascertained during installation. The development of this type of detection is by no means complete, experiments have actually taken place with rotating beams for long-range outdoor coverage where there is obvious scope if faults can be eradicated; under some conditions the very sensitivity of this type of equipment can negative its possible application.

Proximity detectors

The full potential of these may not yet have been fully utilised. The principle is that when an object is brought near to an area connected with a proximity detector, or touches it, a capacitance change takes place which operates the alarm on the circuit of the detector. It is essential that the proximity detector be linked with a metal surface so the use is restricted to safes, filing cabinets, and articles which, somehow or other, are connected to metallic surfaces. This is done in the case of paintings by placing aluminium foil over the rear of the frames and similar steps can be taken in connection with showcases by placing strips of foil over or under the articles to be protected.

The disadvantage of this system is that its limited detection range restricts its application. Efforts to utilise the principle on a wider scale have so far been impeded by the liability to false alarms.

Trip wires

These are the most elementary types of warning devices but they can be effective because they are unexpected in modern society. The drawback is that an intruder will almost certainly become aware that he has been detected and will make his escape. Not only that, if he wishes to renew his attempt at a later date he will be aware of the barrier and can easily avoid it.

Nylon trip lines are probably the most effective and least likely to be noticed. Where used externally false alarms must be expected from animals and possibly children. They could be of considerable value inside compounds and building sites when connected to instantaneous alarms.

A variation of these could be used for the protection of windows and roofs by arranging that they could be under constant tension which if varied in any way would set off an electronic alarm. This would enable almost invisible steel wires to be stretched behind windows as an alternative to closed-circuit wiring, thereby facilitating the use of the windows for ventilation. A similar arrangement, with suitable intervals between wires could be put on the underside of roofs. A period of settling down would have to be allowed before these could be wired into the system to avoid false alarms caused by initial stretching or straightening of the wire.

Developments

Whilst these are the more commonly encountered devices, there is a constant development being carried on by the larger companies to improve and miniaturise their equipment. Much of this is in the ultra high frequency field and will be most difficult to counteract since the presence of the intruder will be noted before he can possibly interfere with any part of the system. This, coupled with a radio link, could well produce the ultimate in protection. It will not be cheap and users might even consider dispensing with insurance coverage if no substantial reductions in premium ensue.

PRECAUTIONS AGAINST INTERFERENCE AND LIMITATION OF FAULTS

Expert thieves are fully alive to the potential danger to them from burglar-alarm systems. They also realise that premises equipped with burglar alarms must also contain property of attractive value. It can be anticipated therefore that these will offer a challenge to a certain class of criminal who will apply expert knowledge to overcoming them. It is not proposed to discourse on how this can be attempted, but only on precautions that can be taken to prevent any success being achieved.

The normal Post Office 999 line is most vulnerable. The following steps can be taken which will enhance its value:

1 Have the transmitter coupled to an ex-directory line which can be made 'out-going only' on application to the Post Office. It is possible for the Post Office to arrange that this shall be a normal two-way line for certain periods of the day and out-going only after a fixed time. This will involve an extra payment.

2 Where an ex-directory line is used remove the disc from the centre of the telephone showing the number.
3 Ensure that the line connected to the telephone exchange is in the form of an underground cable.
4 Site the transmitting apparatus in such a position that any intruder, even if he knows its whereabouts, must pass through a form of trap protection and be unable to reach the transmitter before the message has been sent out.
5 Get the alarm company to fit a 'line sense unit' which, should the Post Office lines be non-operational, automatically brings into action external alarm bells which are usually self-contained with their own battery supply and encased against interference.

Power supplies

Most systems are mains operated, but technical developments are in hand which may make them self-contained. Rectifiers and transformers are included in the circuitry to produce a direct current output of 6 or 12 volts. It follows that a mains failure means that the system becomes non-operational unless there is an alternative source of supply. Alarm companies therefore install wet or dry batteries and the testing of these is part of the periodic servicing carried out by them. These offer a stand-by of adequate duration in case of failure of the power supply.

Control panels

These are a necessity to test the state of the wiring and the state of the power supply. Multiple control panels can be installed whereby pre-arranged sections of the premises may be made subject to the alarm circuits, allowing others to continue to be used for production purposes. More specialised panels can combine this facility with one of fault location whereby, should the control panel show a fault on test, it is possible to identify at once the offending contact. All control panels now incorporate locking devices to prevent the system being switched on unless all the triggering devices are set correctly.

Operating procedure

The normal procedure is that the operator tests the circuit and switches on when he has found that it is in order. He then goes to the main exit door possibly via other contacted doors and either the last of these or the main exit door itself will be fitted with a switch lock which he will turn to incorporate those doors into the alarm circuit, including the one he is locking. These switch locks are usually of multiple lever type, the keys to which cannot be duplicated other than by the alarm company who can be anticipated to be very circumspect about to whom they issue them. Maintaining the security of the switch lock keys is obviously of vital importance and strict control must be exercised on the number of persons who have access to them.

Bells

If premises are insured, the insurers will have to approve the location, number and operation of any bells incorporated in the alarm system. Their main insistence is likely to be on a minimum time delay between the alarm being triggered off and their operation; some agreement will be necessary with the police to achieve a compromise. Police practices vary but they do not normally want external bells openly displayed on premises where there are direct line facilities into a police station.

Where the major reason in installing a system is that of damage prevention there is a case for immediate bells to disturb the intruders. This is currently particularly true of the lowlands of Scotland where damage by children and youths has become excessive.

Installation costs

Normal practice of alarm companies is to make an installation charge and thereafter a quarterly maintenance fee. This will vary with the extent of the installation and also whether rays, which involve more detailed maintenance and the replacement of bulbs, are included in the system. Costs will also be affected by the proximity of the firm's nearest branch for installation and servicing. Where engineers have to travel a distance and perhaps incur lodging allowances to effect the installation, these can be calculated to add to the installation charges. One of the considerations in deciding which firm's services to employ is that of the availability of local servicing facilities. It is of little use having an installation which might be out of order for anything up to forty-eight hours pending the arrival of an engineer to rectify the defect. The availability of servicing facilities is one in which the crime prevention officers will again be of great assistance—the alarm company can hardly be expected to admit to deficiencies in this respect.

They have taken action on this which should result eventually in much higher standards of workmanship in installations and possibly stop the unrestricted growth of inefficient firms. Their Association has sponsored and financed an independent body to ensure that alarms conform with British Standards and to investigate complaints. 'The National Supervisory Council for Security Alarm Systems' should be very much to the advantage of users.

It must be appreciated that one has to pay for protection; ninety-five per cent protection might cost £200; one hundred per cent protection of the same premises might cost £800 to £1000. This will be a matter for the customer to decide on the risks he is prepared to incur. In all instances, the customer should not pay the final account until he is satisfied that the installation is entirely to his satisfaction and working correctly. It should be such that it can be easily covered during redecoration of the premises so that the wiring is not obtrusive to the eye. If it can be built into new premises so much the better.

Where there are recurrent faults it must be noted that the chief constables of the area concerned can take a very strong line and will refuse attendance if these become too recurrent. In these circumstances, pressure must be put on the alarm company to rectify the mistakes as soon as possible, failing which they should be told to remove the equipment. Alarm companies cannot stand publicity from such instances and are very sensitive in this respect. There is one unavoidable fault which is accepted by the police—that of interference with the alarm systems due to sudden surges of current or electrical breakdowns. These can occur during the course of electrical storms where it would not be unusual for all the alarms in a given area to function at once. It is in these instances that stand-by batteries are an absolute essential. Expect a limited number of faults initially, but after the first day or so these should cease.

30

Radio and television

As yet these facilities have by no means been utilised to their full potential in industrial security work. There are reasons for this, at least as far as closed-circuit (CCTV) goes, other than the obvious one of expense, but these do not apply to VHF radio-telephone installations, nor is the expense item in that case by any means as severe.

RADIO

Many large firms with numerous security officers and large areas to protect have already appreciated the enhanced value of their patrolmen when they have means of instantaneous communication with a base station. Observation on intruders is made easy, when there is no need to break off and perhaps lose contact with them when making a telephone call for assistance. A most satisfying experience for one security officer was to sit on the top of a tower watching thieves struggle with heavy slabs of metal while the police assembled in accordance with his commentary. There was, of course, no hurry and the exhausted thieves had not the energy to run when the police eventually closed in.

Despite the fact that he is guarding premises a security officer may be overpowered and left out of action for a considerable time before his colleagues are aware that thre is anything wrong. Where there are no regulations concerning 'ringing out' procedures from various points on his round this period could be almost the full duration of the round, leaving the thieves with ample opportunity to do as they wish. There is less chance of this happening where a portable set is carried; at least one modern version is equipped with an 'attack' button to give an automatic alarm if the carrier is suddenly assaulted. This modification, at some considerable expense, can be accompanied by a computerised addition to the base station, which will indicate exactly where the incident has taken place.

The fact that there is radio contact can also be a protection for the base station, which will almost certainly be sited in the office at the entrance to the premises. Numerous instances are on record of attacks being made on the gate office before anything is attempted elsewhere in the premises, as by eliminating the man in that office the main means of contact with the outside world can be cut off. Hence the recommendations to keep the office door locked or bolted at all times after dark, to have an inspection hatch through which visitors can be seen before letting them have access, and to have an alarm siren, which is preferably linked to the factory siren, and can quickly be operated by a push button inside the office.

Licensing—a necessity

The Ministry of Posts and Telecommunications is the sole authority in the United Kingdom for regulating the use of radio and broadcasting equipment, and for the granting of licences for their use. The Ministry is also responsible for the allocation of radio frequencies and for honouring obligations binding on Great Britain as a member of the International Telecommunication Union. The Ministry has a special responsibility for tracing and measuring interference to radio reception, especially in suggesting methods of suppressing interference, and for encouraging the responsible use of equipment—whether for business or pleasure.

Radio frequencies available for use in this country are overcrowded and must be shared between many services, such as aeronautical, defence, maritime and land mobile services, radio-navigation, radiolocation and communications satellites.

To ensure that maximum benefit from the frequencies is obtained and that they are operated in an orderly manner with the minimum interference, any use of radio in the UK must be licensed.

There are certain kinds of transmitters, however, for which the Ministry cannot issue licences because they are liable to cause interference to authorised radio services.

These include 'walkie-talkies' of foreign manufacture designed to operate on frequencies around 27 MHz. The use of these without authority may result in prosecution.

Would-be purchasers of radio transmitting equipment are advised to confirm that the equipment has been approved by the appropriate authority of the country in which it will operate.

Call-signs

Specified call-signs must be rigidly adhered to. In any individual case the operative name will be the same for both base and portable sets; for example, base station will be 'Yorkmet Base,' the first portable 'Yorkmet Alpha', the second 'Yorkmet Bravo', and so on in sequence, using the phonetic alphabet.

Only spoken messages are to be used and only authorised persons may operate the stations. An onus is laid upon the licensee to ensure the latter provision is strictly carried out and that those engaged in operating the stations will observe the terms, provisions, and limitations of the licence at all times. The apparatus must be designed, constructed, and used so as to avoid causing interference with any other wireless telegraphy. The call-signs must precede every message and be repeated at the end, though the two identifications are not mandatory if the transmission is less than one minute in duration. Messages should be brief, clearly phrased and spoken, and those making them should never forget that freak conditions may lead to their words being heard over a wide area. A senior police officer was once very embarrassed to find that an installation which went wrong at the crucial stage of a ticklish operation, and which he was addressing in terms which were certainly not of endearment, was transmitting all he was saying but was not receiving the frantic appeals addressed to him by the base station.

Limitations

Both stations and licence must be available at all reasonable times for inspection by duly authorised officers of the Post Office, at whose request the station must close down. The licence is not transferable and can be revoked or varied as desired by the Minister of Posts and Telecommunications, who must be notified of any proposed changes to be made. Amongst a series of conditions and limitations are those that stations shall not be linked to the public telephone exchange system; near aerodromes and power lines restrictions have to be complied with concerning the transmitting aerial; and transmission of certain misleading messages is forbidden.

Uses

Modern portable sets are small and efficient. They are being progressively improved and no longer impede those who have to carry them. The range that can be covered will naturally be governed by the terrain and buildings, but inter-communication between portables should be possible up to about 4 kilometres and considerably further with the base station. In fact, the range will be in excess of requirements in most cases, thus care must be taken with matter that is transmitted. The installing firm will advise on the siting and type of aerial to be used; remote control facilities can be provided when it is inconvenient to operate the fixed station itself, or, for instance, to reduce the length of down lead from aerial to set in the case of high buildings.

Full use must be made of such equipment where it is available. For security purposes this will mainly be after normal working hours, but there is no reason at all why, in a large factory, it should not be used to maintain contact with important managerial personnel or visitors touring the site.

Regular contact with base station should be kept during patrols by positional reports made every fifteen minutes or so.

This enables the base station to know almost exactly where their man is at any given time, and if no message is forthcoming his whereabouts can be pinpointed with a reasonable accuracy. Where vehicles are in use, or dog patrols maintained, the additional facility of wireless inter-communication enables greater use to be made of them. From all points of view this is an expenditure which can well be justified and there are few running or maintenance difficulties to be encountered.

VHF 'POCKET PAGING' SYSTEMS

These are better known as 'bleeper' systems, the name being derived from the warning sound which is put out by the pocket receiver carried by the person who is wanted. They provide a means of locating key members of staff almost instantly, wherever they may happen to be. The person it is wished to contact is called individually, without disturbing the other carriers of these pocket receivers, by a transmission to which only his will respond. Dependent upon the system that is employed, or the type of equipment that he is carrying, the person will either ring the telephone switchboard to receive a message or listen to the message direct from his base station on his personal set. A 'talk-back' version may also be used which will give certain key personnel two-way speech facilities anywhere on the premises. The more sophisticated the installation, naturally, the greater the cost. Though normally used inside very large buildings this type of communication is equally effective externally.

Equipment can be designed to cater for a large number of 'bleepers' to be covered by the system; modifications can provide for units of 2, 6, or 12 individuals to be called simultaneously as teams. Receivers are very small and easily carried in a pocket quite unobtrusively. The system is particularly useful for office blocks, large shops, and hospitals.

CLOSED-CIRCUIT TELEVISION

This equipment may be obtained by direct purchase, or on a

leasing rental basis. It is already in use in many places where high security risks are incurred, but its cost and a reluctance on the part of employees to undergo continual surveillance has possibly limited its development for all the situations where it could be of value. In some instances where it is in use there is a tacit agreement between the management and unions that matters concerning minor breaches of factory regulations observed by this method will not be acted upon. The attitude of some large retail shops may be governed by the thought that a percentage of their wealthier customers might have objections to being 'televised' in this way.

The basic installation is that of a fixed television camera linked to a single television screen or monitor. This can be indefinitely extended by the use of additional accessories and equipment, the purpose-designed monitor gives better results than the ordinary TV set. Fixed cameras are of main value in watching a single vulnerable point with no necessity to obtain closer viewing. A wages office, a corridor, and a small compound would be suitable objectives. By linking several fixed cameras and using a simple switching arrangement a single viewer can inspect a whole series of vulnerable points from a remote position; alternatively, a battery of monitor screens could be used so they could be seen simultaneously.

Accessories

The value of individual cameras can be very much increased by accessories to control the rotation and up and down movements of the camera—'pan' and 'tilt'. In addition, protective housings can be fitted with screen wipers to offset the effects of condensation; cameras can have thermostatically controlled demisters for operational use out of doors in all conditions; 'zoom' lenses can be used under remote control. However, it must be appreciated that the cost of these accessories, plus their means of activation, may be individually in excess of the cost of the original installation.

The quality of picture received depends on light and shade in the 'target area' which will of course vary at different times of the day and season. Automatic lens adjustment can be provided on the cameras to ensure an even brightness of pictures under varying conditions of scene illumination. Artificial lighting can be quite satisfactory internally and existing lighting on roads may provide usable pictures. Recent use has been made of special types of camera tube which are sensitive to infra-red radiation and have proved effective in conditions of apparent total darkness where the area under inspection has been lit by infra-red lamps. These vidicon tubes are about three times the price of the ordinary kind but have a similar life of some 2000 hours. In all cases where television cameras are to be used for security purposes it must be made certain that there is always adequate lighting in the areas to be surveyed for reasonable pictures to be obtained.

A further development has been the use of the video-tape recorders in conjunction with closed-circuit television; these will record TV pictures for future reference, but again expense could be prohibitive except in highest security risks. The price range here is extremely wide and specialist advice should be sought by interested parties; foreign equipment, of relatively low cost, can be quite satisfactory.

Remote control of gates

Economies in manpower, or constant supervision of an uncontrolled access can be effected by a combination of closed-circuit television and an electrically operated gate release mechanism. Loudspeaker arrangements can be incorporated to challenge individuals, pass instructions, or answer enquiries. The cameras used are normally the fixed variety, though there is no reason why the pan and tilt facilities should not be fitted—other than price. At dispersed sites the extent of cable runs for the more expensive variety will probably cost them out of acceptance.

Optimum circumstances are where it is not economic to constantly man a gate or turnstile which is needed for a matter of minutes at shift changes only and used by a limited number of personnel. The system does not provide perfect security, except where the only property to be stolen is so bulky that it cannot avoid being seen, spot checks are really a necessary adjunct. Even where grounds of suspicion do exist, the egress of the persons concerned may be difficult to prevent until enquiries have been made. The siting of a fixed camera so as to observe both persons entering and leaving sufficiently to be able to identify is an added complication.

The same type of installation of course can be used at the entrance doors to high security risk buildings and to control the entry of vehicles into a restricted area. A successful TV procedure in respect of these has been to allow access on visual inspection and verbal identification into an outer compound, itself sealed off, to await supervision of unloading or collection in the restricted risk area.

The combination of outer and inner gates remotely controlled eliminates constant manning and ensures the driver cannot do anything overt until it is possible to directly supervise his actions.

Maintenance

When quotations are being asked for, prospective users should satisfy themselves that the firms concerned can provide adequate maintenance facilities.

Retail store supervision

This is an additional application in which television not only provides a means of catching shoplifters but also constitutes a monumental deterrent where it is known to be in use in a store. Cameras have to be sighted above normal head height which allows a much wider field of view than would be available to the normal store detective. The observer therefore has a much better chance of seeing shoplifters in action without bringing to notice his own presence. Multiple viewing points can be installed in shop ceilings which will give a very wide field of coverage and these can be fitted to use sequential photography for a permanent record. These viewing points are hemispherical and have five 'eyes' giving an all-round field of vision; as these become better known the detrimental value of 'dummies' for future consideration needs no comment.

Apart from the pure security applications, sight should not be lost of the potential of closed-circuit television for the remote observation and inspection of dangerous processes, where it also could have a very important safety application (see Chapter 21).

31

Services of security companies

The growing need to protect property against loss from various causes has led to the formation of an increasing number of firms who will provide such protection at a price.

The larger and well-known companies have joined together to form a trade association called the British Security Industry Association, whose secretary is Allen Baldry Holman and Best, 36 New Broad Street, London EC2M 1NX.

1 To promote and encourage high standards of ethics, service, and equipment throughout the trade or business of manufacturers of and traders in intruder-detection devices and fire-alarms and of positive protection devices such as safes, locks, and allied products; and also the trade or business of providing security services such as protected transport of cash or valuables and security guard or patrol services of all kind; and also to promote and encourage the practice of security methods and generally to increase the awareness of the public to security considerations.

2 To provide a negotiating body to meet with government, insurance and police representatives and other interested parties so as to facilitate the achievements of the objects of the Association.

TRAINING

The companies which form the membership of the Association have the experience and facilities to provide their employees with formal training in their duties. There are, however, a great number of smaller firms supplying guard and patrol services who do not enjoy those advantages and, consequently, before engaging their services it is recommended that the quality of training their employees receive is assessed. Some of these firms use the training courses of the Industrial Police and Security Association (see Chapter 5).

SERVICES

The Association is divided into four sections:

Safe and lock
Alarms and appliances
Transport—carrying
Guard and patrol—watching

Safes and locks; alarms and appliances

The supplying of these types of security equipment will be dealt with in Chapter 32.

Transport—carrying

This service concerns taking money to or from a bank. The service can be extended to the money collected being made up into wage packets for delivery to the customer or the distribution of them at pay-out points. A further use of the carrying service is the collection and delivery of computer data. The transport of valuable property other than money is also undertaken by special arrangement and contract. An example of this service is the transfer by overnight carriers of urgent or confidential information where normal postal facilities are unsuitable.

When a contract is arranged for the transport of money from a bank to premises, it must specify with some detail the place at which the delivery is to be made. It is not sufficient to give just the address: the precise location of the transfer of the money must be given; for example, the cashier's office on the second floor. This leaves no doubt when the performance of the contract has been completed. It also affects the decision under whose insurance policy a claim is made should a theft of the money occur on the premises. Recent information from two of the largest companies engaged in the transport of cash shows that out of fifteen attacks by thieves on security personnel delivering money six were inside the premises concerned. See Chapter 15 for examples.

The security personnel who are to be directly concerned must be introduced by an accredited representative of their employers to members of the firm or the person engaging their services. The customer will then introduce the security personnel to the representative of the bank concerned or anyone else from whom other valuable property is to be collected. Photographs and specimen signatures of the security personnel must be supplied to the customer and the bank for retention. Frequent changes in the security personnel are highly undesirable and if they continue it is recommended that the services of another company be substituted.

The names of the members of the customer firm authorised to accept delivery of the money must be supplied to the security company with specimens of their signatures. Sufficient names must be included to take account of holidays and sickness. An acknowledgement of the names must be received in writing from the security company. If money were afterwards handed to any other person and was stolen the responsibility to replace it would be that of the security company.

If an unknown person purporting to be from the security company attended premises to collect cash or anything valuable his identity documents must be inspected and his employers telephoned to confirm his instructions. His documents might have been stolen or forged.

It is the responsibility of the security company to deliver the money by the time it is required. If instances occur where it is received at a later time the company should be informed.

Insurance—Before contracting for the transport of cash or valuable goods it is advisable to require proof of the quality of the insurance cover for any loss which might occur whilst the property is in the possession of the transporting company.

Efficiency—The efficiency of the service which is contracted for must not be taken for granted but examined critically. If the men supplied to carry out the duties do not satisfy reasonable requirements replacements must be asked for. What is not recommended is the employment particularly on site security of inexperienced temporary personnel such as university or pre-university students on vacation or part-timers performing work which is secondary to their main occupation, often undertaken without the knowledge of their primary employer who might object if he knew.

Guard and patrol—watching

These services include:

1 Supplying uniformed security officers to carrying out security duties at premises with or without guard dogs (see Chapter 32);
   (*a*) continuously over twenty-four hours, seven days a week,
   (*b*) for specified shorter periods, or
   (*c*) by security visits at irregular intervals.
2 Supplying uniformed security officers to supplement a company's own security staff on special occasions or at times when security coverage is reduced through such staff being on holiday or absent because of sickness.

Site security—clocking points—The security companies will supply and install clocking points free of charge on request for use either by the patrolling or visiting security officers. It is recommended these be requested and fixed with the requirement that greater attention be given to the places or areas most vulnerable to loss. A copy of the times of the visits to the clocking points should be asked to be supplied with the invoice for the services rendered. Chapter 4 deals more fully with information on the use of clocks.

SPECIALIST SERVICES

Consultants

When it is decided that an assessment of the state of security in a concern should be obtained this can be done by employing a security consultant. He can be an individual who is self-employed or a member of a security services company. Before employing the latter it should be borne in mind that he has a vested interest to encourage the engagement of the services of his company. Therefore an opinion based on good professional experience by someone independent of such companies has certain advantages. There are no criteria of the capability required to warrant the description of 'consultant' and personal knowledge or recommendation is desirable.

Investigators and store detectives

The employment of such persons can be for a specific inquiry to find out the causes of losses of goods or cash in circumstances outside the responsibilities of the police. Store detectives can be hired hourly, weekly, or for longer periods for duty in one or more premises belonging to the hirer. The employer of the security company personnel must satisfy himself he is indemnified in writing by the company against a claim of civil damages which may result from their action. Chapter 21 refers to their employment in retail stores.

OTHER SERVICES IN RELATION TO PREMISES

Keyholders

To reduce the inconvenience of having to attended premises at night and weekends in consequence of something occurring which requires attention, usually brought to notice by police, security services companies will hold the keys and become registered with the police. Should their visit to premises result in no further action being necessary the instance is subsequently reported to the customer. If it is of a nature that attendance of a representative of the owners of the premises is immediately necessary the previously nominated person is informed at once.

Protecting specially valuable property

Security personnel in uniform or plain clothes can be supplied to protect such property on display at, for example, trade exhibitions.

Security custodians of premises

Persons employed to carry out a security function on premises which are otherwise unattended are in danger of being attacked by intruders or of being taken ill or sustaining incapacitating injury and lying without assistance for some hours. A mutual security aid scheme which has been referred to in Chapter 4 will take account of that situation. Where for any reason such an arrangement is not made the services of a security company are available to receive telephone calls at their operations room from the security officer or watchman at prescribed times. If a call is not received within the agreed tolerance period the premises are telephoned and if the call is not answered they are then visited. To provide for an occasion, which is far from rare, when the security officer may make a call under duress from intruders an innocuous phrase should be decided on which will indicate to the receiver the circumstances in which the call is being made.

32

Guard dogs

It is only in recent years that the recognised qualities of guard dogs have been applied to police and security purposes. Even now industrial firms do not use their full potential, though the police and the commercial security organisations use them on an increasing scale.

The alsatian is generally recognised as one of the most intelligent breeds of dog in the world. This is the essential quality to ensure that, in addition to being fearless without being aggressive, they can be eventually trained to act as companion dogs, guide dogs, guard dogs, or working police and security dogs. They have an appearance which is consistent with the general conception of what a guard dog should look like: sturdy build, combined with a physical ability and strength to work hard in all types of weather. Other types of dog have been used for the same purpose but these have become increasingly restricted in numbers; the only other one which now is in general use is the dobermann which is widely used on the Continent but has found little favour in this country; breeding is less likely to guarantee planned development of required characteristics and it is accepted to be less predictable in behaviour. For these reasons it should be assumed that when guard dogs are referred to in this book they are alsatians.

ALSATIANS' INSTINCTS

There are characteristics of the alsatian which render them to be of particular value to users; these are of a hereditary nature arising from their origin in pack life—hence the alternative name, wolf-hound. They have a real sense of self-preservation, quick reaction to danger with an alertness to every sight and sound; loyalty to those with whom they are in regular contact coupled with a suspicion of strangers, and a quickness of supple movement. They are comparatively short sighted but their range of hearing covers seven octaves above the human ear and is infinitely more acute. Their power of scent is even more greatly developed in comparison—it can warn them of the presence of intruders or strangers even when they are motionless and cannot be seen or heard. Having an ancestry associated with cattle and sheep herding, under all conditions, they can pass instantly from sleep to alertness. It is no hardship for them to work up to fourteen hours a day if this should be necessary, although for humane reasons they should not be worked for periods of this duration. While they may be used every day, they will reflect the benefit if one clear night's rest each week can be allowed. There is no difference whatsoever between dogs and bitches in security usage—they are equally efficient.

ACQUISITION OF GUARD DOGS

Even police forces and the commercial security firms with their large numerical requirements rarely find it either advisable or economical to breed their own dogs. This is a specialised field and there are kennels owned and manned by persons who have spent a lifetime in this work and can supply dogs to meet all needs. Over a period of years these breed to eliminate undesirable characteristics and to select those which are of value for particular purposes. Dogs, like human beings, vary considerably in their mental attitude to training and in their ability to assimilate it; the more experienced their handler the earlier any shortcomings in the dogs will be observed and the less time invested in training a dog which lacks essential qualities. To acquire a dog of unknown background is to run a risk of complications and unforeseen behaviour which is best avoided.

SELECTION OF DOGS

No breeder of repute will risk selling a dog which may reflect unfavourably upon his standards; the more reputable the establishment the more certain it is that any dog purchased from it will conform to requirements.

An expert will have detailed and minute points for a dog to match up to, but the main general characteristics that are desirable are:

1. An appearance of alert watchfulness to both sight and to sound, an interest in everything around him, and an immediate and unafraid suspicion in the presence of strangers.
2. Well proportioned, strongly boned, muscular without being too heavy, smooth and supple in movement.
3. A long lean head in proportion to the body size, black nose, and a long strong muzzle.
4. Ears rather large, pointed, carried erect giving an alert expression; eyes normal—not bulging.
5. Muscular broad-backed body with straight legs viewed from either front or back.
6. Feet should be round with strong toes, firm pads, and short strong claws.
7. The outer coat should be straight, hard, and lying flat, the inner coat woolly in texture, thick, and close. The tail should normally hang down in a slight curve.

These are general points from which a would-be purchaser may gauge whether he is being correctly informed by the breeder.

USES OF DOGS

The value of a dog lies in the full utilisation of its basic instincts. However intelligent they cannot replace the presence of man, but correctly trained and handled there are many instances in which a dog and handler are more valuable than two men. In any case the view has been expressed by a handler, accustomed to work in a dangerous area, that whatever else his dog did it gave him the confidence to go anywhere at any time. The main uses of dogs are:

1. As an aid to more thorough and effective patrolling by using their organs of sense, smell, and hearing to detect the presence of strangers.
2. As a means of apprehending intruders. Few will run in the presence of a dog and they will have little chance of escape if they do.
3. As a protection to the handler. Their appearance and, perhaps undeserved, reputation makes them a formal deterrent to the use of force against their handler.
4. To search premises in darkness or with an involved layout and find intruders quickly. This is of particular value in complicated premises where a person can easily hide himself from the security officer but be detected by the dog's sense of smell.
5. Utilising their sense of smell to track intruders from the scene of an incident.
6. Searching for small articles lost over a wide area.
7. As a deterrent where there are signs that a situation of potential violence is building up.

So much importance is attached to the presence of dogs that merely the display of a notice warning 'danger, guard dogs' has a marked effect in keeping intruders away from premises.

THE LAW

Following as it does several well-publicised incidents involving guard dogs, the Animals Act 1971 has great relevance to their use in security. The Act does not apply to Northern Ireland or Scotland.

All previous common law and statutory liabilities regarding damage done by animals ferocious by nature or of known or presumed vicious or mischievous propensities are repealed (section 1).

It is unlikely that the provisions relating to 'dangerous species' will have any bearing on industrial or commercial practice and guard dogs are not included in this category. They come under the provisions of section 2(2) which makes the keeper of 'an animal not belonging to the dangerous species' liable for damage (injury) if:

1 The damage was 'of a kind which the animal, if not restrained, was likely to cause, or which, if caused by the animal was likely to be severe' *and*
2 the likelihood of the damage or its severity was due to characteristics of the individual animal which were not normal to its type other than at particular times or circumstances *and*
3 these characteristics *were known* to the keeper or the person in charge of the animal at the time as the keeper's servant.

In other words, if a dog is known to be vicious, and does inflict a severe bite, the *keeper* is liable unless he can avail himself of statutory exceptions under section 5. These are:

1 A person is not liable for any damage which is wholly due to the person suffering it.
3 A person is not liable under the circumstances shown in 1, 2 and 3 above if the damage is suffered by someone who has voluntarily accepted the risk.
3 A person is similarly not liable if the damage is caused by an animal kept on any premises or structure to a trespasser if it is proved *either:*
(*a*) The animal was not kept there for the protection of persons or property.
(*b*) If it was so kept, then to do so was not *unreasonable.*

The exception which will apply to guard dogs is (*b*). This only applies to trespassers. Liability will extend to bona fide visitors who may be bitten. So far as an employee is concerned, when he incurs such a risk incidental to his employment he will *not* be regarded as doing so voluntarily. 'Unreasonable' is not expanded on and would in law be related to the environment in which kept and all the attendant circumstances. To use a dog of really savage nature which might even kill could be deemed 'unreasonable'.

Children are not specifically mentioned but it would be dangerous to assume a court would be prepared to accept them as 'trespassers' in the event of injury on premises to which they had easy access.

The use of good guard dogs under supervision has considerable advantages. If it is proposed to allow them to 'run loose' this practice has fundamental liability dangers and it would be essential to ensure the elimination of the slightest risk of child entry into the area of use.

HANDLERS

It must be clearly understood that where a dog is used with more than one handler its efficiency will not be that of a dog which has only one master. This is inevitable since no two persons can have the same attitude towards a dog which is not their sole property, nor will they place the same emphasis on their words of command. The main use of guard dogs is, of course, at night and it therefore follows that should a decision be reached to utilise one dog with one handler, that person is committed to permanent night duty unless the staff is big enough to employ sufficient dogs and handlers to have round-the-clock cover. Though special financial allowances will have to be made for this unappealing perpetual tour of duty, experience has shown that firms with smaller security staffs have difficulty in keeping a dog handler for this task for any length of time.

It is possibly this limitation, which has been widely publicised, that prevents more firms making use of guard dogs. Whilst it must be reiterated that a multi-handler dog cannot be as efficient as one with one permanent handler it still has a massive value, sufficient to justify its keeping. War-time experience showed that a dog need not be considered expendable purely because its handler was killed or wounded; it could be readily transferred to another handler and it is a short step from that practice to accustoming a dog to a series of handlers—multi-handling.

TRAINING: DOG AND HANDLER

Serious training of dogs starts at the age of adolescence. This can vary between different dogs from the age of seven months to fourteen months, according to whether the dog is a slow developer or not. It should already have been given some elementary obedience training by a professional. This is the make-or-break stage with a dog; if it is not trained properly then, bad tendencies are likely to persist later. Similarly, there is no specific age at which a dog becomes suitable for practical usage. This will depend upon the expertise of the trainer, in addition to the age of adolescence which is indicated by the dog suddenly becoming aggressive towards other dogs in its presence. Generally speaking, a dog which is purchased from a training school will have been given basic obedience training and will have had its most undesirable traits corrected. Most important, it will have become accustomed to receiving instructions.

The training of dogs requires a clear mind and the ability to observe and study each dog individually to diagnose the degrees of loyalty, tenacity, cunning, stubbornness, wilfulness, ability to absorb instructions, and other desirable qualities. This is where the sheer professionalism of dog training shows its real value, for unconventional methods may be needed with a particular dog which would be beyond the capabilities of a novice handler. Transformation of raw material into a trained dog cannot be achieved overnight. There must be systematic and planned periods allocated to each day and patience must be exercised by the trainer who, by his experience, will know when a dog has reached the point of frustration. He will then divert it to an exercise with which it is fully conversant, then flatter it and return it to its kennel.

Even when a dog will have to be multi-handled, when it is transferred from the kennels to its permanent home, it is advisable to associate it with one particular handler. Training men for handling is at least as vital as the dog's own training—or even more so. If one particular individual can be designated he should spend preferably two weeks at the kennels with the dog before it is brought back to the firm. The first week would be spent mainly in developing an association between the dog and the handler and giving the handler instruction; the second week would be basic training with dog and handler working together. During this period the handler should obtain expert advice from the kennel supervisor, not only on the future training, but on health, housing, and any particular problems or tasks which the dog will subsequently be asked to perform.

If a man is to be trained solely for duties as a dog handler the first essential is that he must be already a sound and experienced security officer, even-tempered, patient, and mentally alert. If he is a cheerful individual this will be reflected in the behaviour of the dog, and if he has a strong character and persistence to achieve his objectives the training of the dog will be much more thorough. An irritable and bad-tempered handler will soon ruin a good dog.

It is those latter characteristics of handlers which have to be watched for where a dog is being multi-handled. It is easy to ensure that a standardisation is achieved in words of command that are used; it is much more difficult to see that the dog is treated equitably and reasonably by all who have its handling. A dog, like a child, will soon realise and react to variations in the degree of control which is imposed upon it, particularly if one individual acts maliciously or sadistically towards it.

Where a dog is used by a number of handlers it seems even more essential to display appreciation of good behaviour and good work by making a fuss of it. It is in the human sense 'deprived' by not having a single person to look up to and, in this sense, the wearing of a security uniform can be an advantage as it can tie the dog's affections to a particular body of men more easily. One of the virtues of alsatians is that they are a suspicious breed, which makes it all the more essential to assure them that they are regarded with affection and to keep them amply fed and warm.

Before a dog is taken into full use at a firm it must be given a period of familiarisation with all the sounds, sights, and smells of the place where it is to operate. This time should also be utilised for training in the particular tasks which are to be required of it. The individual who has already had experience in its handling can, during this period, introduce to the dog those of his colleagues who are to assist in its handling and ensure that they are fully conversant with the correct words of command, in effect passing on the knowledge which has been imparted to him by the professional who has supplied the dog.

The training must not stop when the dog is taken into use for the first time; a specific period should be allocated each day during which the dog should be given obedience training and progressively more advanced training. Since this will probably be done by persons who are not specifically professional trainers not only will the process be lengthy but certain faults may creep in. For this reason it is advisable, after the dog has become throughly acclimatised, for it to be returned for a period, say a week, to the kennels from which it came for a form of professional check-over, comment on how the training has been carried out, and the formulation of a further programme.

Obedience

An excellent handbook, *Police Dogs, Training and Care,* is published by HM Stationery Office, price 50p, and deals at length with obedience training; reference should be made to this book for the detailed manner in which it should be carried out.

Standardised words of command are used throughout and obedience is brought about by serious habit-forming exercise to specific orders. Patience is required, for it must not be expected that a dog, no matter how intelligent, will become immediately aware of the meaning of a new command. It must be shown what is needed and there must be repetition, until it automatically associates the command with a required action.

The training must be progressive and care must be taken to avoid boring both dog and handler. Single words of command must be used wherever possible and in the early stages of training it is better that the dog should be on a lead so that it can be guided into the action that is required. Disobedience to or disregard of commands must not be allowed, but physical punishment should not be used; use of the choke chain and a reprimanding tone of voice show disapproval. Praise should always be given for good performance. At the conclusion of any exercise period one of the simpler procedures should be carried out so that the dog can be praised and encouraged before being put away.

In multi-handling, each person who has the dog should be taken through the full sequence of obedience training to ensure that he is familiar with it, and the dog accepts him as its master. Every time it is taken out it should first have a ten-minute spell of obedience training to accustom it to accepting commands from the current handler.

Tracking and searching

There is wide variation in the abilities of different dogs in both these fields. The sense used is that of scent; an intruder may leave a trace on either the ground or in the air of his presence or passage.

Ground scent—This is caused by disturbance of the ground—movement of the soil or crushing of vegetable life—causing the emission of a scent which to a degree adheres to the outerwear of the intruder.

Wind-borne scent—This is the scent given off by the person of a particular individual and his clothing. A dog can make use of either of these in his efforts to track.

Climatic conditions have a great effect upon the work of a the dog for these purposes; thus a scent is persistent:

1 In dull mild weather with little wind.
2 In low-lying areas where there is little air movement.
3 At night time when the temperature of the ground is higher than that of the air.

Conversely, scent is soon dispersed in:

1 Hot sunshine.
2 Winds.
3 Heavy rainfall.

There is a limit to ability to track on footpaths and road, pedes trian and motor traffic quickly break up any scent and ground disturbance is negligible. The age of a track is a major factor and too much must not be expected of a dog which has not been thoroughly trained specifically for tracking.

A dog's obedience training includes teaching him to 'speak', that is to bark on particular occasions. This is of importance in searching for intruders—there is no desire that they should be bitten and it is adequate for a dog to stop their escape and indicate their presence by barking. This, in itself, will have the desired effect in rendering the criminal submissive enough to surrender readily.

Attack

A dog must be throughly advanced in other aspects of its training before this is taught, otherwise serious injury may be caused out of all proportion to the offence which is being committed. Moreover, the dog must not anticipate that every time it sees a person running he is doing something wrong—this could have complications at any factory when employees are leaving. With the use of a protective sleeve a dog should be taught how to attack the forearm only, and to let go immediately on the word of command 'finish'.

Police Dogs, Training and Care deals with all matters of training at considerable length, with suitable diagrams.

KENNELLING

The best position for a kennel is near a works entrance where it is in full view of everyone and within full sight of the gate office so that employees have no opportunity to bait the dog and supervision can be given to ensure that no poisonous material is thrown into the run. Whilst warning notices have got a preventive value, the obvious physical presence of a dog is even more beneficial.

The kennel itself should either be a permanent or semi-permanent structure. Whichever it is it should have certain characteristics. It must be:

1 Waterproof and windproof.
2 Correctly ventilated so that there are no draughts but yet a reasonable admission of fresh air.
3 Of sufficient size not only to allow the dog to lie down in comfort but also to allow the entry of a handler to clean out the bedding and interior.
4 Fitted with a removable floor, a few inches off the ground for warmth and easy cleaning.
5 Connected with an outer run for exercise purposes.
6 So placed that rain water and moisture will run away from underneath the kennel and the enclosed run.

For both prestige and psychological reasons the kennel and run should be well made and substantial in appearance if they are in the public view.

Either sawdust or straw should be used for bedding purposes, inspected and shaken-up regularly and swept out for cleanliness as and when required.

The dog should be thoroughly groomed each day with a brush and comb. The coat should first be brushed against and then with the direction of the hair growth, then combed to remove dead and tangled hair. Apart from keeping the coat and skin healthy and clean this helps to keep up the appearance of the dog which, like the dress of a security officer, reflects upon the efficiency of the department. Bathing is seldom necessary and may remove the natural grease in skin and coat, thereby being detrimental to health and appearance. If it is necessary to wash the dog, soap must not be allowed to enter the eyes and ears and it must be thoroughly rinsed off before the dog is dried with towels.

FEEDING

Adult dogs need to be fed only once a day. Opinions differ

somewhat as to the best time to do this but once a time is fixed it should be made a routine to avoid disturbance of the dog's digestive processes. Where a dog is used overnight it seems best to feed it in the morning immediately before putting it back in its kennel to sleep. An adequate supply of clean water is essential for a dog and this should be available at all times.

A full-grown working alsatian needs about 3 pounds (1.4 kg) of mixed meat and biscuits, or their equivalent, to supply the energy it requires and the necessary protein, vitamins, and minerals. Meat and/or fish should supply about half the total, either raw or cooked. If bones are given they must be raw and only the larger type of beef bones that will not splinter. Food should be prepared as near the time it is to be eaten as possible. The dog should be left to consume it in peace, without being diverted by grooming or anything else. Frozen meat must be thoroughly thawed before being mixed.

HEALTH

A sick dog, suffering discomfort or pain may act totally out of character, which, in an industrial environment, could lead to considerable trouble with personnel. Handlers must recognise the symptoms of illness since a dog, naturally, has no means of telling anyone about them. If there is any suspicion of it being ill, rest, quiet, and warmth will go a long way towards remedying the condition; if any illness appears to be serious or there is doubt as to what it is, a veterinary surgeon should be called in without delay.

Usual minor illnesses are those of constipation and diarrhoea, both are due to an improper diet, possibly coupled with a chill of the stomach. If these are recurrent, checks should be made that the kennel is not draughty and causing a chill to develop. Worms are also a common cause of complaint; there are on the market proprietary drugs to deal with all these conditions and instructions are clearly shown on the packets. Regular inspection should be made of the dog's ears for excessive wax or irritation due to foreign bodies; great care should be used in dealing with the ears and again veterinary treatment may be advisable.

Canker is a fairly common ear complaint in some breeds but is fortunately rare amongst alsatians; this has a foul smell and causes the dog to scratch and shake its head—it definitely needs the early attention of the veterinary surgeon. Feet and eyes are other points which require close periodic inspection.

POINTS IN USAGE

Different environments will require different techniques in the general use of dogs. Whenever there are employees about the dog should be kept on a lead, unless it is one hundred per cent trustworthy. There should be a gradual build-up for a new dog in the peculiarities of its new environment. This includes getting used to employees in mass without displaying signs of nervousness or aggression by 'giving voice'. If the dog does misbehave it must not be struck under any circumstances, except in an absolute emergency. A well-trained dog will probably attack if it sees a raised foot. Adequate reprimand is by scolding and using the choke chain. The handler should be on the alert at all times and ready with the command 'finish' to prevent unnecessary action by the dog.

On normal patrol the dog should be kept on a lead to ensure that it remains close to the handler and also to prevent incidents with employees. It should be trained to walk quietly to heel on or off the lead, unless it is ordered to do otherwise. Its natural reaction, if the handler is attacked, is to counter-attack and it will evade kicks and blows by natural instinct.

The dog can be left in a vehicle or chained to something that it is required to protect; it must be remembered that doing so restricts the dog's ability to defend itself and as much latitude to move as possible should be allowed. It is possible for a dog to operate over a considerable area by stretching a wire between two posts and fitting the running line to this so that the dog has a field of movement at either side of the line. However, this also gives more opportunity to attack the dog by throwing poison or missiles at it; if it is used in this fashion it should be visited at regular intervals. The same applies to dogs which are turned loose in buildings or compounds during the night.

Extra care should be observed where the dog has to be used in an environment where there are acids or metal which could cause severe injury to its pads. Similarly, where (as in many industrial premises) rat poison is put out the dog should be kept on tight leash otherwise it might die in agony. Under no circumstances should it be turned loose for guarding purposes in any premises where any of these dangers exist.

CONCLUSION

To get the maximum value from a dog it is necessary to use one handler only if this is an economical proposition. Training must be thorough initially and continual thereafter to maintain the sense of obedience in the dog and progressively extend its intelligence. If it is engaged in incidents which cause complaint these must be thoroughly investigated to ascertain the cause and prevent recurrences. If the dog is at fault this fact must be admitted and no blame laid on the person who is objecting to its behaviour—but it must be borne in mind that the allegations may be malicious or for an ulterior motive. Above all, where a dog is multi-handled, there must be absolute uniformity in orders and treatment by all concerned. Firmness and affection to the dog will be rewarded in like measure.

33

Locks, keys, safes, strongrooms

Despite sad experience, there still exists a tendency to regard anything given the magic name of 'lock' as adequate protection against any form of intrusion. As in everything else, the true value of what the buyer gets is reflected by what he has to pay. The price range of locks and safes in general is wide, but there is little variation between the cost of those of equivalent strength produced by reputable manufacturers.

LOCKS

In effect, locks can be regarded as falling into two broad categories: those that provide privacy and those that provide security; it is incidental that the latter also give privacy, but the former only give security in the imagination of the user or the advertisement of the maker. Needless to say, locks of the first type are the simpler and also the cheaper.

Night latch or pin-tumbler rim lock

These locks are fastened to the face or 'rim' of the door and the locking device is a spring-loaded bolt, which engages in the staple or bolt-housing when the door is closed. The principle is that the correct key will align a series of pins and tumblers, revolve the cylinder, and withdraw the bolt. These locks have the advantages of being low in cost, easy to fit by any amateur handyman, unobtrusive in appearance, and usually trouble-free in use. Moreover, their keys are small and can have an immense number of variations or 'differs'. These are quite easily the most commonly fitted locks in Britain for internal and external house doors.

Unfortunately, such locks rarely give any trouble to a determined thief either and their limitations are no secret. If a partial list of the means whereby they can be defeated is given, it is to be hoped that this may deter persons from placing too much trust in them.

1 The door or an adjacent window will almost invariably provide a breakable pane of glass from which the operating knob inside can be reached and turned.
2 Both lock and staple are surface mounted and the staple is held by only two or three screws. A firm kick will remove these with the minimum of effort or expertise. The body of the lock itself is probably only secured by four similar screws.
3 The bolt is spring loaded and, unless the locking catch is down, if a thin sheet of mica, or other flexible strip, can be worked between the door edge and the staple, sufficient pressure can often be exerted on the head of the bolt to force it back against the spring. The more forcible intrusion of a screwdriver at the same point can give identical results.
4 By clamping a pipe wrench on the outside edge of the key cylinder it can be turned bodily thereby opening the door.
5 The key mechanism can be drilled out or the entire cylinder pulled out by force.
6 Duplicate keys are readily available to anyone who knows the number and they can also be easily made if possession of the key is obtained for even a limited time.
7 Most of these locks can be picked with patience and some experience—though this is rarely necessary in the light of easier alternatives.

Manufacturers have overcome some of these weaknesses by various steps, all of which, of course, increase the price of the lock. The main improvement is that of dead-locking the latch and securing the operating handle on the inside, by either an extra turn of the key or other means according to the pattern of the lock. Nevertheless, it is not liked by insurance companies and, while it gives a degree of privacy against the casual caller who hopes to find an unlocked door, it gives little real protection.

Box locks

These too are fastened to the face of the door and the bolt housing to the face of the door jamb. Again, these easily give way under bodily pressure since the usual means of fixing is by screws. Moreover, they usually operate on the 'ward' principle; obstructing plates are placed between the keyway and the operating cam, and the key is cut to miss the obstructions. The result is that any key which is cut away simply to leave the operating part will provide an effective 'skeleton' key. Extra strength can be given by putting strapping over the lock and the housing on the door jamb. With the older types of this lock it is even possible to get hold of the end of the key with a sharp pair of pliers and turn it from the outside.

The two above-mentioned forms of lock offer a low security value but even with the better types of lock which are available, it is essential that the door itself should be of robust construction. It is useless putting a most expensive mortise lock on a flimsy door with a weak and narrow jamb. If a door is faulty, in that the edge is not close to the locking plate or staple, it will be possible to spring out the bolt by means of a jemmy.

Mortise locks

When installed in a suitable door a good quality mortise lock eliminates many defects of those previously mentioned. Both lock and the housing for the bolt are fitted inside the woodwork of the door and jamb and will therefore stand considerably greater pressures. These locks operate on the 'lever' principle: they contain thin strips of notched metal held in position by springs and the correctly cut key brings all the levers into alignment and allows the bolt to be moved. It follows that any increase in the number of levers brings additional security out of proportion to the multiplicity of key variations that become possible. If these locks are made to operate from both sides of the door—'double sided'—this reduces the maximum number of basically different keys; in the case of the four-lever mortise lock this is only nine for a particular type, though additional key variations may be obtained by the use of wards on the inside of the lock case. Nothing less than a five-lever lock is normally regarded as a security lock under present conditions; if this is operative from the one side only a very high number of differs are afforded—some 3000 as opposed to 125 with a double-sided lock.

A striking plate of 8 to 10 inches (20 to 25 cm) in length is desirable to receive the bolt; the best types include a small box of strong metal behind the opening for the bolt to prevent the use of end pressure upon it to force it back into the lock case—this would be possible if there were purely a wood surround to the bolt end.

Where these locks are used in doors where it is possible to attack the bolt by means of a hack-saw—in double non-rebated doors—better quality locks must be fitted which incorporate hardened steel rollers in the bolt. These are becoming standard in this type of lock; when they are reached by the hack-saw blade the rollers revolve freely giving no bite to the saw edge.

The number of levers in these locks can be progressively extended. At least one firm operates a series of locks with no fewer than ten levers providing almost limitless combinations for keys. The question of 'master keys' will be dealt with later.

Combination or code locks

The substantial forms of these are usually fitted to safes and strongroom doors. They have a number of discs each cut with a small notch. When the correct code is set on the dial, the notches are in alignment and the bolt can be withdrawn. The number of different combinations which are possible is of course dependent upon the number of discs in use. To give an example, if two discs only are utilised in conjunction with 100 different markings on the dial, the possible combinations are 100 multiplied by 100; these will increase proportionately as each additional disc is added.

These are keyless locks but the dial itself can be fitted with a key operation to prevent it being removed. The value of this type of lock is in the security of the combination number and it is essential to ensure that the person who is setting the mechanism should not be overlooked while he is doing so.

Time lock

This is a special mechanism used in conjunction with safes, and strongrooms which operates independently of any key or a combination lock. It is set to go off at a specific hour and the door cannot be opened until this time is reached, even if any other locks have been unlocked.

Padlocks

Padlocks really come into a separate category. The locks themselves can be extremely good; a well designed one should be made in one piece with the shackle close enough to the body of the lock to prevent an instrument being inserted to force it; the whole should be of hardened steel to prevent cutting. At least five and preferably more levers or pin tumblers should be contained in padlocks used for security purposes.

The main practical fault is one of sheer carelessness, in that the hasps and staples or the locking bars may be subject to cutting or be insufficiently fixed to the door or jamb. Like the padlock itself, the locking bar should be of hardened steel with no external rivets. It should be bolted through the door to the inside and through a backing plate with the bolt ends burred over. Manufacturers sell padlocks and locking bars complete.

It is a good practice periodically to change the locations of padlocks in use. This will go towards defeating anyone who has an unauthorised copy of a key. A spare set of locks is recommended which can be incorporated in the rotation plan.

Master key 'suites' or 'systems'

These have been designed to reduce to a minimum the number of individual keys that need to be carried by senior executives; for the same reason these will be of use to security staff. The best way to illustrate the function of such a system is to quote what will happen if it is installed in a works or office block. Locks from a planned grouping of locks will be fitted to all doors. Each of these locks will be different yet there will be one key which will open all of them—this will be the key of the managing director or senior executive on the site, the 'master' key.

In each subsidiary department there will again be one key which will open all the doors of that department but which will not fit the doors of any other department. These keys are 'sub-master' keys and will be held by the departmental heads. Finally there will be one key which will only open one lock and this will be held by each individual who has responsibility for the enclosure behind that particular door.

This system will pin-point responsibility for keys and reduce the number that need to be in existence. Each person will only carry one key and this will give him access to as many—or as few—doors as he has the right to enter. None of the inferior keys in such a system can be converted into a master key and special care is taken by all manufacturers to ensure no extra keys can be obtained other than by properly authorised persons. Manufacturers keep a full record of the firms to whom such systems are sold to ensure that maximum security of keys is obtained.

Colour coding of key tabs

It is possible, if it is thought necessary, to obtain keys with identifying coloured end-tabs. This can be of assistance in security offices where large numbers of keys are kept for a variety of departments. Each department will have a special identifying colour and a key would only be given to a member of that department presenting a similarly coloured numbered disc. Manufacturers will no doubt readily cooperate in supplying the needs for particular circumstances or this can be applied to a series of master key systems.

Conclusion

Remember that the security of the most highly priced lock is no better than that which is accorded to its key. If the key is misplaced by rank carelessness the lock can be valueless. Where security is of any importance the making of duplicate keys must be rigidly suppressed and the strictest records kept of the issue and return of keys.

SAFES

In a similar way to locks, many safes are not worthy of the name. This is particularly true of the older varieties where the backs, and with the oldest models even the sides, are riveted rather than welded together. In fact they are little more than strongboxes and can easily be ripped or cut open, especially if the back is readily accessible. Even the metal from which they have been manufactured is occasionally suspect and it has been known for a large safe to virtually disintegrate from a blow with a 14 pound (6.4 kg) hammer. Few of these old safes even live up to their fire resistant claims due to the deterioration of the fire resistant material they contain.

Modern safes really came into being with the First World War when the use of welding equipment became commonplace. The dangers of attack by explosives were recognised at an early stage and devices were introduced to combat this. However, only the most important safes were fitted with them and it is reasonable to assume that any safe manufactured before 1945 will not be explosive-resistant. The usual anti-explosive device consists of a spring-loaded steel plunger which is activated when an explosive charge is detonated in the safe door; this prevents movements of the safe bolts and the safe then requires factory attention to open it. Resistance to cutting by oxyacetylene burners has also only been developed since about 1945.

It can be difficult to assess the value of a really old but substantial-looking safe. Tests can be made by striking the top with knuckles—if there is any hollow sound or feel the safe is virtually useless. The safe's key, too, may give some indication as to whether this has been originally a superior type. The cutting on the key should be clean and sharp, there should be a minimum of seven levers, and if the key is 'double bitted' (cut at both sides) this is an indication of a good lock.

The ideal would be to progressively down-grade safes to storage of records and finally to scrap them, replacing them by more modern types. Regrettably, while new premises appear to merit new and strong safes, economic considerations are frequently quoted as reasons for not replacing them elsewhere, despite the fact that they can no longer fulfil their modern function.

If there is reluctance to replace a safe the alternative is to shield the metalwork from direct attack, so far as possible, leaving only the stronger part, the door, exposed. The best way to do this is to encase it in reinforced concrete—ordinary concrete could be peeled off and bricks are of little use. The same precaution can be taken in respect of any portable safe; to place any reliance in a safe which the criminals could take away, to open at leisure, is ludicrous. If setting in concrete is not practicable it should at least be possible to fix the safe's base into reinforced concrete by means of bolts.

Even a manufacturer of the most modern safes would be unwise to claim that his products are impregnable. If a thief has adequate time at his disposal and the most modern means—and it is not divulging a criminal trade secret to quote the thermic lance as a typical example—he will eventually cut his way into that safe. However, the longer the thief is delayed in attaining his objective the more likely he is to abandon the project and, if a well-designed safe is accorded further protection by well-devised alarm systems and periodic inspection, it should be adequately resistant to all the normal attacks.

There are few problems attached to the choice of the modern safe; production of these has been going on for many years and those firms that sold inferior products have long since become extinct. All manufacturers supply adequate literature describing their products with details of their capacity and resistance to types of attack. Insurance companies too, will advise on a type of safe commensurate with the risk that is being incurred.

Insurance requirements

Special data has been compiled by insurance companies equating each make of safe with a cash figure representing the risk, against which they are prepared to insure it. Some of these amounts are surprisingly low to the layman. It is no use at all going ahead with a purchase which appears to fit all requirements, and then find the insurers are not prepared to cover the amount of money it is desired to lodge in it. Non-compliance with their wishes will simply result in subsequent claims being refused. It is obvious therefore that insurers' advice is all important and must be followed if they are involved.

Siting of safes

The idea of safes being placed in out-of-the-way places where they may evade being found is obsolete (except for small wall safes—see below). Wherever possible safes should be put where they are visible to patrolling police officers, security staff, or even members of the public. Time and the opportunity to work unseen are the factors that the safe-breaker desires. The safe should be illuminated to facilitate inspection and the absence of the light will at once attract attention. A double-filament bulb is recommended. Observation points may be through peepholes in doors or walls and mirrors can be used to allow inspection round corners and where direct viewing is impossible.

Anchorage of safes

Any safe which weighs less than a ton and is sited at ground floor level should be secured against being removed bodily. Above ground floor level suitable measures depend on the facilities available for removal—such as lifts and hoists. A further limiting factor would be what the fabric of the building will carry, which also might control the size of the safe.

Practically all the larger modern types have anchorage points so that the safe can be bedded firmly into reinforced concrete. Where this is not so, substantial rag-bolts can be sunk into the floor and grouted in, while corresponding holes are drilled in the base of the safe; it is then set over the threaded ends and mild steel locking plates, linking each pair of bolts, are fitted before the nuts are put on—the object of this is to spread the force and prevent the safe being levered over the nuts. Similar means can be used to attach the safe to walls by the sides or back instead of the base.

Combination safe locks

The workings of these locks have been described earlier in this chapter; they are ideal where there are apt to be frequent changes of staff since the combination can be altered to suit the person then responsible for the security of the safe. Moreover there is no possibility of the key being copied by a previous holder as could happen with normal locks. A further attribute lies in the difficulty of applying explosive substances, since there is no key-hole whereby they can be inserted in the door.

Having established a code for a combination lock the user should deposit a copy of it at some normally inaccessible point to obviate the gross inconvenience that might occur if he were to forget it or if the safe needed opening in his absence. The ideal place is at his bank. Any patrolling security officer, or person with security responsibilities, should occasionally glance at the side of the safe to ensure that no one has been so stupid as to write the code number there for convenience—this is not a flight of fancy, either!

Apart from the normally accepted free-standing kinds of safe there are smaller types which are used for specific purposes; the most important of these are described below.

Wall safes

These are used for safe keeping of sums of money and small valuable personal belongings. They can be obtained in a variety of sizes from approximately that of one brick upwards. Their resistance against attack may be limited and this type of safe should be concealed in some out-of-the-way place. If they are to contain valuables of consequence they should be drill resistant, fitted with anti-explosive devices, and firmly fitted into the wall in a similar manner to which a safe is set into the floor. A substantial wall safe is of little use in an ordinary brick wall from which it can easily be prised. Smaller safes have key locks only but the larger ones may have either key or combination locks. The essential point is that of concealment.

Floor safes

These are becoming increasingly common and are particularly valuable for guarding cash. Relatively speaking, they are much stronger than wall safes; they are easy to set in reinforced concrete; the removable tops are tapered and therefore cannot be knocked in; they are normally fitted with anti-explosive bolts; and they can easily be concealed beneath the floor tiles. In a variation money can be dropped into a floor safe by way of a chute, without a key being in the possession of the person making a deposit.

Night safes and trap deposits

These are simply means of enabling money to be put into a safe without opening it and without substantially reducing its security by the method employed. The form taken is usually that of having a rotary trap linked to the safe interior by a chute. Apart from use at banks, for the convenience of shopkeepers and businessmen in general, such safes can be of value at petrol stations and the like in conjunction with normal safes. The main advantage is that substantial sums of money need not be kept on the premises under conditions of needless risk after normal working hours.

STRONGROOMS

Erection of strongrooms is a job for the expert and should not be given to the ordinary building contractor. Current criminal practice is to attack strongrooms by way of the walls, floors, or ceilings—frequently from adjacent premises.

As with safes, strongrooms should ideally be sited where they can be regularly inspected, and there certainly should be some form of alarm system fitted to them. In new constructions special consideration should be given to the siting of the strongroom at the drawingboard stage. Extremely resistant heavy doors and grilles are available but the other potential areas of attack, through doors, walls, and ceilings, must not be overlooked and these weaknesses countered by the use of strongly reinforced concrete or other methods.

PART EIGHT

PROFESSIONAL BODIES

34

The Industrial Police and Security Association

This association was formed in 1958 to be the professional and representative organisation for those persons employed on security duties. It was the first and remains the only such body. It is non-political in character with no trade union affiliations or aspirations.

The objectives of the association are :

1 To establish, promote and encourage the science and professional practice of industrial and commercial security, and all operations and expedients connected therewith.
2 To promote, and make more effective security measures in industry and commerce and to improve the status of the individual by providing a close liaison between all members of the profession, thus making possible an exchange of ideas, knowledge, information and experience, between members and others in all matters of common interest and mutual concern.
3 To establish, foster and encourage ethical and professional standards of work and conduct for members of the security profession and to take all such steps as are necessary and expedient in this connection.
4 To provide and to promote a means of maintaining a centralised and representative body of the industrial security profes-

sion and to collect, collate, coordinate and distribute by any means deemed advisable: data, information, ideas, knowledge, methods and techniques for the benefit and improvement of the industrial security profession and its members.

5 To make surveys and studies, hold conferences, forums and training courses and arrange for the presentation of lectures and the reading of papers on matters and problems of interest to members of the Association; to foster, promote, encourage and facilitate discussion, study and research on matters and problems of all kinds connected with industrial security; and generally to collect and disseminate in any manner or by any means deemed appropriate, information of service or interest to Association members or to the public at large.

6 To establish and to operate for the benefit of industry, commerce, the industrial security profession in general and the Association's members in particular, an Appointments Bureau.

7 To administer an 'Institution of Industrial Security' with categories of Graduate, Member and Fellow. To arrange for the appointment of an Examinations Board, to set and to hold annual examinations in such centres as may be decided and to award certificates of qualification to successful students passing these examinations. (The Institution will be dealt with in Chapter 35.)

MEMBERSHIP CATEGORIES

Member

This is available to members of a security organisation employed in a whole- or part-time capacity by an industrial or commercial establishment, government department, nationalised industry, or public utility company. The membership fee is 75p a year which entitles the member to participate in all regional activities, to attend any training course organised by the Association (on payment of the appropriate fee), to free advice and assistance on any security problem, and to use the services of the Appointments Bureau free of charge.

Corporate

This category of membership is available to chief or senior security officers, directors, personnel managers, plant engineers, officers, officials and executives who employ and supervise staff engaged in the industrial and commercial security profession. The fees are £3 a year. In addition to the privileges of general membership this grade of member may attend special studies, conferences, and forums which are arranged by the National Council of the Association from time to time to suit their requirements.

Associate

This category of membership is available to individuals in executive, professional, supervisory, and administrative positions in organisations engaged in manufacturing or supplying security materials, equipment, or services to industry and commerce. The fees of this membership are £7.50 a year. Such membership will provide those persons so employed with an opportunity of meeting together to discuss matters of common interest and concern. It also permits participation in all regional activities, attendance at training courses, and free advice on any matter connected with industrial security.

Group member

Membership in this category is available to industrial and commercial companies, corporations, nationalised industries, public utility undertakings, banks, government departments and local authorities. The annual fees are £10.50 a year. The membership entitles such organisations to send any members of their staff to training courses, studies, conferences, or forums organised by the Association on payment of the course fee. Membership of the Association is normally necessary to attend training courses but employees of group members may attend without being members in their own right. Free use of the Appointments Bureau is also available.

This type of membership provides managements with an opportunity to give tangible expression to their desire to assist the Association in its work for the benefit and for the improvement of industrial and commercial security measures.

ADMINISTRATION

National

The governing body of the Association is the National Council which consists of the Chairmen of the Regional Councils, which will be described, and a National Secretary/Treasurer.

Regional

The Regional Councils referred to are the bodies representing members in the following geographical areas of the UK. Overseas members and (at present) members in Ireland have no direct representation on the National Council. For administrative purposes the country is divided into Regions each with its Council. They are as follows:

London and South East Region. The counties of Bedfordshire, Buckinghamshire, Essex, Hertfordshire, Kent, Surrey, Sussex, and Greater London. Headquarters are in London.

Midland Region. The counties of Herefordshire, Shropshire, Staffordshire, Warwickshire, and Worcestershire. Headquarters are in Birmingham.

Northern Region. The counties of Cumberland, Durham, Northumberland and Westmorland. Headquarters are in Newcastle-on-Tyne.

North East Region. The East and West Ridings of Yorkshire. Headquarters are in Leeds.

North Midland Region. The counties of Derbyshire (part), Leicestershire, Lincolnshire, Nottinghamshire, Northamptonshire, and Rutland. Headquarters are in Nottingham.

North West Region. The counties of Cheshire, Derbyshire (part), and Lancashire. Headquarters are in Manchester.

Southern Region. The counties of Berkshire, Dorset, Hampshire (including the Isle of Wight), and Oxfordshire. Headquarters are temporarily in London.

South West Region. The counties of Cornwall, Devon, Gloucestershire, Somerset, and Wiltshire. Headquarters are in Swindon.
Welsh Region (North). Counties of Anglesey, Caernarvon, Flint, Merioneth, Montgomery, and parts of north Cardigan and Radnor. Headquarters at Wrexham.
Welsh Region (South). All other counties of Wales, and Monmouthshire. Headquarters are at Port Talbot.
Eastern Region. The counties of Cambridge, Lincoln, Norfolk, and Suffolk. Headquarters at Norwich.
Scottish Region. All counties of Scotland. Headquarters at Edinburgh.

TRAINING

One of the main activities of the Association is the holding of training courses in security and associated duties for security personnel of all categories. These are described in Chapter 5.

Meetings of members are arranged frequently at regional and sub-regional levels where talks are given by well-known authorities on security subjects and demonstrations made of new security equipment and techniques.

INSTITUTION OF INDUSTRIAL SECURITY

A qualifying period of membership of the Association, with other qualifications required by the Governors, is the basis of an application for membership through examinations of the Institution of Industrial Security, the by-laws of which are described in the following chapter.

JOURNAL

The journal of the Association and the Institution is *Security and Protection*. It is published monthly, free to corporate, associate and group members, from Pembroke House, Campsbourne Road, Hornsey, London N8.

OTHER PROFESSIONAL BODIES

The following organisations represent practitioners in security in other parts of the world:

The American Society for Industrial Security
404 Nada Building
2000 K Street Northwest
Washington, DC 20006
USA

The Canadian Society for Industrial Security
336 Whimby Avenue
St Lambert
Province of Quebec
Canada

The Industrial and Commercial Security Association of South Africa Ltd
PO Box 175
Mobeni
Durban
Natal
South Africa

The Institute of Commercial and Industrial Security Executives
Mascot
Australia

Interested persons can obtain further information from those addresses.

35

The Institution of Industrial Security

In common with members of other professions with institutes or institutions, the security profession has its own body. Such institutions award coveted membership on proof through examination of a high level of professional qualification, experience, and suitability.

The establishment of the Institution of Industrial Security (note that 'industrial' includes 'commercial') followed years of preparation and membership is awarded to members of the Industrial Police and Security Association who satisfy the requirements of the Governors, who include prominent businessmen and lawyers. Exceptions are made where membership is awarded to other persons who have achieved eminence in their particular fields which would be beneficial to the Institution.

Membership of the I.I.Sec., as of any other professional institution, shows the possession of knowledge and experience, recognised by all persons in the particular profession, and in return requires a high standard of performance and ethical behaviour. Failure to maintain these standards usually leads to the cancellation of membership which can have serious consequences and therefore the possibility has a controlling influence on members.

An extract of the relevant by-laws follows.

BY-LAWS

1 *The Institution of Industrial Security,* hereinafter called the Institution, shall be governed and administered by the National Council of the Industrial Police and Security Association who shall have the power to delegate certain responsibilities hereinafter defined to a Board of Governors.

2 The President, Chairman and Secretary of the National Council for the time being, shall be members of the Board of Governors for as long as they shall hold office, and thereafter by appointment. The remaining Governors shall be appointed by the National Council of the Association, who shall also decide any cases of doubt or difficulty as to eligibility for or retirement from office.

6 The Board of Governors of the Institution shall have the power to appoint committees chosen from their own body, and where special circumstances prevail or make it advisable and with the agreement of the National Council of the Association, such committees may include co-opted members.

7 *Membership of the Institution of Industrial Security.* The Institute shall consist of—

Fellows
Members
Graduates

Fellows and Members of the Institution must have been Corporate or Associate Members of the Industrial Police and Security Association for not less than three years. Graduates must have been Members of the Association for not less than two years.

9 *Abbreviated Titles and Description of Membership.* A Fellow of the Institution shall be entitled to the exclusive use after his name of the initials F.I.I.Sec.; a Member of the initials M.I.I.Sec.; a Graduate of the initials Grad.I.I.Sec.

10 *Certificates.* Subject to such regulations and on payment of such fees as the bylaws of the Institution may from time to time prescribe, the Institution shall issue to any member of any class a certificate showing the class to which he belongs. Every such certificate shall remain the property of, and shall on demand be returned to, the Institution.

11 *Fellows.* Every candidate for election or transfer to the class of Fellow shall satisfy the Board of Governors either—

(*a*) that he—

i Is at least thirty-five years of age and has been a Member of the Institution for at least two years;

ii Has satisfied the requirements of the Examinations and Training Regulations;

iii Has had at least five years' experience in a senior or supervisory capacity in industrial or commercial security.

(*b*) or that he—

Has such knowledge of industrial and commercial security and has acquired such eminence in the profession that his admission as a Fellow would in the opinion of the Governors conduce to the interests of the Institution and the industrial security profession.

12 *Members.* Every candidate for election or transfer to the class of Member shall satisfy the Board of Governors either—

(*a*) that he—

i Is at least thirty years of age and has been a Corporate or Associate Member of the Industrial Police and Security Association for at least three years;

ii Has satisfied the requirements of the Examinations and Training Regulations;

iii Has had appropriate experience and has held rank in a civil police force including Colonial and HM Forces' Police or Fire Brigade, plus a period of not less than three years' experience in a senior or supervisory capacity as an industrial security officer, or has had at least seven years' experience in industrial and commercial security, plus a period of not less than three years' experience in a senior or supervisory capacity as an industrial and commercial security officer.

(*b*) or that he—

Has had such experience and has acquired such knowledge in the industrial security profession that his admission as a Member would in the opinion of the Governors conduce to the interests of the Institution and the industrial security profession.

The Board of Governors of the Institution may require any

candidate for Membership to attend an interview conducted by them on their behalf in order that he may better satisfy them that he possesses the requisite qualifications and experience.

13 *Graduates.* Every candidate for election to the class of Graduate shall have attained the age of twenty-five years and shall satisfy the Board of Governors—

(*a*) That he has attained such a standard of general education and has received such training in general security subjects as satisfy the requirements of the Examinations and Training Regulations.

(*b*) That he possesses such knowledge and has acquired such experience in the industrial security profession as to satisfy the requirements of the Board of Governors, and that he is employed in a full-time capacity in the industrial security profession.

(*c*) That it is his intention to remain in the profession and to seek Corporate Membership of the Industrial Police and Security Association as soon as appropriate.

14 *Annual subscriptions.* The following annual subscriptions shall be payable by Fellows, Members, and Graduates—

Fellows £3.15
Members £2.10
Graduates £1.05

(In addition to the relevant membership subscription of the Association.)

15 The annual subscriptions shall be due on 1 January each year for the year commencing on that day.

16 Members of any class elected before 1 July in any year shall pay the annual subscription for that calendar year, and those elected on or after 1 July in any year shall for that calendar year pay half such annual subscription.

Provided that in the case of any member elected in the last two months of any calendar year who shall elect to pay his first subscription at the full rate, such subscription shall cover the remainder of the year of his election and the next succeeding year.

17 Any member of any class whose annual subscription remains unpaid after 30 June shall not be entitled to attend or to

take part in any meeting of the Institution that may be held, or to receive any notice or publication of the Institution that may be issued before he has paid his subscription in full.

18 Any member of any class whose annual subscription remains unpaid for one year may by resolution of the Governors be excluded from the Institution, and he shall thereupon cease to be a member and his name shall be removed from the Register; but such removal shall not relieve him from his liability for the payment of the arrears of subscription due from him calculated up to 31 December preceding his exclusion.

20 The Board of Governors may readmit to membership in the class to which he formerly belonged any person whose membership has terminated from any cause, provided he satisfies the Board of Governors that he is worthy of readmission and pays such amounts in respect of arrears of subscription as the Board of Governors may determine.

In the event of the Board of Governors deciding to refuse readmission in any particular case, they may do so without assigning any reason thereto.

21 *Expulsion.* In case any member of any class shall be convicted of any criminal offence, the Board of Governors may decide that his name shall be removed from the Register of the Institution, and the Secretary shall communicate such decision to such member forthwith.

22 Any member of any class who shall act in contravention of any by-law or who shall refuse or wilfully neglect to comply with any of the by-laws or who shall be alleged to have been elected or transferred to any class, or to have procured election by the Board of Governors as a result of false representation, or who shall in any other respect have been guilty of such conduct as in the opinion of the Board of Governors either shall have rendered him unfit to remain a member of the Institution or shall be injurious to the Institution shall be liable to have his name removed from the Register of the Institution if the Board of Governors shall so decide, after taking the matter into consideration and giving such member an opportunity of defending himself and meeting any accusation that may be made against him, either in person

before the Board of Governors or in writing. The Secretary shall communicate such decision to such member forthwith.

23 *Professional conduct.* Every member of any class is required at all times so to order his conduct as to uphold the established traditions of the Institution and the dignity of the industrial security profession.

Any alleged breach of this by-law which may be brought before the Board of Governors, properly vouched for and supported by sufficient evidence, shall be investigated and if proved shall be dealt with by the Board of Governors either by expulsion of the offender from the Institution under the procedure of by-law 22 as far as it applies, or in such other manner as the Board of Governors, with the approval of the National Council of the Industrial Police and Security Association, may think fit.

25 *Notices.* Any notice may be served or any communication may be sent by the Board of Governors or by the Secretary of the Institution upon or to any member of any class either personally or by sending it prepaid through the post addressed to such person at his address as registered in the books of the Institution. Any notice or communication, if served or sent by post, shall be deemed to have been served or delivered on the day following that on which the same is posted; and in proving such service or sending it shall be sufficient to prove that the notice or communication was properly addressed and posted.

28 The Board of Governors shall have the power to add to, alter or vary any of these by-laws, always provided that such additions, alterations, or variations shall first have been considered and approved by the National Council of the Association.

APPENDICES

I

Chief security officer: job description

It is suggested that the chief security officer's position and responsibility be on a par with other middle line management positions, as shown in the following diagram. His colleagues of similar status might include the stores manager, the maintenance manager, the work study manager, etc. Titles and relative status are obviously different in every organisation.

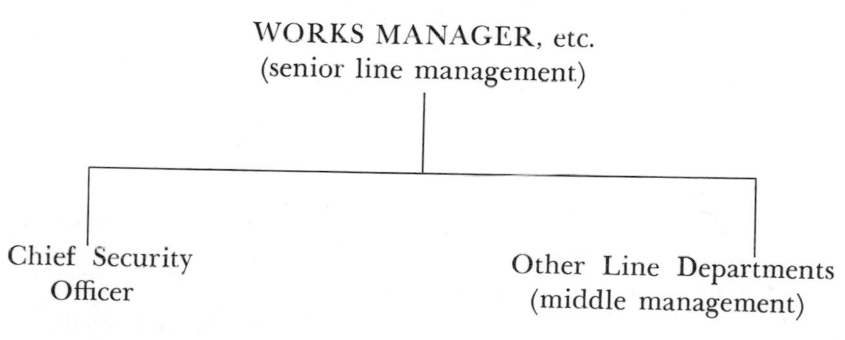

WORK INVOLVED

1 To organise the systematic supervision and patrolling of all factory boundary fences, storage areas, works and office buildings, etc., to ensure the safekeeping of company property.
2 To make recommendations to management on all matters where he considers the security of the company's plant, buildings, materials, and other property can be improved. He will have similar responsibility in connection with the property of any company employee, but, where there is no legal liability on the company for such matters, this will be only in an advisory capacity to help the individual without legal obligation.
3 To be responsible for the recruitment and day-to-day administration of the security officers, their instruction in the various aspects of their duties and ensuring that they main-

tain a suitable standard of efficiency and deportment. He will also take steps to provide them with facilities to participate in periodic first aid and security training to keep their abilities in this field to a satisfactory level consistent with the requirements of the Factories Act.

4 To ensure that the security staff are fully conversant with the operation of all fire equipment in the works; that such equipment is fully and adequately maintained; that the satisfactory liaison is created and maintained at a high level with the local fire prevention officer and fire brigade; and that all fires occurring on the company property are fully investigated and reported upon.

5 To organise duties at all gatehouses, to ensure that the company's rules and regulations relating to the entry and exits of employees, contractors' employees, visitors, and vehicles belonging to company and other parties are observed. This will include the discretionary right to search persons and vehicles.

6 To cause such books to be kept in the main gatehouse as are necessary to ensure the permanent recording of commercial vehicle movement in and out of the works, with notation of purpose or load. He will also cause a day-to-day diary to be kept for reference purposes and such other records as he deems essential for the efficient functioning of the department.

7 To be responsible for the guarding of wages after receipt from the bank and durıng distribution to employees.

8 To inquire as necessary into and report upon any thefts within the works and use his discretion in connection with any of these matters he deems should be reported to the civil police.

9 To maintain the best possible cooperation at high level with the police and fire authorities of adjacent areas and be responsible for dealing with any inquiries from them.

10 To arrange the allocation of staff to carry out any miscellaneous duties which may be required in the company's interests; these will include:

(*a*) Control and direction of all commercial and private traffic entering and leaving the works and office areas.

(*b*) Control of the parking of cars and motor cycles of the company's employees at the works.
(*c*) Supervision of the car parks reserved for visitors.
(*d*) Assisting the local police when necessary to control and expedite the exit of traffic from the works area into adjoining roads.

11 To prepare annual estimates of the expenditure to be incurred on the upkeep of the security staff, installations and equipment and submit these for inclusion in the annual budget.

12 To carry out such miscellaneous duties and inquiries as may be required of him in the company's interest.

13 To keep in touch with developments in mechanical and other aids to security; and in techniques by maintaining contacts with persons in parallel positions in other companies and professional associations.

Requirements

The position calls for a thorough training and experience in the administration of the law concerned with the prevention of crime and in the detection of offenders. The ability to train, organise, and supervise the duties of a security staff is also required.

Initiative

He will be expected to advise the company on matters within his experience and to perform his duties without immediate supervision and to make decisions. In his dealings with all levels of management, staff, and general workforce, he will be expected to maintain an amicable and cooperative relationship.

His normal hours of duty will be the same as those required of other members of management on Monday to Friday inclusive. He will have authority at his discretion to vary those hours in accordance with the needs of economy.

Responsibility

He will be directly responsible to the works manager for the proper performance of the duties described under 'Work Involved'.

Alterations

No alterations or additions to the duties of the chief security officer will be made without authority.

Security department organisation chart

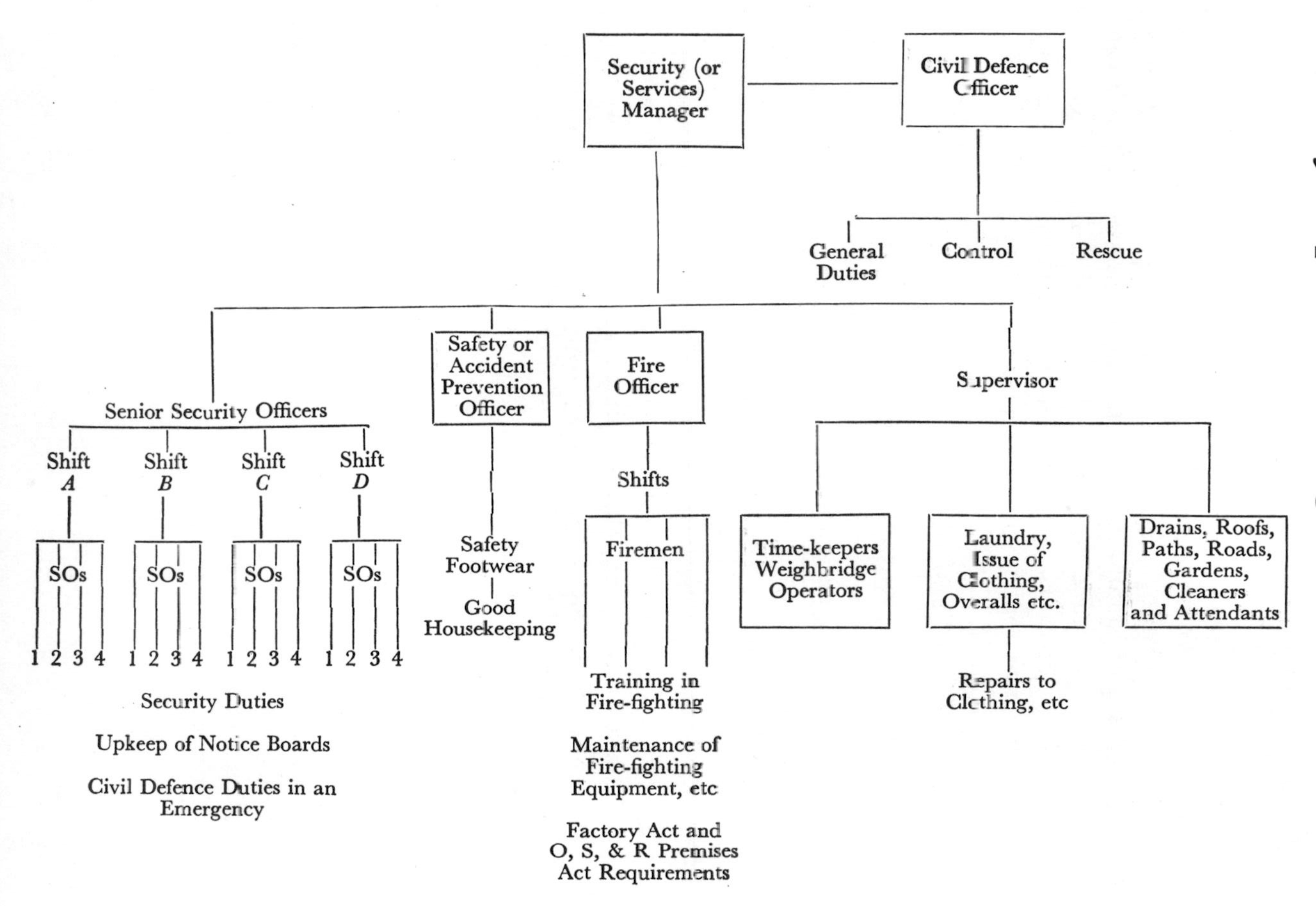

3

Confidential standing orders to security staff

To assist management of premises where uniformed security officers are employed and where no consolidated form of written instructions regarding their duties has been issued, the following suggested list of 'standing orders' has been drawn up. These can be elaborated to take account of local conditions and requirements. They, of course, supplement the usual conditions of employment which apply to all employees.

DUTIES OF THE SECURITY STAFF

1 They will be punctual in commencing their periods of duty, as instructed by the chief security officer (CSO), and be in a fit and proper condition. They will sign the attendance register at the commencement and the completion of their periods of duty. They will not go off duty without special permission until their duties are taken over by another member of the security staff.

2 They will perform duty in the uniform provided by the company including shirts, collars, and ties, unless wearing plain or civilian clothes has been specially authorised. Black footwear will be worn. Black or dark blue scarves only may be worn during inclement weather. Shirt sleeve order will be adopted only by authority of the CSO or other authorised person. Medal ribbons should be worn. Where two sets of uniform are held the older garments will be worn for night duty.

3 Uniforms, which remain the property of the company, will be kept in a clean and tidy condition. The wearing of uniform trousers to and from duty is permitted but uniform jackets and caps will not be worn outside the perimeter of company property unless on direction.

4 Should any article of uniform become unfit for wear because of damage it is to be handed to the CSO, where appointed, or other official responsible for the supervision of the staff, with a report in writing as to the cause.

5 If such clothing be damaged through any wilful act or omission or carelessness on the part of the person to whom issued, the company reserves the right to assess that person the cost of the replacement after taking into consideration the length of time the garment has been worn or of the cost of repair.

6 On a member leaving the security department for any reason he will be required to return his complete uniform in an acceptable condition, taking into consideration fair wear and tear, otherwise the company may exercise the right described in paragraph 5.

7 Instructions to the security staff will be given only by the person responsible for their supervision.

8 Members of the security staff are expected to keep themselves informed of all instructions affecting their duties and of any amendments issued by proper authority and where in writing to append their signatures to such orders as directed.

9 Security measures which will be introduced from time to time and all instructions to the security staff are confidential and are not to be discussed with anyone outside the Security Department. Any breach of this order will be dealt with as a serious offence against discipline.

10 Members of the security staff are expected to make themselves proficient in first aid by the passing of the relevant primary and refresher examinations set by the Orders of St John or St Andrew or the Red Cross Association.

11 They are similarly expected to become proficient in fire fighting and attend training periods as required.

12 They will be expected to carry out searches of company employees and their vehicles to detect the unauthorised removal of company property within the terms of the conditions of employment of the company.

13 They will also be expected to carry out searches of personnel and vehicles of contractors working on the premises subject to their previous agreement being obtained to comply with the conditions of employment of company personnel.

14 The principal duties of security officers are as follows:

(*a*) To protect the property of the company at all times against theft both from inside and outside the premises. [Here would follow detailed instructions regarding action on discovering the unauthorised removal of company property.]

(*b*) To protect the buildings of the company and their contents from damage by fire or water or from the effect of bad weather conditions, and prevent waste of materials.

(*c*) To ensure all safety regulations, including smoking restrictions, of the company are observed and to assist as directed in the prevention of accidents.

(*d*) To ensure that all the regulations of the company, affecting the security of its property and the property of employees, are carried out.

(*e*) To pay attention to all water, steam, gas, and electrical installations, to detect breakdowns and wastage and to take action where necessary in accordance with instructions.

(*f*) To see that all fire-fighting equipment is in the designated locations and not treated in such a manner as to affect its immediate and efficient use.

(*g*) To report in the manner directed all accidents affecting employees or property.

(*h*) To ensure that no unauthorised persons or vehicles enter company premises and to see that all persons properly seeking admission are courteously received and assisted. To record as directed the entry and exit of all motor vehicles and visitors.

(*i*) To ensure that no employee, or employee of a contractor, or any vehicle leaves the premises in any irregular manner.

(*j*) To require the production of authorisations for the removal of company property, including borrowed tools, and to deal with as directed.

(*k*) To ensure that any employee below supervisor's grade does not leave the premises at irregular times without a written note of authority. To deal with as directed any pass out surrendered.

(*l*) To take charge of all personal property found on company premises and to record in the relevant register a description of the property together with details of the time, date and place of finding. The entry should be signed by the finder of the property in agreement of its correctness.

(*m*) To record in the relevant register all complaints of the loss of personal property within the perimeter of the premises together with all necessary information.

(*n*) To take charge of keys and to issue them to authorised persons only in accordance with local instructions.

(*o*) To attend time clocks to prevent and detect irregularities.

(*p*) To render all proper assistance to employees in any emergency such as accepting and communicating messages received on their behalf.

(*q*) To give all proper assistance to the police.

(*r*) To record all occurrences concerning the security department in the log book for the information of management and other interested persons. It is essential that full and accurate information be recorded.

(*s*) To regulate traffic and organise and control the parking of vehicles or storage of bicycles on company property.

(*t*) To act as weighbridge operators when required.

(*u*) To attend any burning or welding operation with fire-fighting equipment.

(*v*) To observe recording instruments and, within reasonable limits, experiments in laboratories.

(*w*) To maintain fire-fighting equipment.

(*x*) To act as escorts for the carrying of cash.

15 These instructions do not touch on all circumstances which may call for the attention of the security staff. Where a situation arises and no specific instructions have been issued which apply to it the members of the security staff will be expected to use intelligence, imagination, and discretion to ensure it is dealt with satisfactorily.

16 If breaches of industrial discipline are observed by the security staff their action will be confined to reporting the matter to the chief security officer or to whom they are directly responsible. If immediate action is required the matter will first be reported to the supervisor of the department or section concerned.

17 Security officers should discuss any problem affecting their duties they may have with the CSO and when in doubt regarding their action confer with him for directions.

18 Additions or amendments to these orders will be made only in writing and by the chief security officer or other official in charge of the security staff.

19 On appointment a security officer will be issued with a personal copy of these orders, and any subsequent additions or amendments, which will be returned by him on his relinquishing his position.

20 A copy of these orders, and additions and amendments, will be kept at the security department for ready reference when required by any member of the security staff.

Date................................ Signed...........................

4

Duty rosters

CONTINUOUS WORKING—THREE SHIFT, FOUR GANG

4 × 7 SYSTEM

Cycle starting Monday, 21 days working 7 days rest per month

	DAY	06.00 to 14.00	14.00 to 22.00	22.00 to 06.00	REST	TAKING GROUP 'A' ONLY		
						Rest Period	Rest Hours	Hours Worked
1	M	A	B	C	D			
2	T	A	B	C	D			
3	W	A	B	C	D			
4	T	A	B	C	D	NIL	NIL	56
5	F	A	B	B	C			
6	S	A	B	C	D			
7	SUN	A	B	C	D			
8	M	D	A	B	C			
9	T	D	A	B	C			
10	W	D	A	B	D			
11	T	D	A	B	C	NIL	NIL	56
12	F	D	A	B	C			
13	S	D	A	B	Y			
14	SUN	D	A	B	C			
15	M	C	D	A	B			
16	T	C	D	A	B			
17	W	C	D	A	B			
18	T	C	D	A	B	NIL	NIL	56
19	F	C	D	A	B			
20	S	C	D	A	B			
21	SUN	C	D	A	B			
22	M	B	C	D	A			
23	T	B	C	D	A	6am MON	168	NIL
24	W	B	C	D	A	to		
25	T	B	C	D	A	6am MON	NO PAY	
26	F	B	C	D	A	(Pay usually averaged per week		
27	S	B	C	D	A	over duration of rota)		
28	SUN	B	C	D	A			

CONTINUOUS WORKING—THREE SHIFT, FOUR GANG

CONVENTIONAL SEVEN-SHIFT CYCLE

		Day	06.00 to 14.00	14.00 to 22.00	22.00 to 06.00	Rest	Rest period	Rest hours
	1	S	A	C	B	D	2pm Sat to	72
	2	M	A	C	B	D	2pm Tues	
	3	T	A	D	B	C	10pm Mon to	72
1	4	W	A	D	B	C	10pm Thurs.	
	5	T	A	D	C	B	6am Thurs to	
	6	F	A	D	C	B	6am Sun.	72
	7	Sat	A	D	C	B		
	8	S	B	D	C	A	2pm Sat to	
	9	M	B	D	C	A	2pm Tues	72
	10	T	B	A	C	D	10pm Mon to	
2	11	W	B	A	C	D	10pm Thurs	72
	12	T	B	A	D	C	6am Thurs to	
	13	F	B	A	D	C	6am Sun	72
	14	Sat	B	A	D	C		
	15	S	C	A	D	B	2pm Sat to	72
	16	M	C	A	D	B	2pm Tues	
	17	T	C	B	D	A	10pm Mon to	72
3	18	W	C	B	D	A	10pm Thurs	
	19	T	C	B	A	D	6am Thurs	
	20	F	C	B	A	D	6am Sun	72
	21	Sat	C	B	A	D		
	22	S	D	B	A	C	2pm Sat to	72
	23	M	D	B	A	C	2pm Tues	
	24	T	D	C	A	B	10pm Mon to	72
4	25	W	D	C	A	B	10pm Thurs	
	26	T	D	C	B	A	6am Thurs to	
	27	F	D	C	B	A	6am Sun	72
	28	Sat	D	C	B	A		

CONTINUOUS WORKING—THREE SHIFT, FOUR GANG

3 × 2 × 2 'CONTINENTAL' SYSTEM

		Day	06.00 to 14.00	14.00 to 22.00	22.00 to 06.00	Rest	Rest period	Rest hours
	1	S	A	B	C	D	6am Sun to 6am Wed	72
	2	M	A	B	C	D		
	3	T	A	B	C	D		
1	4	W	D	A	B	C	6am Wed to 6am Fri	48
	5	T	D	A	B	C		
	6	F	C	D	A	B	6am Fri to 6am Sat	48
	7	Sat	C	D	A	B		
	8	S	B	C	D	A	6am Sun to 6am Wed	72
	9	M	B	C	D	A		
	10	T	B	C	D	A		
2	11	W	A	B	C	D	6am Wed to 6am Fri	48
	12	T	A	B	C	D		
	13	F	D	A	B	C	6am Fri to 6am Sun	48
	14	Sat	D	A	B	C		
	15	S	C	D	A	B	6am Sun to 6am Wed	72
	16	M	C	D	A	B		
	17	T	C	D	A	B		
3	18	W	B	C	D	A	6am Wed to 6am Fri	48
	19	T	B	C	D	A		
	20	F	A	B	C	D	6am Fri to 6am Sun	48
	21	Sat	A	B	C	D		
	22	S	D	A	B	C	6am Sun to 6am Wed	72
	23	M	D	A	B	C		
	24	T	D	A	B	C		
4	25	W	C	D	A	B	6am Wed to 6am Fri	48
	26	T	C	D	A	B		
	27	F	B	C	D	A	6am Fri to 6am Sun	48
	28	Sat	B	C	D	A		

FOUR CREWS, THREE SHIFTS

W=hours worked.

SHIFTS	Week 1	Week 2	Week 3	Week 4
	S M T W T F S	S M T W T F S	S M T W T F S	S M T W T F S
NIGHTS	*A A A A A A B*	*B B B B B C C*	*C C C C D D D*	*D D D A A A A*
AFTERNOONS	*C C D D D D D*	*D A A A A A A*	*B B B B B B C*	*C C C C C D D*
MORNINGS	*B B B B C C C*	*C C C D D D D*	*D D A A A A A*	*A B B B B B B*

CREWS	W	W	W	W
	A 48	*A* 48	*A* 40	*A* 40
	B 40	*B* 40	*B* 48	*B* 48
	C 40	*C* 40	*C* 40	*C* 40
	D 40	*D* 40	*D* 40	*D* 40

DUTY	Week 9	Week 10	Week 11	Week 12
	S M T W T F S	S M T W T F S	S M T W T F S	S M T W T F S
NIGHTS	*B B B B C C C*	*C C C D D D D*	*D D A A A A A*	*A B B B B B B*
AFTERNOONS	*A A A A A A B*	*B B B B B C C*	*C C C C D D D*	*D D D A A A A*
MORNINGS	*C C D D D D D*	*D A A A A A A*	*B B B B B B C*	*C C C C C D D*

CREWS	W	W	W	W
	A 48	*A* 48	*A* 40	*A* 40
	B 40	*B* 40	*B* 48	*B* 48
	C 40	*C* 40	*C* 40	*C* 40
	D 40	*D* 40	*D* 40	*D* 40

DUTY	Week 17	Week 18	Week 19	Week 20
	S M T W T F S	S M T W T F S	S M T W T F S	S M T W T F S
NIGHTS	*C C* D *D D D D*	*D A A A A A A*	*B B B B B B C*	*C C C C C D D*
AFTERNOONS	*B B B B C C C*	*C C C D D D D*	*D D A A A A A*	*A B B B B B B*
MORNINGS	*A A A A A A B*	*B B B B B C C*	*C C C C D D D*	*D D D A A A A*

CREWS	W	W	W	W
	A 48	*A* 48	*A* 40	*A* 40
	B 40	*B* 40	*B* 48	*B* 48
	C 40	*C* 40	*C* 40	*C* 40
	D 40	*D* 40	*D* 40	*D* 40

SEVEN-DAY WEEK

Repeat at Week 25

Week 5	Week 6	Week 7	Week 8	
S M T W T F S	S M T W T F S	S M T W T F S	S M T W T F S	
A A B B B B B	*B C C C C C C*	*D D D D D D A*	*A A A A A B B*	
D D D D A A A	*A A A B B B B*	*B B C C C C C*	*C D D D D D D*	
C C C C C C D	*D D D D D A A*	*A A A A B B B*	*B B B C C C C*	
W	W	W	W	Hours
A 40	*A* 40	*A* 40	*A* 40	=336÷8=42
B 40	*B* 40	*B* 40	*B* 40	336÷8=42
C 48	*C* 48	*C* 40	*C* 40	336÷8=42
D 40	*D* 40	*D* 48	*D* 48	336÷8=42

Week 13	Week 14	Week 15	Week 16	
S M T W T F S	S M T W T F S	S M T W T F S	S M T W T F S	
C C C C C C D	*D D D D D A A*	*A A A A B B B*	*B B B C C C C*	
A A B B B B B	*B C C C C C C*	*D D D D D D A*	*A A A A A B B*	
D D D D A A A	*A A A B B B B*	*B B C C C C C*	*C D D D D D D*	
W	W	W	W	
A 40	*A* 40	*A* 40	*A* 40	=336÷8=42
B 40	*B* 40	*B* 40	*B* 40	336÷8=42
C 48	*C* 48	*C* 40	*C* 40	336÷8=42
D 40	*D* 40	*D* 48	*D* 48	336÷8=42

Week 21	Week 22	Week 23	Week 24	
S M T W T F S	S M T W T F S	S M T W T F S	S M T W T F S	
D D D D A A A	*A A A B B B B*	*B B C C C C C*	*C D D D D D D*	
C C C C C C D	*D D D D D A A*	*A A A A B B B*	*B B B C C C C*	
A A B B B B B	*B C C C C C C*	*D D D D D D A*	*A A A A A B B*	
W	W	W	W	
A 40	*A* 40	*A* 40	*A* 40	=336÷8=42
B 40	*B* 40	*B* 40	*B* 40	336÷8=42
C 48	*C* 48	*C* 40	*C* 40	336÷8=42
D 40	*D* 40	*D* 48	*D* 48	336÷8=42

TEN-MAN ROTA

Example of 10-man rota designed to provide extra cover daily for the period 6 p.m. to 2 a.m., plus an additional man for wage payment on Thursdays and Fridays.

WEEK	1	2	3	4	5	6	7	8	9	10
SUNDAY	7 a.m.	L	10 p.m.	2 p.m.	7 a.m.	L	10 p.m.	2 p.m.	6 p.m.	L
MONDAY	7 a.m.	L	10 p.m.	2 p.m.	7 a.m.	L	10 p.m.	2 p.m.	6 p.m.	L
TUESDAYS	7 a.m.	10 p.m.	L	2 p.m.	7 a.m.	10 p.m.	L	2 p.m.	L	6 p.m.
WEDNESDAY	7 a.m.	10 p.m.	L	2 p.m.	7 a.m.	10 p.m.	L	2 p.m.	L	6 p.m.
THURSDAY	7 a.m.	10 p.m.	2 p.m.	L	7 a.m.	10 p.m.	2 p.m.	L	10 a.m.	6 p.m.
FRIDAY	7 a.m.	10 p.m.	2 p.m.	L	7 a.m.	10 p.m.	2 p.m.	L	10 a.m.	6 p.m.
SATURDAY	L	10 p.m.	2 p.m.	7 a.m.	L	10 p.m.	2 p.m.	6 p.m.	7 a.m.	L

NB. 2 p.m. denotes the period 2 p.m. to 10 p.m. 10 p.m.—10 p.m./7 a.m. 7 a.m.—7 a.m./2 p.m.
Average 40.8 hours per week.

SEVEN-MAN ROTA—48-HOUR WEEK

WEEK	1	2	3	4	5	6	7
SUNDAY	2 p.m.	6 a.m.	10 p.m.	2 p.m.	6 a.m.	L	10 p.m.
MONDAY	2 p.m.	6 a.m.	10 p.m.	2 p.m.	6 a.m.	10 p.m.	L
TUESDAY	L	6 a.m.	10 p.m.	2 p.m.	6 p.m.	10 p.m.	2 p.m.
WEDNESDAY	6 a.m.	L	10 p.m.	2 p.m.	6 a.m.	10 p.m.	2 p.m.
THURSDAY	6 a.m.	10 p.m.	L	2 p.m.	6 a.m.	10 p.m.	2 p.m.
FRIDAY	6 a.m.	10 p.m.	2 p.m.	L	6 a.m.	10 p.m.	2 p.m.
SATURDAY	6 a.m.	10 p.m.	2 p.m.	6 a.m.	L	10 p.m.	2 p.m.

Lost property report form

NAME AND ADDRESS OF COMPANY

SECURITY DEPARTMENT

Report of Property Lost or Suspected Stolen

Form Serial Number

Full details of loser, or, if company property, of person reporting.	Full Name .... Address .... Home phone number .... Work phone ext. .... Dept. ....
Full description of property, including make, colour, size, shape, and any distinguishing marks by which it could be identified. If money, how made up.	....
Date, time, and place when property last known to have been in loser's possession.	Date .... Time .... hours Place ....
Date, time, and place when loss discovered.	Date .... Time .... hours Place ....
If lost from the person, routes taken by loser since property last known to have been in possession.	....
Has loser any further information which may assist recovery of the property?	YES/NO (Delete as required) ....
Date, time, and to whom reported.	Date .... Time .... hours To whom reported .... Log Book page .... refers.

[*Action taken by the Security Department should be shown on the back of the form*]

6

Found property report form

NAME AND ADDRESS OF COMPANY

SECURITY DEPARTMENT: Report of Found Property

Form Serial Number

Time, date, and place ....
where found ....
Name, Address, and Department ....
and phone extension of finder

....

Time and date of report ....
Description of property ....

....

Initials of finder ....
Report received by .... Log Book page.......... refers

DISPOSAL OF PROPERTY

Restored to loser   Received by ....
Address ....
Department.............................. Date..............................

Finder informed of disposal and by whom ....
Returned to finder

I, the undersigned, hereby acknowledge having received the above described property and undertake to indemnify (name of company) in respect of any claim which may be made against the company in connection with this property.

Signature ....
Witness.............................. Date..............................

[*Action taken by the Security Department should be shown on the back of the form*]

7

Found property label

(showing both sides)

NOTE	DESCRIPTION OF PROPERTY
Property will be	....
handed to finder	....
if not claimed	....
within	....
three months	Received by..............................

Security Department	542
542	FOUND PROPERTY
FOUND PROPERTY	Found by
Found by	....
Department	....
Date Found	....
Where Found	....
Date Handed in	Date ....

8

Pass out: contractors' materials

Copy of a form showing (in capitals) how it should be completed by contractors before removing their property from premises where employed.

Serial Number 136

To the Security Department From THOMPSON BROS. LTD
J Bloggs and Sons Ltd DEPTFORD, LONDON SE
Loamville Time 10 00 Date 2.4.68

Please allow to leave the premises motor vehicle registered number XPY 123 E Driver HARRISON with the following items:

1 LADDER

2 cwt BAGS OF CEMENT

1 WHEELBARROW

10 LENGTHS OF STEEL SCAFFOLDING

Total items 14

H. Harrison
Driver's signature
J. E. Thompson
Authorised Representative

Received

Time 10 15 Date 2.4.68

J. White
Signature of Security Officer

9

Supervisors' security course

The following items could be included in a curriculum:

1. *Current position in country.* Describe the fire losses experienced stressing that the published figures do not take account of consequential loss and firms forced out of business. Emphasise increases in crime by 500 per cent in the last twenty years coupled with a general lowering of moral standards.

2. *Current position in own firm.* Describe losses that have become known, how committed and how to prevent, and unaccountable losses which have occurred and how it is possible for them to have been concealed.

3. *Sources of loss—police—courts.* Define what constitutes stealing in law; refer to 'perks' with the probability of their escalating to become a serious drain on a company if not stopped; taking away and non-return of tools; clocking offences; how and when to obtain police assistance; court procedure.

4. *Firm's policy on prosecutions.* Give details of the policy and the procedures that are to be followed in the event of any contravention by (*a*) outsiders, (*b*) contractors and (*c*) own employees.

5. *Loyalty to firm—responsibilities of carrying authority.* The consequences of deliberately turning a blind eye on misconduct. The construction placed upon negative action. Emphasise that dishonest employees constitute a very small anti-social minority who will try to influence relationships to their own advantage.

6. *Conclusion.* Underline the value of loyalty, cooperation with management generally and the security department where this exists and any system of notification to that department to remove personal involvement.

10

Contracts: security clauses

In addition to any other conditions of the contract concerning work to be done on company premises the following should be included:

FACTORY REGULATIONS

The contractor shall require all his employees to observe the factory rules and regulations applying to the company's employees, especially relating to safety, health, canteens, security, and smoking.

SECURITY CONTROL OF VEHICLES

The drivers of contractors' vehicles leaving company premises must be provided with an authorised pass out in respect of any load. The company or its agent shall have the right to search such vehicles, including the drivers and any passengers, and the right to require the vehicle to be unloaded in continuation of the search.

IDENTIFICATION PASS

Contractors should obtain from the company identification passes for all their employees entering company premises and shall require them to observe the conditions printed on them. These should bear the name of the employee, his employer's name, and the date of issue and expiry. On the pass should be the information that the holder is required to comply with the conditions of employment of the company concerning safety, health, canteen, security, and smoking which are exhibited on official notice boards on the premises.

11

Radioactive materials: storage

PLACE OF STORAGE

1 When not in use, radioactive sources should be kept in a place of storage assigned for this purpose only.
2 The place of storage should be adequately shielded.
3 Only authorised personnel should be allowed to introduce or remove sources from the place of storage which should be secure against tampering.
4 The place of storage should be in a room provided with a suitable means of exit that can be operated from the inside.
5 The place of storage should be chosen so as to minimise risk from fire.
6 The places where sources are stored should be inspected regularly and checked for possible contamination.

CONDITIONS OF STORAGE

1 All radioactive sources should be clearly labelled, giving information on the activity and nature. It may be found desirable to include the name of the person who is responsible for the source. In the event that a number of sources are normally in use in a fume-hood or other working area, as, for example, in analytical work, the marking might be of a general nature to apply to the whole working area. Any source involving hazards greater than those listed in the general warning should be specially marked.
2 The containers for beta-emitting isotopes should have adequate thickness to reduce the primary radiation to a safe level. Considerable bremsstrahlung may arise from high intensity sources and additional shielding should be provided if necessary.

3 Gamma-emitting sources should be stored in such a way as to limit the radiation exposure from other sources when any one source is being handled.
4 When either sealed or unsealed sources are liable to release a radioactive gas, their place of storage should be efficiently vented to the open air by mechanical means before it is opened.
5 Special equipment should be provided for storing unsealed sources of radioactive substances to prevent not only external irradiation hazards, but also radioactive contamination hazards.
6 In type *C* working places the sources may be stored in special cupboards providing adequate protection; in type *B* working places it is better to use a special secure receptacle which provides adequate protection and could be ventilated if necessary. (Type *C* working places are good chemical laboratories and type *B* are radioisotope laboratories.)

STORAGE OPERATIONS

1 Records should be kept of all stored radioactive sources. The records should give clear information on type of source, activity, and time of removal and return as well as the name of the person responsible for the source during its absence from the store.
2 Periodic inventories should be performed.
3 The removal of sources from the store and the time for which they are removed should be checked to provide adequate control.
4 Thermally unstable solutions containing radioactive materials in nitric acid or other oxidising solutions containing even traces of organic material and stable solutions with alpha-activity in excess of 5 mCi, or beta-activity in excess of 50 mCi, should always be stored in vented vessels.
5 Bottles and containers which open easily should be chosen.
6 Solutions having a high alpha-activity in excess of 1 mCi/ml should not be stored in thin-walled glass bottles, since irradiation might weaken the glass. All glass vessels must be expected to fail without apparent cause.

7 Bottles containing radioactive liquids should be placed in vessels large enough to hold the entire contents of the bottles in case of breakage.
8 Special precautions are required when opening vessels containing radioactive liquids liable to catch fire, explode, or froth.

TRANSPORTATION OF RADIOACTIVE MATERIALS

Transportation within an establishment

1 The amount of radioactive material moved should be limited to that required.
2 Transportation should be done in adequately shielded and closed containers. The containers should be constructed to prevent accidental release of the source material in case of upset.
3 If radioactive material in liquid or gaseous form or in powder, or other dispersible form, is in a shatterable container it should be transported in an outer non-shatterable container. With liquid sources the container should be provided with absorbing material able to retain all the liquid in case of breakage.
4 Suitable means should be provided for the transfer of the source to and from the transport container.
5 The transport container should be clearly marked with warning signs.
6 Containers in transit should bear a transportation tag showing necessary information for safety such as:
(*a*) Nature of contents.
(*b*) Physical condition.
(*c*) Activity in curies.
(*d*) Dose rate of radiation at contact of the outer surface of the container.
(*e*) Dose rate of radiation at a specific distance.
(*f*) Kind of packing (when applicable).

7 In case of unsealed sources, the transportation tag should, in addition, certify that the outsides of the container and carrier are free from contamination.

8 It is recommended that the transportation tag should be disposed of only when the source is in the charge and under the complete physical control of a person who is aware of the nature of the radioactive material and of the radiation hazards involved.

9 Emergency procedures should be planned to cover accidents to radioactive material in transit.

10 Any loss of radioactive materials during transport should at once be reported to the radiological health and safety officer.

11 Suitably trained workers should be in charge of all transportation of hazardous quantities of radioactive material inside an establishment.

Transportation outside an establishment

1 The route and method of shipment should be ascertained in sufficient detail to permit compliance with the rules and regulations established by all authorities through whose charge the shipment will pass. It should not be assumed in the case of trans-shipment that, because a shipment meets the requirements of the initial carrier, it will meet all requirements.

2 The recipient should be notified about shipment and receive all significant information in time to make any necessary preparations for receiving the shipment. Such information should at least include the method of shipment and estimated time of arrival. The notification should include any special storage instructions and details of safe opening techniques for special shipping containers.

Unless the rules and regulations have different provisions, the following general recommendations should be applied:

3 Packing should be such that the dose rate of radiation outside the package and the foreseeable duration and con-

ditions of handling, transport, and storage is unlikely to result in any person receiving radiation in excess of that permitted for non-occupational exposure. The limitation of the number of packages acceptable in one vehicle or storage location should depend upon the same considerations.

4 The outer surface of the package should not be significantly contaminated. Returnable containers should be free from significant external contamination.

5 Packaging should be adequate to prevent any loss of the radioactive material under normal conditions of transport, and any dispersal of radioactive material as a result of accidents that can reasonably be expected with the form of transport used. The packaging should be resistant to shocks, fire, water, and in addition the following factors should be considered:

(*a*) Possible corrosion of container.

(*b*) Effects of changes in outside temperature and barometric pressure.

(*c*) The degree of self-heating by large sources.

(*d*) Possibility of gas formation and pressure build-up.

(*e*) The possibility of the escape of gases produced and the effects to be expected from any radioactive gases released.

(*f*) The possibility of increased activity after packing due to build-up of daughter products.

6 The shipment should be clearly marked. The marking should:

(*a*) Indicate the presence of radioactive danger. This marking should remain legible under adverse conditions.

(*b*) Indicate the nature and quantity of radioactive material, the dose rate at the surface of the package and at a specified distance.

(*c*) Indicate that no person should stand by needlessly and that undeveloped photographic films should be kept at a specified safe distance.

(*d*) Be adequate to prevent any loss or misplacement of the shipment if damaged in transit, and provide instructions for safe disposal if delivery cannot be effected.

7 Radioactive material should not be loaded together with dangerous substances such as explosive, inflammable, oxidising, or corrosive substances.
8 The person responsible for transportation should be notified in writing of all particulars and of all other necessary instructions for safe transport, handling, or storage of the shipment.

Reprinted from Safe Handling of Radioisotopes (*Number* 1 *Safety Series*), *by permission of the International Atomic Energy Agency, Vienna.*

12

Liquid petroleum gas: storage

SIMPLE PRECAUTIONS THAT CAN FORESTALL TROUBLE

Cylinders should always be kept upright. Partly this is to help protect the valve assemblies, but there is another important reason. A full LPG cylinder always contains a volume of gas above the liquid level. This means that if a leak occurs at the valve when the cylinder is upright, only gas will escape. If, however, the cylinder is stored in any other position and the same thing happens, it may be the liquid that will spill out—and this will result in a more voluminous and rapid loss of contents. The same conditions apply when cylinders are in use—if they are not kept upright, liquid rather than gas may reach the burner. So if a cylinder is found to have a crumpled base ring it should not be used; it must be put to one side and the suppliers should be informed immediately.

Care is also needed in the way cylinders are attached to equipment, particularly if flexible hose or tubing is used. All connections with flexibles should be made by crimping or by using jubilee clips; merely pushing hozes onto nozzles is bad practice because they can easily be tugged off again, and wiring-on is also to be avoided because it may puncture the tubing. Gas pressure regulators should always be used so that the flow of gas is controlled.

When operating LPG heating equipment from a cylinder, always make sure that the two are kept well apart. It is best if cylinder and heater can be kept apart by a brick wall or similar heat-resisting barrier. Alternatively, there should be at least 10 feet (3 m) between the two; this space should not be used as a passageway unless the tubing is well protected—by ramps, for example. Appliances that are remote from the gas vessel should be fitted with their own taps or valves, but the supply should also be shut off at source when they are not in use. It is not a good idea to run too much equipment off one cylinder. Icing-up provides a rough check as to whether a cylinder is being made to discharge its contents too rapidly, and if this occurs the cylinder should be allowed to defrost and it can then either be changed for a new one or used in parallel with a further supply.

LPG storage cylinders should preferably be kept outside buildings. If they are connected to a fixed installation they will require a housing round them, but neither this nor the way they are sited must be allowed to interfere with access to them. Cylinders should never be sited where floor-level ventilation is bad, and they must be kept well clear of cellars, drains, and hollows where escaping gas might collect. When indoors they should, of course, be kept well away from heating appliances. British Standard Codes of Practice CP 338 (1957) and CP 339 (1956)[1] give further information on fixed domestic installations, while it is mainly American Standards[2] that cover the larger equipment.

Where LPG burners are used in buildings—especially for heating—adequate ventilation should always be provided. Windows alone cannot always be relied on to provide this since they will almost inevitably be closed in cold weather when the equipment is most likely to be used. An inadequate air supply can result in incomplete combustion of the gas so that carbon monoxide is produced.

INVESTIGATING LEAKS IN EQUIPMENT

There are a few more safety points that are sometimes overlooked. When lighting LPG equipment, for example, it is always best to bring the light to the gas outlet before turning on the fuel—the reverse procedure may cause an accident. If leaks are suspected—the gases can usually be smelled—the faulty section

1 'British Standard Code of Practice' 388 (1957). Domestic Propane-Gas-Burning Installations Permanent Dwellings. 'British Standard Code of Practice' 339 (1956). Domestic Butane-Gas-Burning Installations (issued in 3 parts). British Standards Institution.

2 'National Fire Protection Association's Standard Number 58' Liquefied Petroleum Gases and Number 59 Storage and Handling of LP Gases at Utility Gas Plants, published by NFPA, 60 Batterymarch Street, Boston, Massachusetts 02110, USA.

may give itself away by icing up; but this is not always so, and soap and water may have to be used to trace the fault. Never, of course, use a naked flame to find a leak, and always switch off the supply as soon as possible.

Reprinted with permission from the Fire Protection Journal, *October* 1967.

13

Flammable liquids: storage in main building

The store for bulk quantities of flammable liquids should be provided in accordance with the following recommendations:

1 *Location.* Preferably the siting of the store should be so arranged that there are two external walls.

2 *Construction.* The building should be single storey. The walls should be non-combustible and of fire-resisting construction (9 inch (23 cm) brickwork or equivalent).

The roof should be of substantial reinforced concrete construction. Window openings should have wired fire-resisting glazing fitted in fixed frames of equal fire resistance.

Doors should open outwards, be fire resisting and preferably non-combustible.

The floor should be of concrete with sills or ramps provided at door openings to contain the liquid contents of the store plus a 10 per cent margin of safety; the height of this sill should not exceed 2 feet (61 cm).

3 *Lighting equipment.* Lighting equipment, fixtures and fittings should be either flameproof or intrinsically safe for use in flammable atmospheres. The installation should be in accordance with the Wiring Rules of the Institute of Electrical Engineers.

4 *Heating.* The store should be unheated, but if heating is essential it should be by low-pressure hot water, low-pressure steam, or other approved safe form of heating.

5 *Ventilation.* Ventilation should be by natural means. In addition to ventilation at roof level there should be ventilation as near ground level as possible, protected with fine gauze of copper (or other non-corroding metal) of a mesh not less than twenty-eight to the linear inch (25 mm). Pipe ventilators or other vent pipes should be similarly protected

and should terminate at a height sufficient to ensure adequate dissipation of vapours. The total area of ventilation should be not less than $2\frac{1}{2}$ per cent of the floor area.

'Explosion vents' should be provided. This can be achieved by providing top-hung windows, glazed with fire-resisting glazing, and retained by either (*a*) counterbalanced weights, or (*b*) tension spring.

6 *Sources of ignition.* Smoking or the use of naked lights or other sources of ignition should be prohibited inside the store and within a reasonable area around the store.

Sparks from metallic objects may cause the ignition of flammable vapours and all possible steps should be taken to avoid them. Footwear or tools, from which sparks are likely to be struck should not be used. The sparking hazard of a floor is reduced by covering the concrete with a cement screed.

7 *Liquid containers.* Liquid containers should be protected against mechanical damage. Drums should be stored hung uppermost and not more than two high. Open containers should not be used.

Transfer of liquid from one container to another should be by means of pumps.

8 *Means of escape.* Adequate means of escape should be provided, and the layout should be such that the distance of travel to an exit does not exceed 10 feet (3.05 m).

Exit doors should open outwards and be kept open when the store is occupied.

9 *Fire extinguisher equipment.* 7 pound (3.2 kg) size carbon dioxide extinguisher(s) and two buckets of dry sand (three-quarters filled) should be provided outside the store. This equipment should be suitably protected from the weather.

10 *General.* Care should be taken to ensure that unauthorised persons cannot enter the store.

The store should be prominently marked 'danger—highly flammable liquids stored' and notices affixed 'no smoking or naked lights within 20 feet (6.10 m) of this building.'

14

Flammable liquids: storage in isolated building

The store for bulk quantities of flammable liquids should be provided in accordance with the following recommendations:

1 *Location*. The site of the store should be on level ground and as far away from any other building as possible. The store should not be less than 20 feet (6.10 m) from any other building, especially one in which a naked flame or any other possible source of ignition is present.

2 *Construction*. The building should be single storey. The walls should be of non-combustible and of fire-resisting construction (9 inch (23 cm) brickwork or equivalent).

The roof should be constructed of light non-combustible material which is easily shattered to minimise the effect of an explosion. Asbestos cement sheeting may be used for this purpose, but other suitable materials may also be available. Window openings should have wired fire-resisting glazing fitted in fixed frames of equal fire resistance.

Doors should open outwards and be fire resisting and preferably non-combustible.

The floor should be of concrete with sills or ramps provided at door openings to contain the liquid contents of the store plus a 10 per cent margin of safety; the height of this sill should not exceed 2 feet (61 cm).

3 *Lighting equipment*. Lighting equipment, fixtures and fittings should be either flameproof or intrinsically safe for use in flammable atmospheres. The installation should be in accordance with the Wiring Rules of the Institute of Electrical Engineers.

4 *Heating*. The store should be unheated, but if heating is essential it should be by low-pressure hot water, low-pressure steam, or other approved safe form of heating.

5 *Ventilation.* Ventilation should be by natural means. In addition to ventilation at roof level there should be ventilation as near ground level as possible, protected with fine gauze of copper (or other non-corroding metal) of a mesh not less than twenty eight to the linear inch (25 mm). Pipe ventilators or other vent pipes should be similarly protected and should terminate at a height sufficient to ensure adequate dissipation of vapours. The total area of ventilation should be not less than 2½ per cent of the floor area.

6 *Sources of ignition.* Smoking or the use of naked lights or other sources of ignition should be prohibited inside the store and within a reasonable area around the store.

Sparks from metallic objects may cause the ignition of flammable vapours and all possible steps should be taken to avoid them. Footwear or tools from which sparks are likely to be struck should not be used. The sparking hazard of a floor is reduced by covering the concrete with a cement screed.

7 *Liquid containers.* Liquid containers should be protected against mechanical damage. Drums should be stored hung uppermost and not more than two high. Open containers should not be used.

Transfer of liquid from one container to another should be by means of pumps.

8 *Means of escape.* Adequate means of escape should be provided, and the layout should be such that the distance of travel to an exit does not exceed 10 feet (3.05 m).

Exit doors should open outwards and be kept open when the store is occupied.

9 *Fire extinguishing equipment.* 7 pound (3.2 kg) size carbon dioxide extinguisher(s) and two buckets of dry sand (three-quarters filled) should be provided outside the store. This equipment should be suitably protected from the weather.

10. *General.* Care should be taken to ensure that unauthorised persons cannot enter the store.

The store should be prominently marked 'danger—highly flammable liquids stored' and notices affixed 'no smoking or naked lights within 20 feet (6.10 m) of this building'.

15

Fire instructions

THE PERSON DISCOVERING A FIRE SHOULD:

Give the alarm by telephoning the company's telephone operator stating location of fire.

Use the fire extinguishers and, if necessary, the fire hose reels provided (see plan). Only use a hose reel in event of a major fire.

Smother burning clothing and similar small fires with coats, curtains, carpets, etc.

Switch off current before using the fire extinguishers or fire hose reels on electrical apparatus.

EXIT AND MEANS OF ESCAPE:

Employees should make themselves acquainted with all the means of exit from the premises. Staircases, landings, ordinary and emergency exits should be kept clear and unobstructed.

ON RECEIVING FIRE WARNING:

Close all doors and windows.

Act on the instructions of the floor fire stewards who will be responsible for the safety of the staff during the period of the fire. Instructions should be carried out immediately and without question.

DURING A FIRE:

Do not evacuate unless ordered by a fire steward. If evacuation is ordered:

Do not use lifts.

Maintain silence.

Do not rush.

Do not attempt to pass others.

Do not panic.

YOUR FLOOR FIRE STEWARDS ARE ....

Fire report form

To ....Company Secretary   Date..
                                                      Copies to..

..............................Factory Manager   ....

A Location and extent of fire..
....
....

B Fire found at.......................... hours on.......................... Date by..

C Area affected last seen in order.................... hours on....................date by....................

D Public Fire Brigade
Called at...hours by.........................
Appliances attending (1)................................ arrived............ hours left............hours
(2)................................ arrived............ hours left............hours
(3)................................ arrived............ hours left............hours
Officer in charge.. Station........................ Tel No............
Cause to which outbreak attributed (if given)...
....
....

E Factory Personnel Notified

....	Chief Engineer	at....................hours
....	Production Supervisor	at....................hours
....		at....................hours
....		at....................hours
....	Safety Officer	at....................hours
....	Chief Security Officer	at....................hours
*..	Fire Assessor at..................	hours..................date

* This is responsibility of Senior Engineer/Production Supervisor

---

F Brief description of outbreak and details of any injuries

Person Reporting

G INVESTIGATING OFFICER'S REPORT
Appraisal of damage and likely effect on production

H Comments and recommendations

(Other sheets attached as required)   Investigating Officer's
Signature..

17

Static (dry) fire drill

NOTICE TO MANAGEMENT

A practice fire drill will be held at .... hours on .... to test the fire alarm system, and provide training for fire wardens and fire parties; the .... fire brigade will be co-operating.

It is *not* intended that staff shall carry out an evacuation when the alarm is sounded; with the exception of those with specific fire duties everyone should continue to work normally.

Fire wardens, already nominated, will visit all parts of their areas of responsibility, remind staff therein of evacuation instructions, including exits to be used, their assembly points outside, and answer any queries that may be raised.

The commencement of the drill will be preceded by an announcement over the public-address system and will be similarly terminated.

MEMORANDUM TO HEADS OF DEPARTMENTS

A practice fire drill be held at .... hours on .... to test the fire alarm system and train personnel. The sounding of the alarm will be preceded by a public address announcement, but you are requested to ensure that your staff have prior notice that this exercise will take place.

Staff *will not carry out an evacuation,* but will continue with normal working. Departmental fire wardens are being instructed to visit all sections and personnel for whom they are responsible; a copy of the memorandum which is being circulated to them is attached for your information, and you are asked to arrange their duties so that they are available to participate in this exercise.

A fire will be presumed to have broken out in the .... area; the security staff and fire parties will take immediate action to extinguish it; .... fire brigade has signified willingness to participate and the arrival of their vehicles may be expected soon after the alarm.

The practice will be terminated by a further public address announcement, and progress reports may be passed by this means during its course.

Copy of memorandum to fire wardens attached.

MEMORANDUM TO FIRE WARDENS

A practice fire drill will be held at .... hours on .... in the interests of the safety of employees and the premises. In your capacity of fire warden, on the giving of the alarm, you are to visit all sections within your sphere of responsibility and ensure that the staff therein are conversant with evacuation arrangements including their means of leaving the building and their assembly point. *It is not intended that they should leave their places of work* and there should be minimum interference with normal business.

Reference should be made to your written emergency instructions and advantage taken of the opportunity to check the nominal roll of the personnel for whom you are responsible.

The practice will be preceded by a public-address announcement and similarly terminated.

You are asked to report back to the chief security officer on any points arising from the practice which you think open avenues for improvement.

EMERGENCY INSTRUCTIONS

For fire wardens and others in the case of fire or other emergency necessitating evacuation.

Name.................................. *Department*.......................

1 *Means of alarm*—continuous sounding of the fire siren, ringing of handbells or the calling of a general alarm over the public-address system.
2 *All* persons, except fire wardens and persons having specific emergency duties, *must immediately leave* the bulding. They should *walk* not run.
3 Those leaving should not delay to collect personal belongings or for any other reason. They must use the exits laid down for them to use and assemble at the prearranged point.
4 Where it is possible to do so instantaneously, electric motors and machinery should be stopped, or switched off.

Fire wardens

5 Ensure that everyone in your department is instructed in their primary and secondary means of escape from where they work and the action they should take when an alarm is sounded.
6 Keep an up-to-date list of all persons in the department for the purposes of a roll call.
7 Nominate persons to take specific action necessary to effect evacuation as quickly and efficiently as possible.
8 In the event of fire, raise the alarm, attack the site with available apparatus and periodically check its serviceability, see that fire points are not obstructed.
9 See that all doors and windows that can be closed are closed, to limit the spread of the fire.
10 Do not allow persons to go into cloakrooms to collect personal belongings.
11 Arrange for the cloakrooms and lavatories to be searched to ensure no staff are left behind.
12 Switch off machinery and motors if possible, but do not take unnecessary risks.
13 Report subsequently on usage of extinguishers that should be refilled or replaced.
14 If the fire is quickly put out, report immediately to minimise inconvenience.

PUBLIC ADDRESS ANNOUNCEMENTS

First announcement

Attention please. Attention please.
A practice fire exercise is about to take place to test the fire alarm system and the arrangements for your safety.
I repeat, a practice fire exercise is about to take place to test the fire alarm system and the arrangements for your safety.
The fire siren will be sounded as soon as this broadcast ends.
Do not, I repeat, 'do not' leave your place of work unless you have specific duties in connection with the exercise to carry out.
No further action is required by you, fire wardens will visit all departments and will answer any queries.
A further announcement will be made on the termination of the exercise.
End of broadcast.

Second announcement

Attention please. Attention please.
The practice fire exercise is now over.
I repeat, the practice fire exercise is now over.
Thank you for your cooperation.
End of broadcast.

Accident report form

DATE and TIME		Clock Number	SURNAME	PLACE OF ACCIDENT	LOCATION OF INJURIES
of Report	of Injury				

STATED CAUSE ....

FOR USE OF SUPERVISION

OCCUPATION	SHIFT	From / To	ACTION AFTER ACCIDENT—DELETE AS NECESSARY OFF WORK/LIGHT WORK/OWN JOB

What was the accident and how did it happen? ....

....

....

....

....

IF DUE TO MACHINERY STATE—
1. Part of m/c causing injury.
2. In motion by mechanical power. YES/NO
3. Type of lifting m/c or crane.

WITNESSES 1. Clock Number
2. Clock Number

EXACT LOCATION

ACTION TAKEN TO PREVENT RECURRENCE ....

....

....

....

....

....

....

Signature   Date

SUPERVISOR

MANAGEMENT COMMENT AND ACTION ....

....

....

....

....

Signature   Date

19

Driver application form

CONFIDENTIAL

Complete in ink
Block capitals please
Applicant must provide driving licence at interview
No approach will be made to present employer without applicant's prior consent

1 **Full Name** ....
....

Date of birth ....
Nationality ....
Married/single ....
Number of children ....

2 **Permanent address** ....

3 **Next of kin**
Name .... Relationship ....
Address ....

4 **Health**
(*a*) Details of illness or disablement in last 5 years
....
....

(*b*) Serious illness, injury or operation prior to last 5 years
....
....

(*c*) Is your hearing/vision impaired? YES NO. If yes give details
....

5 **Prior employment**
Give details of present and previous employments, most recent first, include military service

Name and address of employer	Job	From—to	Reason for leaving
....	....	....	....
....	....	....	....
....	....	....	....
....	....	....	....
....	....	....	....

6 **Driving ability and record**

NOTE: disclosure of a conviction does *not* mean you will automatically be barred from employment; failure to disclose would, however, mean subsequent dismissal.

(*a*) Have you experience of customer delivery?..........

(*b*) Are you experienced in keeping drivers' records?..........

(*c*) List types of commercial vehicles driven..........

....

(*d*) Give details of any driving convictions..........

....

(*e*) Have you been convicted of any offence involving dishonesty? If so, give offence, date and penalty..........

....

....

(*f*) Has any load, part load or vehicle in your charge ever been stolen? If so, give details ....

7 **Hobbies** ....

8 **Any other matters** ....

....

I apply for employment as a driver and vouch that the preceding details are correct.

I understand that a misleading statement or unsatisfactory reference could lead to my subsequent dismissal.

I accept that my employment may involve overnight stay away from my base.

Signed..........

Date..........

20

Security staff: checklist

Checked by .... Date........................

SECURITY ON INDUSTRIAL PREMISES

Security on

1 Who is in charge of security?
2 Has a job description been prepared?
3 Full- or part-time?
4 If part-time what other duties are performed?
5 Is report on work of security departments submitted quarterly, half yearly, or yearly?
6 Who is the person in charge responsible to?

Security staff (if any)

7 How many are employed?
8 Is the number:
(*a*) under requirements,
(*b*) about right, or
(*c*) over requirements?
9 When was their status and payment reviewed and the result?
10 Is the general standard of morale: excellent, good, fair, or bad?

Training

11 What form of training do they receive in:
(*a*) Security duties,
(*b*) Fire-fighting duties,
(*c*) First aid,
(*d*) Accident prevention, and
(*e*) Civil defence?
12 Does this include attendance at an outside training course? If so, which ones?

Duties

13 Have these been embodied in a form of standing orders? Has each man been issued with a copy?
14 If no standing orders have been prepared are the duties of the security staff been described in writing?
15 When were those duties reviewed, by whom, and with what result?
16 What form of permanent record is kept of matters having a security interest?
17 Who inspects this and how often? Check.
18 What form of supervision of staff is carried out at night, at weekends, and on public holidays and by whom?
19 How is this recorded? Check.

Patrolling

20 When is patrolling of premises carried out?
21 How often?
22 Is this excessive, satisfactory, or not often enough?
23 Are all areas given the same attention or do those more vulnerable to risk of any type receive more frequent attention?
24 Is a patrol man's clock used?
25 Are the clocking points too many, sufficient, or not enough and are they suitably placed?
26 Do the time clock records show satisfactory patrolling?
27 Is the company associated with any other in a mutual security aid scheme?
28 Are searches of company personnel and vehicles carried out? Check.
29 Are these too often, adequate, or not enough?
30 Are searches of contractors' personnel and vehicles carried out? Check.
31 Are these too often, adequate, or not enough?
32 Have any complaints been received? If so, were management informed and what were the results?
33 Are car parking arrangements and discipline of users satisfactory?

21

Gatehouse duties: security checklist

Checked by.................................. Date..............

1 What gates are attended by security staff?
2 Are there other gates, including railway gates?
3 If so, when are they open and in what circumstances?
4 Who is responsible for opening and closing those gates?
5 Where are the keys kept?
6 What action is taken at the gates attended by security staff:

(*a*) When vehicles enter:
to collect goods,
to collect scrap or waste materials,
for any other purpose, for example, delivery of stores, etc.

(*b*) When vehicles leave:
after collecting goods,
after collecting scrap or waste materials,
after making deliveries of stores, etc.

(*c*) When employees wish:
to enter at unusual times,
to leave at unusual times,
to bring in large parcels and cases,
to take borrowed tools away,
to return borrowed tools,
to remove purchases of company products from the premises.

(*d*) When requests are received for keys of premises at:
usual times,
unusual times.

(*e*) When property is reported as having been found on the premises.

(*f*) When property is reported lost on the premises.

(*g*) When a fire is reported to have occurred on the premises.

22

Security checklist: sales of scrap and sundry materials

COLLECTION PROCEDURE

1 The purchaser's motor vehicle arrives.
2 The lorry is weighed at the weighbridge where it is noted whether the driver and any passengers remain on the vehicle.
3 The company's representative (e.g. stores foreman, good housekeeper, chief engineer) is informed from the weighbridge of any instructions obtained.
4 The lorry is directed, under escort if possible, to the loading point.
5 The lorry is loaded under supervision of a representative of the company.
6 The lorry returns to the weighbridge and is weighed under the same conditions as in (2) with respect to driver and any passengers.
7 A four-part set of a materials pass-out, serially numbered, is prepared by an authorised person (such as those mentioned in (3) above), on which is shown the weight of the material and the price and whether the sale is for cash or on account.
(*a*) One copy is given to the driver of the vehicle as his authority to leave the premises, subject to compliance with (*c*) below.
(*b*) One is signed by the driver and retained by the department of the company concerned.
(*c*) If the sale is for cash, the purchaser takes his copy and a second copy of the set to the cash office where, on payment, both copies are stamped as 'paid'. One copy is returned to the driver for use as in (*a*) above. The other is retained for a record of the cash outgoings. If the sale is to be invoiced, the selling deparment, as in (*b*) above, will send the third copy to the accounts department for action.
8 Each month the weighbridge tickets referring to the sale of

scrap and sundry materials will be sent to the accounts department for checking that payment has been received or, if not, that an account is rendered.

23

Cash on premises: security checklist

Checked by .... Date ....

1 Who collects money from bank?

(*a*) Company employees.

(*b*) Cash-in-transit security company.

2 If by the latter:

(*a*) Has the actual point of delivery (for example: desk in wages manager's office on third floor of Blank House) been included in the contract?

(*b*) Has a list of authorised receivers of the money with their specimen signatures been supplied to the security service company and acknowledged by them?

3 If there is a gate to the premises is this locked shut until the money is safely delivered?

4 Where is the money taken on arrival?

(*a*) Petty cash requirements.

(*b*) Wages money.

WAGES MONEY

1 Is the office used for making up locked at that time?

2 Can callers outside the door be seen from inside before the door is open?

3 How many persons are present at the make up?

4 Is any one of those permitted to leave before this is complete?

5 When make up is complete what happens to the packets and how are they protected?

6 How are packets taken to pay-out points?

7 Is a car used where the distance requires this?

8 Are gates of premises locked when transfer of the packets is taking place on company roads?

9 How many persons are concerned in the transferral?

10 What is the physical security of pay-out points, telephone, alarm, etc.?
11 Who is present at pay-out points?
12 What happens to unpaid wage packets?
13 Who is responsible for their security?
14 When there are pay-out times outside the normal office hours, who pays out?
15 Where have the relevant packets been kept up to pay-out time? In a safe?
16 Who holds personally the key to the safe?

SAFES

1 Where are safes located on premises?
2 Are they secured to the floor or wall to prevent their removal?
3 Has any of them a protective thief alarm?
4 Who has the operative key to the safe?
5 Where are any duplicates? In the bank?
6 If safe has a combination lock who has (*a*) code (*b*) copies of it?
7 Who is responsible for switching on the alarm where fitted and retaining key? Preferably this should not be the safe key holder.
8 Who is responsible for the duplicate key and where is it kept?

CASHIERS/WAGES OFFICE

1 What floor is it on?
2 Is it close to stairs or lifts?
3 Can the activities of occupiers be observed from adjacent offices?
4 Can these activities be observed from other buildings?
5 Does it give an exit to a flat roof or fire escape?
6 Has the office an observation window or lens in the door?
7 Has the door a strong slam lock?

8. Has the office a counter with cash drawers fitted with slam locks?
9. Has the counter a grille or armoured glass?
10. Is an alarm fitted and tested daily?
11. Have staff been instructed on the action to be taken on hearing the alarm?

24

Cash in transit: security checklist

Checked by.............................. Date................

1 Is the money carried in pockets or in a bag?
2 When are (*a*) deposits and (*b*) withdrawals made?
3 Do the times display a pattern of irregularity?
4 Who makes (*a*) deposits and (*b*) withdrawals?
5 What escort is provided?
6 Are they dressed in distinctive clothing?
7 Are those persons always the same?
8 Are they supplied with police whistles and any form of self protection against attack?
9 How far away is the bank?
10 What transport, if any, is used?
11 Is it always the same?
12 Is the driver always the same?
13 Is a second vehicle used?
14 Has the cash vehicle any form of alarm?
15 If so, is the alarm tested immediately before going to bank?
16 Where is cash carried in the car?
17 If in the boot, is it locked?
18 If a container is used is it secured to the vehicle in any way?
19 Has the container any alarm fitted?
20 Are the keys of the container or any securing device carried in the car?
21 Can vehicle doors be locked from inside?
22 Is the vehicle a hired car?
23 If the driver is not known what action is taken?
24 If vehicle returning from bank with money is followed in suspicious circumstances what action would be taken?
25 Is a bank night safe used?

26 Are security arrangements satisfactory bearing in mind these observations?
27 Is route to and from bank varied and through busy streets?
28 Who decides route and when?
29 Are police informed of visits to bank?
30 If no vehicle is used do the company employees walk against the traffic and away from the kerb edge?

25

Offices: security checklist

1 Who is responsible for security?

ENTRANCE(S)

2 Describe which:
(*a*) Are attended by a receptionist/commissionaire/security officer.
(*b*) Are unattended and at which entry can be obtained.
3 Can (*b*) entrances be reduced without loss of efficiency?
4 Can they be fitted with springs and slam locks so as to be used for exit only?
5 Where the entrance is attended as above:
(*a*) Are postal deliveries handled correctly?
(*b*) Is there a clear view from receptionist's desk of entrance, stairs and lifts?
(*c*) Is there a notice posted requesting callers to go to receptionist?
(*d*) Have instructions in writing been issued respecting the reception of callers?
(*e*) Where the receptionist is the telephone operator have instructions in writing been issued respecting the action required on the (1) firm alarm (2) thief alarm being heard?
(*f*) Have instructions been issued regarding the admittance of maintenance men, telephone operators, window cleaners, etc.?

SECURITY OFFICERS/COMMISSIONAIRES

6 Where they carry out additional security duties such as patrolling the premises on their closure, have the instruc-

tions been issued in the form of standing orders. Are they up to date?

7 In applicable instances is their concern associated with other(s) in a mutual aid security scheme, or otherwise are any arrangements made for the security officer to communicate at intervals with any security services company?

EMPLOYEES

8 Is there a clause in the conditions of employment respecting the right of search, etc.?

9 Is their cloakroom located in a satisfactory position in relation to any entrances or exits?

10 Is a notice displayed in the cloakrooms disclaiming on behalf of the employer responsibility for losses?

11 Are notices displayed in toilets reminding users against leaving articles such as jewellery on the wash basins?

12 If thefts are being experienced has the assistance of the local crime prevention police officer been sought?

OFFICES

13 Are all duplicate and other keys not in use kept under good security?

14 Has the cutting of extra keys been prohibited unless with proper permission by an authorised person?

15 When a safe is removed from use or transferred to another location are all keys accounted for?

PRESIGNED CHEQUES

16 Are they kept in recommended conditions such as being under the protection of two separate locks, with the keys or codes held by two persons?

17 Are they counted daily by an independent person?

COUPONS

18 Are they cancelled on receipt in a satisfactory manner?
19 Are remittances properly recorded, especially those from unidentifiable senders?
20 Are coupons destroyed in accordance with instructions and correct destruction certificates provided?

CONTRACTORS

21 Are contractors' employees subject to the same searching requirements as own employees?

CONFIDENTIAL DOCUMENTS

22 Is security of vital information adequately protected?

MEETINGS/CONFERENCES

21 Where specially advisable:
(*a*) Are date, place and venue kept secret?
(*b*) Are the times and venues changed at short notice?
(*c*) Can those attending be seen/overheard from outside?
(*d*) Are agendas, blotting paper and pads, and contents of waste paper baskets and ash trays destroyed immediately after the meeting?
(*e*) Is security coverage provided during intervals, e.g. for refreshments?

FIRE

22 Have fire/security stewards been appointed on each floor/department?
23 Have they been issued with written instructions as to their duties?

24 Have replacements been made to account for leavings, transfers of stewards?

25 Are the stewards familiar with handling the fire-fighting equipment?

26 Have notices been displayed instructing employees:

(*a*) What to do on discovering a fire?

(*b*) What to do on hearing the fire alarm?

(*c*) Which is the escape route they should take in event of alarm?

(*d*) Where they should assemble on leaving the building?

27 Are all fire exits correctly marked?

28 Are all fire exits unobstructed?

29 Are those on the ground floor fitted on the inside with breakable bolts?

30 Are they satisfactorily locked when the premises are vacated at close of business?

LOCKING UP PREMISES

31 Who is responsible for locking up the premises after ensuring no security risks exist, e.g. open doors, windows, electrical apparatus left switched on?

32 What happens to key(s)?

33 Who is registered with police as the keyholder? Is this up to date?

34 Have door and safe keyholders been warned about checking the genuineness of requests to attend the premises when they are closed?

SECURITY CHECK

35 When was the last *ad hoc* check-up of security made and by whom?

36 When was a complete check-up of security arrangements and their observation made, by whom and the result?

37 What weaknesses were discovered?

38 What was ordered to be done to improve them?

39 What is the present position?

40 If further weaknesses discovered what instructions have been given?

26

Stores: security checklist

(See also Appendices 11-14)

GENERAL

1 Who holds the keys in daily use and the duplicate(s)?
2 Is the protection of the windows adequate?
3 Is the lock satisfactory and any hasp of sufficient strength?
4 Has the store a service counter?
5 How are items obtained when store closed?
6 Is this under sufficient control?
7 What record is made of such withdrawals?
8 Are requisitions checked after being actioned to detect offences and if so by whom?
9 What are the arrangements for the borrowing of tools by employees?
10 Are company tools of common description given a distinguishing mark?
11 Are the serial numbers, where available, of tools recorded?
12 Where is the burning/welding equipment kept?
13 Are the gas cylinders stored at some distance and near an exit?

CANTEEN

14 What arrangements are made for the security of the keys?
15 Are these satisfactory?
16 Are the arrangements for the security of wines, spirits, cigars and cigarettes satisfactory?

27

Shops and supermarkets: security checklist

STAFF

1 Are all references verified before employees are engaged?
2 Have conditions of employment been prepared?
3 Does every employee on engagement sign and receive a copy?
4 Do the conditions refer to:
(*a*) Bringing parcels or cases onto the premises.
(*b*) Handbags, baskets in the sales area.
(*c*) Not leaving sales area without permission.
(*d*) Searching of handbags, etc.
(*e*) Leaving only by front entrance.
5 Purchasing conditions
Are these outlined in the conditions of employment? Do these deal with:
(*a*) Times of purchase.
(*b*) Dress when shopping.
(*c*) Use of wire baskets.
(*d*) Retention of till slips signed by checker.
(*e*) Sale of goods in short supply.
(*f*) Sale of goods reduced to clear.
(*g*) Amount of discount and how to claim.

KEYS TO FRONT DOOR

6 Is keyholder registered with police? Is information up to date?
7 Who has responsibility for ensuring the premises are secure before being left?

FIRE DOORS

8 Is any device fitted to activate an audible alarm if opened?

REAR OR DELIVERY DOORS

9 Who holds the keys (*a*) during times premises are open (*b*) when closed?
10 Are doors locked between deliveries?
11 Are delivery men admitted to sales area?
12 Are they allowed to enter through front door?
13 If deliveries are made to sales area are they checked there together with any collections, for example, stale bread, cakes, etc.?
14 Are composite delivery notes signed representing multiple deliveries.

DELIVERIES

15
(*a*) How are specially valuable and attractive goods protected?
(*b*) If in a separate compound, who holds key?
(*c*) Does the goods inward checker when tallying hold the delivery note at the same time?
(*d*) Are all over and under deliveries properly recorded?
(*e*) Are deliveries occasionally rechecked by a member of management?
(*f*) Is a goods inward book kept? If so, who make the entries?

RUBBISH AND SALVAGE

16 Who supervises the removal of rubbish and salvage?
17 Are bins examined from time to time to detect stolen goods?

CASH TILLS

18 Can customers see clearly the amount rung up on the till?

19 What pockets are there in the cashiers' overalls?
20 Have cashiers been told they must not serve relatives?
21 Do the cashiers:
(*a*) Check their floats before accepting any more money?
(*b*) Check the date recorder on their till?
(*c*) Check that the total amount previously recorded on the till has not been changed?
(*d*) Have a key to their tills?
(*e*) Take money, handbags, baskets or cigarettes to their check-out point?
(*f*) Ring up on the till each item separately?
(*g*) Shut the drawer of the till after each completed sale?

CASHING UP

22 Is an independent count taken of each till at close of business and reconciled with the registered total?
23 Is a record kept of overs and unders and brought to notice of management?
24 Is the pen record in 22 and 23 signed by the till operator and the clerk concerned?
25 Are notes above requirements collected from time to time from the till and are records made and signed by the respective cashiers?
26 Where is any spare cash and takings as in 25 kept until banked?
27 If in a safe, or other locked and satisfactory permanent container, where is key kept?
28 Are some of the day's cash takings taken to the bank during normal banking hours?
29 Is a bank safe used? (See Appendix 13 for recommendations for security of cash in transit.)

SAFE

30 Where is it located?
31 Can it be seen from outside the premises?

32 Would security of the safe be improved if a light was left burning?

33 Is it secured to the floor?

34 Who holds the key?

35 Where is the duplicate? In the bank?

SHOPLIFTERS AND THEFTS BY STAFF

36 Has a policy for dealing with shoplifters and staff found stealing been decided and circulated?

37 Have 'Rules for the Guidance of Managers concerned with Shoplifters' been circulated to them?

38 Are details of all employees found stealing company property and the result of any prosecutions posted in positions where they can be read by all employees?

39 Are records kept of all shoplifters whether prosecuted or not?

SECURITY STAFF (WHERE EMPLOYED)

40 Are employees told of the responsibilities and authority of such staff and that they are liable to be asked to explain their possession of company property after leaving the premises?

28

Fire: security checklist

1. Has the telephone operator where employed and any deputy been instructed in the action to be taken on fire on the premises being reported?
2. Have the instructions been prepared in writing and are they immediately available to the telephonist and whoever replaces her for meal breaks and other absences?
3. Is there a full-time company fire brigade?
4. If not are there men on each shift who will attend and deal with outbreaks of fire?
5. On an alarm is their notification procedure satisfactory?
6. Has their agreed number been maintained following leavings and transfers?
7. What training do they receive? Check records of attendance.
8. Do they receive special payment for training periods, etc., and is it satisfactory?
9. Who is responsible for fire duties and supervision of any company firemen?
10. How is the fire-fighting equipment maintained?
11. When was it done last? Check records.
12. Is this satisfactory according to recommended standards of maintenance?
13. Has a Certificate of Satisfactory Means of Escape been applied for/been issued?
14. If so have all relevant reconstructions of premises, change of usage since then been reported to the fire authority?
15. What is the type of fire alarm?
16. Can it be heard satisfactorily in every part of the premises?
17. When was the alarm last tested? Any records kept? Where applicable check General Register to ensure these have occurred at least at three monthly intervals.
18. When was a fire/evacuation drill last carried out? Have assembly points been designated.

19 Have supervisors or their equivalent in offices been instructed in their duties in the event of fire being reported?
20 Have the requirements of the fire sections of the Factories Act 1961, such as display of notices, been carried out?

OFFICES AND SIMILAR BUILDINGS

21 As 20 in respect of the Offices, Shops and Railway Premises Act 1963.
22 Have fire wardens/stewards been appointed in sufficient numbers and supplied with a list of their duties?
23 Have they been trained in the use of fire-fighting equipment?
24 Have staff movements of fire wardens/stewards been checked to ensure new appointments have been made where necessary?

29

Bomb threat checklist: telephone procedure

It is recommended that this list be attached to a firm background —e.g. cardboard—and be available immediately to every telephone operator.

INSTRUCTIONS

Be calm. Be courteous. Listen, do not interrupt the caller. If possible notify Supervisor/ Security Officer by pre-arranged signal while caller is on line.
Name of operator..Time....................Date..........

CALLER'S IDENTITY

Sex: ....Male ....Female ....Adult ....Juvenile
Approx. Age: Years..............

ORIGIN OF CALL

..............Local ....Long Distance ....Booth ....Internal
(From within building)

DESCRIPTION

VOICE CHARACTERISTICS			SPEECH		LANGUAGE
......Loud	......Soft	......Fast	......Slow	......Excellent	......Good
......High pitch	......Deep	......Distinct	......Distorted	......Fair	......Poor
	......Pleasant	......Stutter	......Nasal	......Foul	
......Raspy	....	......Slurred	......Lisp	....	Other
......Intoxicated	....	..............Other			

ACCENT	MANNER		BACKGROUND	NOISES
......Local	......Calm	......Angry	......Factory machines	......Trains
......Foreign	......Rational	......Irrational	......Bedlam	......Animals
......Race	......Coherent	......Incoherent	......Music	......Quiet
......Not local	......Deliberate	......Emotional	......Office machines	......Voices
......Region	......Righteous	......Laughing	......Mixed	......Airplanes
			......Street traffic	......Party atmosphere

GETTING INFORMATION

Pretend difficulty with hearing.
Keep caller talking.
If caller seems agreeable to further conversation, ask questions like:

When will it go off?	Certain hour	Time remaining..................
Where is it located?	Building	Floor/Area ....
*What kind of bomb?	....	
Where are you now?	....	

How do you know so much about the bomb?..................
Why has this company been chosen for such action?..................
Has he a grievance?..................
If building is occupied, inform caller that detonation could cause injury or death.
Did caller appear familiar with plant or building by his description of the bomb location?
Write out the message in its entirety and any other comments on a separate sheet of paper and attach to this checklist.
* Explosive, incendiary, etc.

ACTION TO TAKE IMMEDIATELY AFTER CALL

Inform police and fire brigade through the 999 Emergency procedure and members of management in accordance with your instructions. Talk to no one else.

30

Bomb threat checklist: searching

Incorporate in your emergency plans an organisational structure fixing individual responsibility for each building, floor, office or room. Who is more familiar with the surroundings than people who work there already?

Train at least one person in each area who will be prepared to search on receipt of a warning to observe and report any unusual objects or incidents in their sections.

Prepare floor plans of the area and divide into definable sections designated in some way, for example Floor 2—section 4, Production Department—section 2. The more likely areas where an explosive device could be placed—e.g. on the first and ground floors and basement—should be marked on the floor plans and given priority in searching. Have a checklist available to record when reported that an area has been searched with the result and the location of any suspicious objects which are reported.

Inform searchers of the exact information received as well as the predicted location and time of explosion where known.

Searching must stop at least fifteen minutes before and not recommence until at least fifteen minutes after that time during which period all personnel should be evacuated from the floor concerned and in a multiple-storey building at least two floors above. When a building is in multiple tenancy a point of contact should be arranged within the procedure decided so that any measures to be taken can be coordinated.

Building and services personnel must be available with keys to all areas and be prepared to cut off any electricity or other supplies which may appear desirable.

Searchers must report any suspicious objects. These must not be touched or moved. Only explosive experts, who will be called for by police, will attempt to move or dispose of suspected bombs. When searching, all cupboards, lavatories (including cisterns), cloakrooms, or potted decorations should be carefully examined.

Security officer application form

CONFIDENTIAL

Complete in ink, block capitals please
No approach will be made to present employer without applicant's prior consent

1 **Full Name**

Date of birth..........
Nationality..........
Married/Single..........
Number of children..........

2 **Permanent address**..........

Owner-occupier YES/NO
Telephone number..........

3 **Next of kin:** Name ....Relationship ....
Address ....
....

4 **Health**
Have you:
(*a*) Undergone any major operation ....
(*b*) Had any illness causing absence from work in last 5 years.
....
....
(*c*) Any physical disability (e.g. impaired vision/hearing, hernia, slipped disc etc.)
....
....
(*d*) Suffered from dermatitis — YES/NO
asthma/bronchitis — YES/NO
(If YES give details) ....
....

N.B. Successful applicant will have to undergo a full medical examination.

5 **Prior employment:** Give details of present and previous employments, most recent first, include military service. Cover at least last ten years.

Name and address of Employer	Job	From–To	Reason for leaving
....	....	....	....
....	....	....	....
....	....	....	....
....	....	....	....

6 **Experience and record:** N.B. Failure to disclose a conviction would mean immediate dismissal.

(*a*) Have you a current First Aid Certificate? YES/NO
If 'No', list experience, if any ....

(*b*) Experience in combatting fire? ....

....

(*c*) Have you a current driving licence? YES/NO

(*d*) What groups does your licence cover? ....

(*e*) Have you been convicted of any offence involving dishonesty, if so, give offence, date and penalty ....

....

....

(*f*) List security experience and training ....

....

....

7 **Hobbies** ....

8 **Any other relevant information** ....

I apply for employment as a security officer and vouch that the preceding details are correct.

I understand that a misleading statement or unsatisfactory reference could lead to my subsequent dismissal.

I accept that my employment may involve regular shift and weekend working and, in emergency, duty change and/or overtime at short notice.

Signed ....

Date ....

32

Recommended code of practice for security staff

If a security staff is to operate to full efficiency, the department and the members themselves must be held in respect by both management and fellow employees. For this to be achieved, high standards of behaviour, integrity, conscientiousness and deportment are called for and must be maintained.

Failures in performance can have consequences, immediate or long term, more serious to a firm than comparable actions by most of its other employees—timidity and delay, or lack of knowledge of first aid could endanger an injured workman's life; lack of diligence in patrolling and powers of observation could result in gross fire or other damage to property; connivance in theft, or lack of action on information in connection with it, could lead to serious loss; indiscretion, tactlessness or unnecessary exercise of authority could jeopardise the entire industrial relations atmosphere in an organisation and cause labour disputes of indeterminable consequences.

With these factors in mind, and the desirability of professionalising the function to give maximum value to the employer, a more stringent code of practice should be asked of persons engaged in security work than from others. This is reasonable since to do so will react to the benefit of all genuinely interested in their work and will make it easier to weed out the unreliable, untrustworthy undesirables.

The following are given as guidelines and, while the appearance may be that of a disciplinary code, fundamental principles are expressed therein.

1 *Obedience to orders.* No security officer will disobey, or without good and sufficient cause omit or neglect to carry out, any lawful order, written or otherwise, given by a superior to whom he is responsible.

2 *Neglect of duty.* No security officer will:

(*a*) neglect, or without due and sufficient cause omit, promptly and diligently to attend to or carry out anything which it is his duty to do as a security officer,

(*b*) fail to carry out his work in accordance with instructions,

(*c*) leave his place of duty without due permission or sufficient cause,

(*d*) fail to report any matter which it is his duty to report,

(*e*) fail to make any necessary reports and entries in any official document or book kept in the course of his work.

3 *Falsehood.* No security officer shall in connection with his work:

(*a*) Knowingly make or sign any false statement or entry in any official document or book.

(*b*) Wilfully or negligently make any false, misleading, or inaccurate statement.

(*c*) Without due and sufficient cause destroy or mutilate any document or record, or alter or erase any entry therein.

4 *Breach of confidence.* No security officer shall:

(*a*) Divulge any matter which it is his duty to keep secret.

(*b*) Without due authorisation, show any book, record or document in connection with his work, or which is regarded by his employers as being classified as 'confidential' or 'secret', to an unauthorised person.

(*c*) Make any anonymous communications in connection with his work to any person.

5 *Corrupt practice.* No security officer shall corruptly:

(*a*) Solicit or receive any bribe or other consideration from any fellow employee, contractor, or any other person with whom his duty brings him into contact.

(*b*) Improperly use his position in relation to other employees to his private advantage.

(*c*) Fail to account for any found property or monies received by him in his official capacity.

6 *Unnecessary exercise of authority.* No security officer shall:

(*a*) Be uncivil to any fellow employee, visitor, or other person encountered in the course of his work, or use language to such person of a type to which he could reasonably object.

(*b*) Make unnecessary use of his authority in such a manner as to cause reasonable complaint by a fellow employee, visitor, or other person lawfully on the employer's premises.

7 *Discreditable conduct.* No security officer shall act at any time in a manner reasonably likely to bring discredit upon his fellow officers, upon his employers, or upon his fellow employees.

8 *Malingering.* No security officer shall feign or exaggerate any sickness or injury with a view to evading his normal duties thereby causing added burden upon his fellows.

9 *Deportment.* No security officer will wear his uniform other than on his employer's premises, or in such places that he could reasonably be expected to traverse during the course of his work.

No security officer, whilst on duty in uniform, will be dirty or untidy in his personal clothing or in any equipment which he should normally use; nor will he wilfully by carelessness cause any waste, loss, or damage to any clothing or other property issued to him.

10 *Drunkenness.* No security officer will commence duty under the influence of liquor to such an extent as to be incapable of fully carrying out all duties reasonably required of him. During his tour of duty he will not consume drink to such a degree as to impair his efficient performance of such duties.

11 *Conviction for criminal offence.* No security officer, if he is convicted of any criminal offence, shall fail to divulge the fact to his immediate superior.

INDEX

Index